# STUDENT SOLUTIONS MANUAL

**J. Richard Christman**
*Professor Emeritus*
*U.S. Coast Guard Academy*

---

# FUNDAMENTALS OF PHYSICS

**Eighth Edition**

## THE WILEY BICENTENNIAL–KNOWLEDGE FOR GENERATIONS

Each generation has its unique needs and aspirations. When Charles Wiley first opened his small printing shop in lower Manhattan in 1807, it was a generation of boundless potential searching for an identity. And we were there, helping to define a new American literary tradition. Over half a century later, in the midst of the Second Industrial Revolution, it was a generation focused on building the future. Once again, we were there, supplying the critical scientific, technical, and engineering knowledge that helped frame the world. Throughout the 20th Century, and into the new millennium, nations began to reach out beyond their own borders and a new international community was born. Wiley was there, expanding its operations around the world to enable a global exchange of ideas, opinions, and know-how.

For 200 years, Wiley has been an integral part of each generation's journey, enabling the flow of information and understanding necessary to meet their needs and fulfill their aspirations. Today, bold new technologies are changing the way we live and learn. Wiley will be there, providing you the must-have knowledge you need to imagine new worlds, new possibilities, and new opportunities.

Generations come and go, but you can always count on Wiley to provide you the knowledge you need, when and where you need it!

WILLIAM J. PESCE
PRESIDENT AND CHIEF EXECUTIVE OFFICER

PETER BOOTH WILEY
CHAIRMAN OF THE BOARD

# STUDENT SOLUTIONS MANUAL

**J. Richard Christman**
*Professor Emeritus*
*U.S. Coast Guard Academy*

---

# FUNDAMENTALS OF PHYSICS

**Eighth Edition**

**David Halliday**
*University of Pittsburgh*

**Robert Resnick**
*Rensselaer Polytechnic Institute*

**Jearl Walker**
*Cleveland State University*

**John Wiley & Sons, Inc.**

Cover Image: © Eric Heller/Photo Researchers

Bicentennial Logo Design: Richard J. Pacifico



ISBN-13 978- 0-471-77958-2

Printed in the United States of America

10 9 8 7 6 5 4 3 2

Printed and bound by Bind-Rite Graphics.

# PREFACE

This solutions manual is designed for use with the textbook *Fundamentals of Physics*, eighth edition, by David Halliday, Robert Resnick, and Jearl Walker. Its primary purpose is to show students by example how to solve various types of problems given at the ends of chapters in the text.

Most of the solutions start from definitions or fundamental relationships and the final equation is derived. This technique highlights the fundamentals and at the same time gives students the opportunity to review the mathematical steps required to obtain a solution. The mere plugging of numbers into equations derived in the text is avoided for the most part. We hope students will learn to examine any assumptions that are made in setting up and solving each problem.

Problems in this manual were selected by Jearl Walker. Their solutions are the responsibility of the author alone.

The author is extremely grateful to Geraldine Osnato, who oversaw this project, and to her capable assistant Aly Rentrop. For their help and encouragement, special thanks go to the good people of Wiley who saw this manual through production. The author is especially thankful for the dedicated work of Karen Christman, who carefully read and corrected an earlier version of this manual. He is also grateful for the encouragement and strong support of his wife, Mary Ellen Christman.

J. Richard Christman
Professor Emeritus
U.S. Coast Guard Academy
New London, CT 06320

# TABLE OF CONTENTS

# Chapter 1

$$1\,\text{yd} = (0.9144\,\text{m})(10^6\,\mu\text{m/m}) = 9.144 \times 10^5\,\mu\text{m}\,.$$

3

Use the given conversion factors.

(a) The distance $d$ in rods is

$$d = 4.0\,\text{furlongs} = \frac{(4.0\,\text{furlongs})(201.168\,\text{m/furlong})}{5.0292\,\text{m/rod}} = 160\,\text{rods}\,.$$

(b) The distance in chains is

$$d = 4.0\,\text{furlongs} = \frac{(4.0\,\text{furlongs})(201.168\,\text{m/furlong})}{20.17\,\text{m/chain}} = 40\,\text{chains}\,.$$

5

(a) The circumference of a sphere of radius $R$ is given by $2\pi R$. Substitute $R = (6.37 \times 10^6\,\text{m})(10^{-3}\,\text{km/m}) = 6.37 \times 10^3\,\text{km}$. Retain three significant figures in your answer. You should obtain $4.00 \times 10^4\,\text{km}$.

(b) The surface area of a sphere is given by $4\pi R^2$, so the surface area of Earth is $4\pi(6.37 \times 10^3\,\text{km})^2 = 5.10 \times 10^8\,\text{km}^2$.

(c) The volume of a sphere is given by $(4\pi/3)R^3$, so the volume of Earth is $(4\pi/3)(6.37 \times 10^3\,\text{km})^3 = 1.08 \times 10^{12}\,\text{km}^3$.

17

None of the clocks advance by exactly 24 h in a 24-h period but this is not the most important criterion for judging their quality for measuring time intervals. What is important is that the clock advance by the same amount in each 24-h period. The clock reading can then easily be adjusted to give the correct interval. If the clock reading jumps around from one 24-h period to another, it cannot be corrected since it would impossible to tell what the correction should be. The following table gives the corrections (in seconds) that must be applied to the reading on each clock for each 24-h period. The entries were determined by subtracting the clock reading at the end of the interval from the clock reading at the beginning.

| CLOCK | Sun. -Mon. | Mon. -Tues. | Tues. -Wed. | Wed. -Thurs. | Thurs. -Fri. | Fri. -Sat |
|---|---|---|---|---|---|---|
| A | −16 | −16 | −15 | −17 | −15 | −15 |
| B | −3 | +5 | −10 | +5 | +6 | −7 |
| C | −58 | −58 | −58 | −58 | −58 | −58 |
| D | +67 | +67 | +67 | +67 | +67 | +67 |
| E | +70 | +55 | +2 | +20 | +10 | +10 |

Clocks C and D are the most consistent. For each clock the same correction must be applied for each period. The correction for clock C is less than the correction for clock D, so we judge clock C to be the best and clock D to be the next best. The correction that must be applied to clock A is in the range from 15 s to 17s. For clock B it is the range from $-5$ s to $+10$ s, for clock E it is in the range from $-70$ s to $-2$ s. After C and D, A has the smallest range of correction, B has the next smallest range, and E has the greatest range. From best the worst, the ranking of the clocks is C, D, A, B, E.

**21**

(a) Convert grams to kilograms and cubic centimeters to cubic meters: $1\,\mathrm{g} = 1 \times 10^{-3}\,\mathrm{kg}$ and $1\,\mathrm{cm}^3 = (1 \times 10^{-2}\,\mathrm{m})^3 = 1 \times 10^{-6}\,\mathrm{m}^3$. The mass of $1\,\mathrm{cm}^3$ of water is

$$1\,\mathrm{g} = (1\,\mathrm{g})\left(\frac{10^{-3}\,\mathrm{kg}}{\mathrm{g}}\right)\left(\frac{\mathrm{cm}^3}{10^{-6}\,\mathrm{m}^3}\right) = 1 \times 10^3\,\mathrm{kg}\,.$$

(b) Divide the mass (in kilograms) of the water by the time (in seconds) taken to drain it. The mass is the product of the volume of water and its density: $M = (5700\,\mathrm{m}^3)(1 \times 10^3\,\mathrm{kg/m}^3) = 5.70 \times 10^6\,\mathrm{kg}$. The time is $t = (10.0\,\mathrm{h})(3600\,\mathrm{s/h}) = 3.60 \times 10^4\,\mathrm{s}$, so the mass flow rate $R$ is

$$R = \frac{M}{t} = \frac{5.70 \times 10^6\,\mathrm{kg}}{3.60 \times 10^4\,\mathrm{s}} = 158\,\mathrm{kg/s}\,.$$

**35**

(a) The amount of fuel she believes she needs is (750 mi)/(40 mi/gal) = 18.8 gal. This is actually the number of U.K. gallons she needs although she believes it is the number of U.S. gallons.

(b) The ratio of the U.K. gallon to the U.S. gallon is (4.545 963 1L)/(3.785 306 0 L) = 1.201. The number of U.S. gallons she actually needs is

$$(18.8\,\text{U.K. gal})(1.201\,\text{U.S gal/U.K. gal}) = 22.5\,\text{U.S. gal}\,.$$

**39**

The volume of a cord of wood is $V = (8\,\mathrm{ft})(4\,\mathrm{ft})(4\,\mathrm{ft}) = 128\,\mathrm{ft}^3$. Use $1\,\mathrm{ft} = 0.3048\,\mathrm{m}$ (from Appendix D) to obtain $V = 128\,\mathrm{ft}^3)(0.3048\,\mathrm{m/ft})^3 = 3.62\,\mathrm{m}^3$. Thus $1.0\,\mathrm{m}^3$ of wood corresponds to (1/3.62) cord = 0.28 cord.

<u>41</u>

(a) The difference in the total amount between 73 freight tons and 73 displacement tons is

$$\begin{aligned}&(8\text{ barrel bulk}/\text{freight ton})(73\text{ freight ton})\\&\quad-(7\text{ barrel bulk}/\text{displacement ton})(73\text{ displacementton})\\&\quad=73\text{ barrel bulk}.\end{aligned}$$

Now

$$1\text{ barrel bulk}=0.1415\,\text{m}^3=(0.1415\,\text{m}^3)(28.378\text{ U.S. bushel})=4.015\text{ U.S. bushel},$$

so

$$73\text{ barrel bulk}=(73\text{ barrel bulk})(4.015\text{ U.S. bushel}/\text{barrel bulk})=293\text{ U.S. bushel}.$$

(b) The difference in the total amount between 73 register tons and 73 displacement tons is

$$\begin{aligned}&(20\text{ barrel bulk}/\text{register ton})(73\text{ register ton})\\&\quad-(7\text{ barrel bulk}/\text{displacement ton})(73\text{ displacementton})\\&\quad=949\text{ barrel bulk}.\end{aligned}$$

Thus

$$949\text{ barrel bulk}=(949\text{ barrel bulk})(4.015\text{ U.S. bushel}/\text{barrel bulk})=3810\text{ U.S. bushel}.$$

<u>45</u>

You need to convert meters to astronomical units and seconds to minutes. Use $1\text{ m}=1\times10^{-3}\text{ km}$, $1\text{ AU}=1.50\times10^8\text{ km}$, and $60\text{ s}=1\text{ min}$. Thus

$$3.0\times10^8\,\text{m/s}=\left(\frac{3.0\times10^8\,\text{m}}{\text{s}}\right)\left(\frac{10^{-3}\,\text{km}}{\text{m}}\right)\left(\frac{\text{AU}}{1.50\times10^8\,\text{km}}\right)\left(\frac{60\,\text{s}}{\text{min}}\right)=0.12\,\text{AU/min}.$$

<u>57</u>

(a) We want to convert parsecs to astronomical units. The distance between two points on a circle of radius $r$ is $d=2r\sin\theta/2$, where $\theta$ is the angle subtended by the radial lines to the points. See the figure to the right. Thus $r=d/2\sin\theta/2$ and

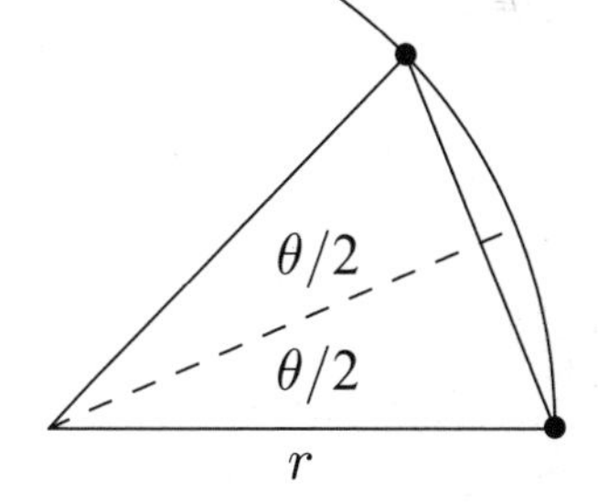


$$1\text{ pc}=\frac{1\text{ AU}}{2\sin(1''/2)}=2.06\times10^5\text{ AU},$$

where $1''=(1/3600)^\circ=(2.78\times10^{-4})^\circ$ was used. Finally

$$1\text{ AU}=(1\text{ AU})/(2.06\times10^5\text{ AU/pc})=4.9\times10^{-6}\text{ pc}.$$

(b) A light year is

$$(1.86\times10^5\text{ mi/s})(1.0\text{ y})(365.3\text{ da/y})(24\text{ h/d})(3600\text{ s/h})=5.87\times10^{12}\text{ mi}$$

and

$$1\text{ AU}=\frac{92.9\times10^6\text{ mi}}{5.87\times10^{12}\text{ mi/ly}}=1.58\times10^{-5}\text{ ly}.$$

# Chapter 2

**1**

(a) The average velocity during any time interval is the displacement during that interval divided by the duration of the interval: $v_{avg} = \Delta x/\Delta t$, where $\Delta x$ is the displacement and $\Delta t$ is the time interval. In this case the interval is divided into two parts. During the first part the displacement is $\Delta x_1 = 40\,\text{km}$ and the time interval is

$$\Delta t_1 = \frac{(40\,\text{km})}{(30\,\text{km/h})} = 1.33\,\text{h}\,.$$

During the second part the displacement is $\Delta x_2 = 40\,\text{km}$ and the time interval is

$$\Delta t_2 = \frac{(40\,\text{km})}{(60\,\text{km/h})} = 0.67\,\text{h}\,.$$

Both displacements are in the same direction, so the total displacement is $\Delta x = \Delta x_1 + \Delta x_2 = 40\,\text{km} + 40\,\text{km} = 80\,\text{km}$. The total time interval is $\Delta t = \Delta t_1 + \Delta t_2 = 1.33\,\text{h} + 0.67\,\text{h} = 2.00\,\text{h}$. The average velocity is

$$v_{avg} = \frac{(80\,\text{km})}{(2.0\,\text{h})} = 40\,\text{km/h}\,.$$

(b) The average speed is the total distance traveled divided by the time. In this case the total distance is the magnitude of the total displacement, so the average speed is 40 km/h.

(c) Assume the automobile passes the origin at time $t = 0$. Then its coordinate as a function of time is as shown as the solid lines on the graph to the right. The average velocity is the slope of the dotted line.

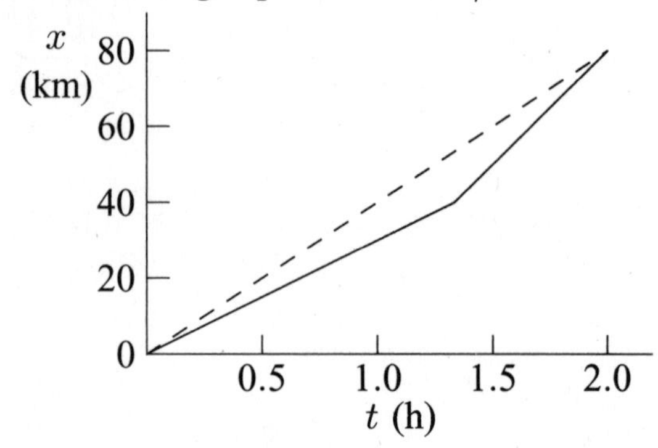


**5**

Substitute, in turn, $t = 1$, 2, 3, and 4 s into the expression $x(t) = 3t - 4t^2 + t^3$, where $x$ is in meters and $t$ is in seconds:

(a) $x(1\,\text{s}) = (3\,\text{m/s})(1\,\text{s}) - (4\,\text{m/s}^2)(1\,\text{s})^2 + (1\,\text{m/s}^3(1\,\text{s})^3 = 0$

(b) $x(2\,\text{s}) = (3\,\text{m/s})(2\,\text{s}) - (4\,\text{m/s}^2)(2\,\text{s})^2 + (1\,\text{m/s}^3)(2\,\text{s})^3 = -2\,\text{m}$

(c) $x(3\,\text{s}) = (3\,\text{m/s})(3\,\text{s}) - (4\,\text{m/s}^2)(3\,\text{s})^2 + (1\,\text{m/s}^3)(3\,\text{s})^3 = 0$

(d) $x(4\,\text{s}) = (3\,\text{m/s})(4\,\text{s}) - (4\,\text{m/s}^2)(4\,\text{s})^2 + (1\,\text{m/s}^3)(4\,\text{s})^3 = 12\,\text{m}\,.$

(e) The displacement during an interval is the coordinate at the end of the interval minus the coordinate at the beginning. For the interval from $t = 0$ to $t = 4\,\text{s}$, the displacement is $\Delta x = x(4\,\text{s}) - x(0) = 12\,\text{m} - 0 = +12\,\text{m}$. The displacement is in the positive $x$ direction.

(f) The average velocity during an interval is defined as the displacement over the interval divided by the duration of the interval: $v_{avg} = \Delta x/\Delta t$. For the interval from $t = 2$ s to $t = 4$ s the displacement is $\Delta x = x(4\,\text{s}) - x(2\,\text{s}) = 12\,\text{m} - (-2\,\text{m}) = 14\,\text{m}$ and the time interval is $\Delta t = 4\,\text{s} - 2\,\text{s} = 2\,\text{s}$. Thus

$$v_{avg} = \frac{\Delta x}{\Delta t} = \frac{14\,\text{m}}{2\,\text{s}} = 7\,\text{m/s}.$$

(d) The solid curve on the graph to the right shows the coordinate $x$ as a function of time. The slope of the dotted line is the average velocity between $t = 2.0$ s and $t = 4.0$ s.

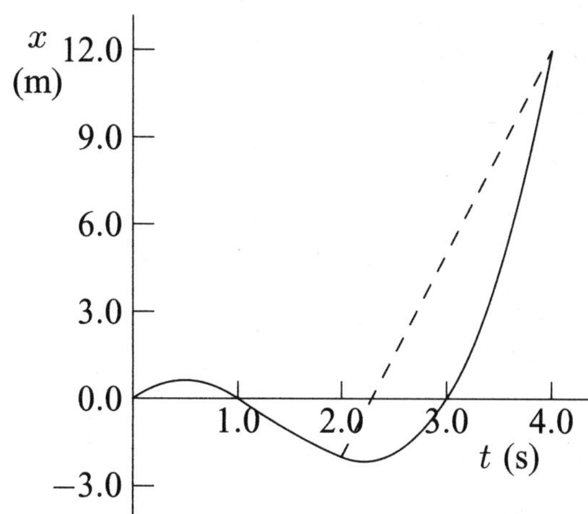


## 19

If $v_1$ is the velocity at the beginning of a time interval (at time $t_1$) and $v_2$ is the velocity at the end (at $t_2$), then the average acceleration in the interval is given by $a_{avg} = (v_2 - v_1)/(t_2 - t_1)$. Take $t_1 = 0$, $v_1 = 18\,\text{m/s}$, $t_2 = 2.4\,\text{s}$, and $v_2 = -30\,\text{m/s}$. Then

$$a_{avg} = \frac{-30\,\text{m/s} - 18\,\text{m/s}}{2.4\,\text{s}} = -20\,\text{m/s}^2.$$

The negative sign indicates that the acceleration is opposite to the original direction of travel.

## 25

(a) Solve $v = v_0 + at$ for $t$: $t = (v - v_0)/a$. Substitute $v = 0.1(3.0 \times 10^8\,\text{m/s}) = 3.0 \times 10^7\,\text{m/s}$, $v_0 = 0$, and $a = 9.8\,\text{m/s}^2$. The result is $t = 3.06 \times 10^6$ s. This is 1.2 months.

(b) Evaluate $x = x_0 + v_0 t + \frac{1}{2}at^2$, with $x_0 = 0$. The result is $x = \frac{1}{2}(9.8\,\text{m/s}^2)(3.06 \times 10^6\,\text{s})^2 = 4.6 \times 10^{13}\,\text{m}$.

## 27

Solve $v^2 = v_0^2 + 2a(x - x_0)$ for $a$. Take $x_0 = 0$. Then $a = (v^2 - v_0^2)/2x$. Use $v_0 = 1.50 \times 10^5\,\text{m/s}$, $v = 5.70 \times 10^6\,\text{m/s}$, and $x = 1.0\,\text{cm} = 0.010\,\text{m}$. The result is

$$a = \frac{(5.70 \times 10^6\,\text{m/s})^2 - (1.50 \times 10^5\,\text{m/s})^2}{2(0.010\,\text{m})} = 1.62 \times 10^{15}\,\text{m/s}^2.$$

## 33

(a) Take $x_0 = 0$, and solve $x = v_0 t + \frac{1}{2}at^2$ for $a$: $a = 2(x - v_0 t)/t^2$. Substitute $x = 24.0\,\text{m}$, $v_0 = 56.0\,\text{km/h} = 15.55\,\text{m/s}$, and $t = 2.00\,\text{s}$. The result is

$$a = \frac{2\left[24.0\,\text{m} - (15.55\,\text{m/s})(2.00\,\text{s})\right]}{(2.00\,\text{s})^2} = -3.56\,\text{m/s}^2.$$

The negative sign indicates that the acceleration is opposite the direction of motion of the car. The car is slowing down.

(b) Evaluate $v = v_0 + at$. You should get $v = 15.55\,\text{m/s} - (3.56\,\text{m/s}^2)(2.00\,\text{s}) = 8.43\,\text{m/s}$ (30.3 km/h).

**45**

(a) Take the $y$ axis to be positive in the upward direction and take $t = 0$ and $y = 0$ at the point from which the wrench was dropped. If $h$ is the height from which it was dropped, then the ground is at $y = -h$. Solve $v^2 = v_0^2 + 2gh$ for $h$:

$$h = \frac{v^2 - v_0^2}{2g}.$$

Substitute $v_0 = 0$, $v = -24\,\text{m/s}$, and $g = 9.8\,\text{m/s}^2$:

$$h = \frac{(24\,\text{m/s})^2}{2(9.8\,\text{m/s}^2)} = 29.4\,\text{m}.$$

(b) Solve $v = v_0 - gt$ for $t$:

$$t = \frac{(v_0 - v)}{g} = \frac{24\,\text{m/s}}{9.8\,\text{m/s}^2} = 2.45\,\text{s}.$$

(c)

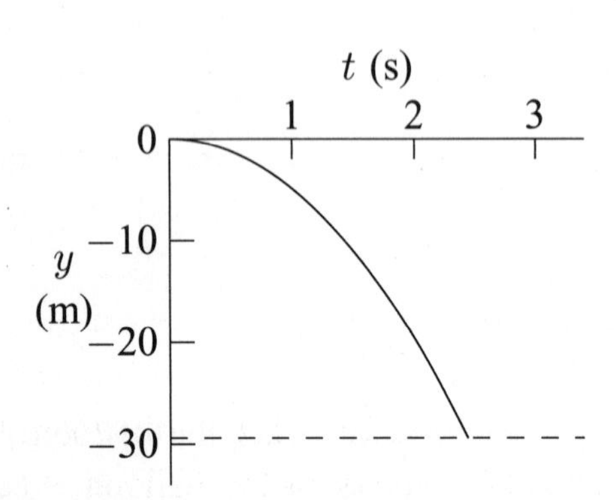


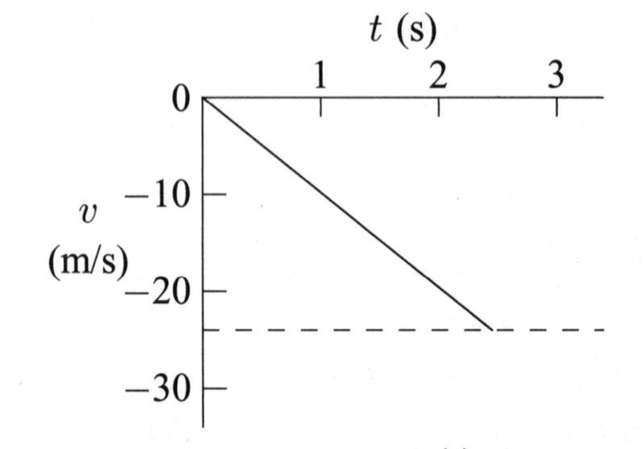


The acceleration is constant until the wrench hits the ground: $a = -9.8\,\text{m/s}^2$. Its graph is as shown on the right.

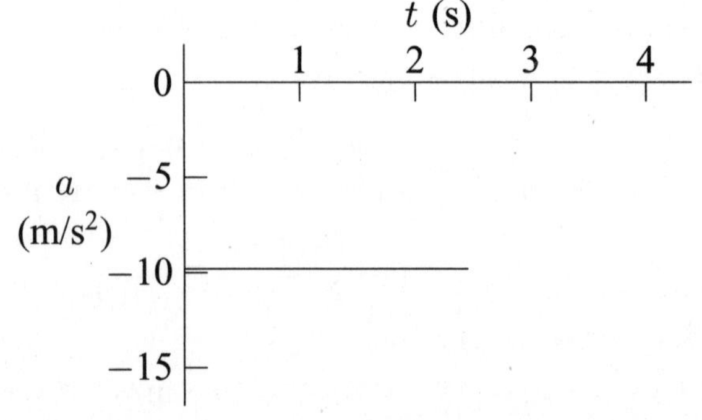


**47**

(a) At the highest point the velocity of the ball is instantaneously zero. Take the $y$ axis to be upward, set $v = 0$ in $v^2 = v_0^2 - 2gy$, and solve for $v_0$: $v_0 = \sqrt{2gy}$. Substitute $g = 9.8\,\text{m/s}^2$ and

$y = 50\,\text{m}$ to get

$$v_0 = \sqrt{2(9.8\,\text{m/s}^2)(50\,\text{m})} = 31\,\text{m/s}\,.$$

(b) It will be in the air until $y = 0$ again. Solve $y = v_0 t - \frac{1}{2}gt^2$ for $t$. Since $y = 0$ the two solutions are $t = 0$ and $t = 2v_0/g$. Reject the first and accept the second:

$$t = \frac{2v_0}{g} = \frac{2(31\,\text{m/s})}{9.8\,\text{m/s}^2} = 6.4\,\text{s}\,.$$

(c)

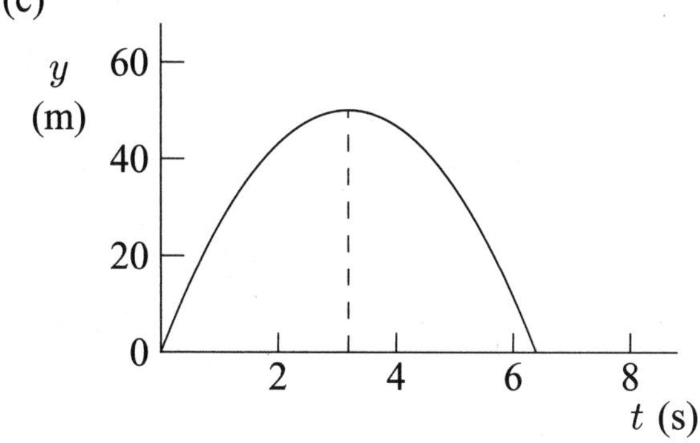


The acceleration is constant while the ball is in flight: $a = -9.8\,\text{m/s}^2$. Its graph is as shown on the right.

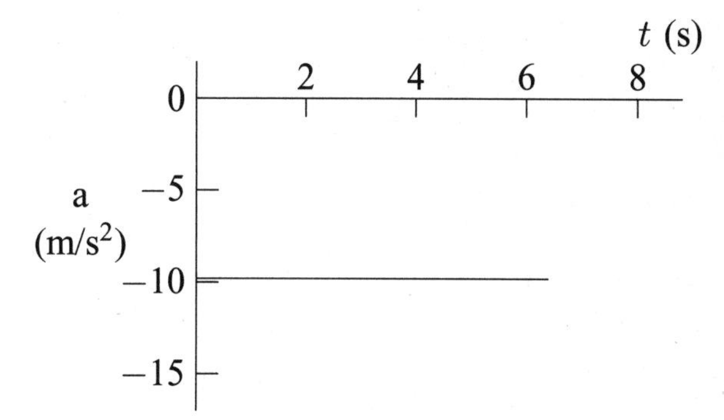


**49**

(a) Take the $y$ axis to be upward and place the origin on the ground, under the balloon. Since the package is dropped, its initial velocity is the same as the velocity of the balloon, $+12\,\text{m/s}$. The initial coordinate of the package is $y_0 = 80\,\text{m}$; when it hits the ground its coordinate is zero. Solve $y = y_0 + v_0 t - \frac{1}{2}gt^2$ for $t$:

$$t = \frac{v_0}{g} \pm \sqrt{\frac{v_0^2}{g^2} + \frac{2y_0}{g}} = \frac{12\,\text{m/s}}{9.8\,\text{m/s}^2} + \sqrt{\frac{(12\,\text{m/s})^2}{(9.8\,\text{m/s}^2)^2} + \frac{2(80\,\text{m})}{9.8\,\text{m/s}^2}} = 5.4\,\text{s}\,,$$

where the positive solution was used. A negative value for $t$ corresponds to a time before the package was dropped.

(b) Use $v = v_0 - gt = 12\,\text{m/s} - (9.8\,\text{m/s}^2)(5.4\,\text{s}) = -41\,\text{m/s}$. Its speed is $41\,\text{m/s}$.

**51**

The speed of the boat is given by $v_b = d/t$, where $d$ is the distance of the boat from the bridge when the key is dropped (12 m) and $t$ is the time the key takes in falling. To calculate $t$, put the origin of the coordinate system at the point where the key is dropped and take the $y$ axis to be positive in the upward direction. Take the time to be zero at the instant the key is dropped. You want to compute the time $t$ when $y = -45\,\text{m}$. Since the initial velocity of the key is zero, the coordinate of the key is given by $y = -\frac{1}{2}gt^2$. Thus

$$t = \sqrt{-\frac{2y}{g}} = \sqrt{-\frac{2(-45\,\text{m})}{9.8\,\text{m/s}^2}} = 3.03\,\text{s}\,.$$

This means

$$v_b = \frac{12\,\text{m}}{3.03\,\text{s}} = 4.0\,\text{m/s}\,.$$

**55**

First find the velocity of the ball just before it hits the ground. During contact with the ground its average acceleration is given by

$$a_{\text{avg}} = \frac{\Delta v}{\Delta t}\,,$$

where $\Delta v$ is the change in its velocity during contact and $\Delta t$ is the time of contact.

To find the velocity just before contact take the $y$ axis to be positive in the upward direction and put the origin at the point where the ball is dropped. Take the time $t$ to be zero when it is dropped. The ball hits the ground when $y = -15.0\,\text{m}$. Its velocity then is found from $v^2 = -2gy$, so

$$v = -\sqrt{-2gy} = -\sqrt{-2(9.8\,\text{m/s}^2)(-15.0\,\text{m})} = -17.1\,\text{m/s}\,.$$

The negative sign is used since the ball is traveling downward at the time of contact.

The average acceleration during contact with the ground is

$$a_{\text{avg}} = \frac{0 - (-17.1\,\text{m/s})}{20.0 \times 10^{-3}\,\text{s}} = 857\,\text{m/s}^2\,.$$

The positive sign indicates it is upward.

**89**

The velocity at time $t$ is given by $v = \int a\,dt = \int 5.0t\,dt = 2.5t^2 + C$, where $C$ is a constant of integration. Use the condition that $v = +17\,\text{m/s}$ at $t = 2.0\,\text{s}$ to obtain $C = v - 2.5t^2 = 17\,\text{m/s} - 2.5(2.0\,\text{s})^2 = 7.0\,\text{m/s}$. The velocity at $t = 4.0\,\text{s} = 7.0\,\text{m/s} + 2.5t^2 = 7.0\,\text{m/s} + 2.5(4.0\,\text{s})^2 = 47\,\text{m/s}$.

**91**

(a) First convert the final velocity to meters per second: $v = (60\,\text{km/h})(1000\,\text{m/km})/(3600\,\text{s/h}) = 16.7\,\text{m/s}$. The average acceleration is $v/t = (16.7\,\text{m/s})/(5.4\,\text{s}) = 3.1\,\text{m/s}^2$.

(b) Since the initial velocity is zero, the distance traveled is $x = \frac{1}{2}at^2 = \frac{1}{2}(3.1\,\text{m/s}^2)(5.4\,\text{s})^2 = 45\,\text{m}$.

(c) Solve $x = \frac{1}{2}at^2$ for $t$. The result is

$$t = \sqrt{\frac{2x}{a}} = \sqrt{\frac{2(0.25 \times 10^3\,\text{m})}{3.1\,\text{m/s}^2}} = 13\,\text{s}\,.$$

**97**

The driving time before the change in speed limit was $t_b = \Delta x/v_b$, where $\Delta x$ is the distance and $v_b$ is the original speed limit. The driving time after the change is $t_a = \Delta x/n_a$, where $v_a$ is the new speed limit. The time saved is

$$t_b - t_a = \Delta x \left(\frac{1}{v_b} - \frac{1}{v_a}\right) = (700\,\text{km})(0.6214\,\text{mi/km}) \left(\frac{1}{55\,\text{mi/h}} - \frac{1}{65\,\text{mi/h}}\right) = 1.2\,\text{h}\,.$$

This is about 1 h and 12 min.

**99**

Let $t$ be the time to reach the highest point and $v_0$ be the initial velocity. The velocity at the highest point is zero, so $0 = v_0 - gt$ and $v_0 = gt$. Thus $H = v_0 t - \frac{1}{2}gt^2 = gt^2 - \frac{1}{2}gt^2 = \frac{1}{2}gt^2$, where the substitution was made for $v_0$. Let $H_2$ be the second height. It is given by $H_2 = \frac{1}{2}g(2t)^2 = 2gt^2 = 4H$. The balls must be thrown to four times the original height.

**107**

(a) Suppose the iceboat has coordinate $y_1$ at time $t_1$ and coordinate $y_2$ at time $t_2$. If $a$ is the acceleration of the iceboat and $v_0$ is its velocity at $t = 0$, then $y_1 = v_0 t_1 + \frac{1}{2}at_1^2$ and $y_2 = v_0 t_2 + \frac{1}{2}t_2^2$. Solve these simultaneously for $a$ and $v_0$. The results are

$$a = \frac{2(y_2 t_1 - y_1 t_2)}{t_1 t_2 (t_2 - t_1)}$$

and

$$v_0 = \frac{t_2^2 y_1 - t_1^2 y_2}{t_1 t_2 (t_2 - t_1)}\,.$$

Take $t_1 = 2.0\,\text{s}$ and $t_2 = 3.0\,\text{s}$. The graph indicates that $y_1 = 16\,\text{m}$ and $y_2 = 27\,\text{m}$. These values yield $a = 2.0\,\text{m/s}^2$ and $v_0 = 6.0\,\text{m/s}$.

(b) The velocity of the iceboat at $t = 3.0\,\text{s}$ is $v = v_0 + at = 6.0\,\text{m/s} + (2.0\,\text{m/s}^2)(3.0\,\text{s}) = 12\,\text{m/s}$.

(c) The coordinate at the end of 3.0 s is $y_2 = 27\,\text{m}$. The coordinate at the end of 6.0 s is $y_3 = v_0 t_3 + \frac{1}{2}at_3^2 = (6.0\,\text{m/s})(6.0\,\text{s}) + \frac{1}{2}(2.0\,\text{m/s}^2)(6.0\,\text{s})^2 = 72\,\text{m}$. The distance traveled during the second 3.0-s interval is $y_3 - y_2 = 72\,\text{m} - 27\,\text{m} = 45\,\text{m}$.

# Chapter 3

**1**

(a) Use $a = \sqrt{a_x^2 + a_y^2}$ to obtain $a = \sqrt{(-25.0\,\text{m})^2 + (+40.0\,\text{m})^2} = 47.2\,\text{m}$.

(b) The tangent of the angle between the vector and the positive $x$ axis is

$$\tan\theta = \frac{a_y}{a_x} = \frac{40.0\,\text{m}}{-25.0\,\text{m}} = -1.6\,.$$

The inverse tangent is $-58.0°$ or $-58.0° + 180° = 122°$. The first angle has a positive cosine and a negative sine. It is not correct. The second angle has a negative cosine and a positive sine. It is correct for a vector with a negative $x$ component and a positive $y$ component.

**3**

The $x$ component is given by $a_x = (7.3\,\text{m})\cos 250° = -2.5\,\text{m}$ and the $y$ component is given by $a_y = (7.3\,)\sin 250° = -6.9\,\text{m}$. Notice that the vector is 70° below the negative $x$ axis, so the components can also be computed using $a_x = -(7.3\,\text{m})\cos 70°$ and $a_y = -(7.3\,\text{m})\sin 70°$. It is also 20° from the negative $y$ axis, so you might also use $a_x = -(7.3\,)\sin 20°$ and $a_y = -(7.3\,\text{m})\cos 20°$. These expressions give the same results.

**7**

(a) The magnitude of the displacement is the distance from one corner to the diametrically opposite corner: $d = \sqrt{(3.00\,\text{m})^2 + (3.70\,\text{m})^2 + (4.30\,\text{m})^2} = 6.42\,\text{m}$. To see this, look at the diagram of the room, with the displacement vector shown. The length of the diagonal across the floor, under the displacement vector, is given by the Pythagorean theorem: $L = \sqrt{\ell^2 + w^2}$, where $\ell$ is the length and $w$ is the width of the room. Now this diagonal and the room height form a right triangle with the displacement vector as the hypotenuse, so the length of the displacement vector is given by

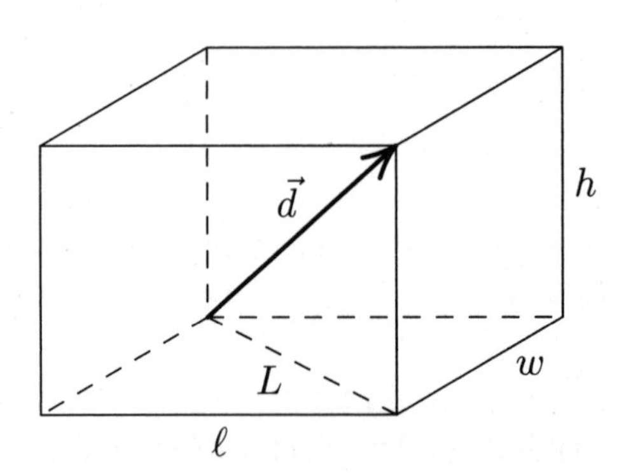


$$d = \sqrt{L^2 + h^2} = \sqrt{\ell^2 + w^2 + h^2}\,.$$

(b), (c), and (d) The displacement vector is along the straight line from the beginning to the end point of the trip. Since a straight line is the shortest distance between two points the length of the path cannot be less than the magnitude of the displacement. It can be greater, however. The fly might, for example, crawl along the edges of the room. Its displacement would be the same but the path length would be $\ell + w + h$. The path length is the same as the magnitude of the displacement if the fly flies along the displacement vector.

(e) Take the $x$ axis to be out of the page, the $y$ axis to be to the right, and the $z$ axis to be upward. Then the $x$ component of the displacement is $w = 3.70\,\text{m}$, the $y$ component of the displacement is $4.30\,\text{m}$, and the $z$ component is $3.00\,\text{m}$. Thus $\vec{d} = (3.70\,\text{m})\,\hat{\imath} + (4.30\,\text{m})\,\hat{\jmath} + (3.00\,\text{m})\,\hat{k}$. You may write an equally correct answer by interchanging the length, width, and height.

(f) Suppose the path of the fly is as shown by the dotted lines on the upper diagram. Pretend there is a hinge where the front wall of the room joins the floor and lay the wall down as shown on the lower diagram. The shortest walking distance between the lower left back of the room and the upper right front corner is the dotted straight line shown on the diagram. Its length is

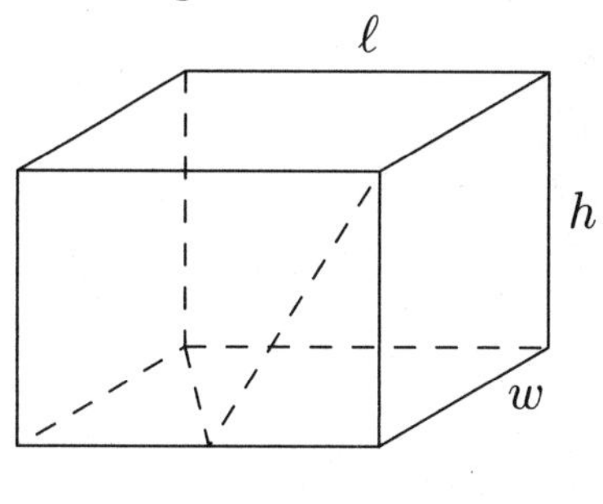


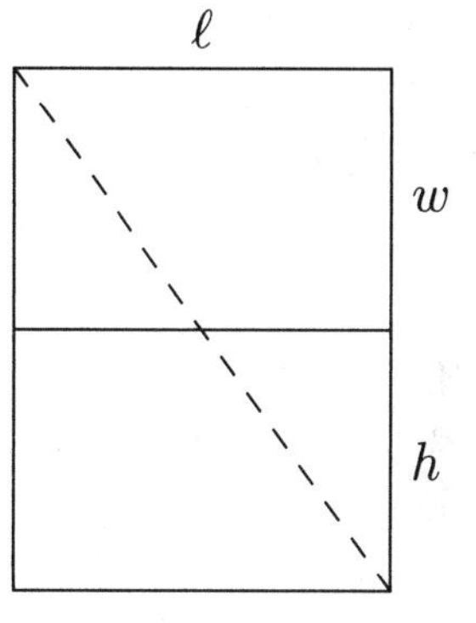


$$L_{\text{min}} = \sqrt{(w+h)^2 + \ell^2}$$

$$= \sqrt{(3.70\,\text{m} + 3.00\,\text{m})^2 + (4.30\,\text{m})^2} = 7.96\,\text{m}\,.$$

<u>9</u>

(a) Let $\vec{r} = \vec{a} + \vec{b}$. Then $r_x = a_x + b_x = 4.0\,\text{m} - 13\,\text{m} = -9.0\,\text{m}$ and $r_y = a_y + b_y = 3.0\,\text{m} + 7.0\,\text{m} = 10\,\text{m}$. Thus $\vec{r} = (-9.0\,\text{m})\,\hat{\imath} + (10\,\text{m})\,\hat{\jmath}$.

(b) The magnitude of the resultant is $r = \sqrt{r_x^2 + r_y^2} = \sqrt{(-9.0\,\text{m})^2 + (10\,\text{m})^2} = 13\,\text{m}$.

(c) The angle $\theta$ between the resultant and the positive $x$ axis is given by $\tan\theta = r_y/r_x = (10\,\text{m})/(-9.0\,\text{m}) = -1.1$. $\theta$ is either $-48^\circ$ or $132^\circ$. The first angle has a positive cosine and a negative sine while the second angle has a negative cosine and positive sine. Since the $x$ component of the resultant is negative and the $y$ component is positive, $\theta = 132^\circ$.

<u>17</u>

(a) and (b) The vector $\vec{a}$ has a magnitude $10.0\,\text{m}$ and makes the angle $30^\circ$ with the positive $x$ axis, so its components are $a_x = (10.0\,\text{m})\cos 30^\circ = 8.67\,\text{m}$ and $a_y = (10.0\,\text{m})\sin 30^\circ = 5.00\,\text{m}$. The vector $\vec{b}$ has a magnitude of $10.0\,\text{m}$ and makes an angle of $135^\circ$ with the positive $x$ axis, so its components are $b_x = (10.0\,\text{m})\cos 135^\circ = -7.07\,\text{m}$ and $b_y = (10.0\,\text{m})\sin 135^\circ = 7.07\,\text{m}$. The components of the sum are $r_x = a_x + b_x = 8.67\,\text{m} - 7.07\,\text{m} = 1.60\,\text{m}$ and $r_y = a_y + b_y = 5.0\,\text{m} + 7.07\,\text{m} = 12.1\,\text{m}$.

(c) The magnitude of $\vec{r}$ is $r = \sqrt{r_x^2 + r_y^2} = \sqrt{(1.60\,\text{m})^2 + (12.1\,\text{m})^2} = 12.2\,\text{m}$.

(d) The tangent of the angle $\theta$ between $\vec{r}$ and the positive $x$ axis is given by $\tan\theta = r_y/r_x = (12.1\,\text{m})/(1.60\,\text{m}) = 7.56$. $\theta$ is either $82.5^\circ$ or $262.5^\circ$. The first angle has a positive cosine and a positive sine and so is the correct answer.

**39**

Since $ab\cos\phi = a_x b_x + a_y b_y + a_z b_z$,

$$\cos\phi = \frac{a_x b_x + a_y b_y + a_z b_z}{ab}.$$

The magnitudes of the vectors given in the problem are $a = \sqrt{(3.0)^2 + (3.0)^2 + (3.0)^2} = 5.2$ and $b = \sqrt{(2.0)^2 + (1.0)^2 + (3.0)^2} = 3.7$. The angle between them is found from

$$\cos\phi = \frac{(3.0)(2.0) + (3.0)(1.0) + (3.0)(3.0)}{(5.2)(3.7)} = 0.926$$

and the angle is $\phi = 22°$.

**43**

(a) and (b) The vector $\vec{a}$ is along the $x$ axis, so its $x$ component is $a_x = 3.00\,\text{m}$ and its $y$ component is zero.

(c) and (d) The $x$ component of $\vec{b}$ is $b_x = b\cos\theta = (4.00\,\text{m})\cos 30.0° = 3.46\,\text{m}$ and the $y$ component is $b_y = b\sin\theta = (4.00\,\text{m})\sin 30.0° = 2.00\,\text{m}$.

(e) and (f) The $x$ component of $\vec{c}$ is $c_x = c\cos(\theta + 90°) = (10.0\,\text{m})\cos 120° = -5.00\,\text{m}$ and the $y$ component is $c_y = c\sin(\theta + 90°) = (10.0\,\text{m})\sin 120° = 8.66\,\text{m}$.

(g) and (h) In terms of components $c_x = pa_x + qb_x$ and $c_y = pa_y + qb_y$. Solve these equations simultaneously for $p$ and $q$. The result is

$$p = \frac{b_x c_y - b_y c_x}{a_y b_x - a_x b_y} = \frac{(3.46\,\text{m})(8.66\,\text{m}) - (2.00\,\text{m})(-5.00\,\text{m})}{-(3.00\,\text{m})(2.00\,\text{m})} = -6.67$$

and

$$q = \frac{a_y c_x - a_x c_y}{a_y b_x - a_x b_y} = \frac{-(3.00\,\text{m})(8.66\,\text{m})}{-(3.00\,\text{m})(2.00\,\text{m})} = 4.34\,.$$

**47**

(a) The scalar product is

$$\vec{a}\cdot\vec{b} = ab\cos\phi = (10)(6.0)\cos 60° = 30\,.$$

(b) The magnitude of the vector product is

$$|\vec{a}\times\vec{b}| = ab\sin\phi = (10)(6.0)\sin 60° = 52\,.$$

**51**

Take the $x$ axis to run west to east and the $y$ axis to run south to north, with the origin at the starting point. Let $\vec{r}_{\text{dest}} = (90.0\,\text{km})\,\hat{\jmath}$ be the position of the destination and $\vec{r}_1 = (50.0\,\text{km})\,\hat{\imath}$ be the position of the sailor after the first leg of his journey and $\vec{r}_2$ be the remaining displacement required to complete the journey. The total journey is the vector sum of the two parts, so $r_{\text{dest}} = \vec{r}_1 + \vec{r}_2$ and

$$\vec{r}_2 = \vec{r}_{\text{dest}} - \vec{r}_2 = (90\,\text{km})\,\hat{\jmath} - (50.0\,\text{km})\,\hat{\imath}\,.$$

The magnitude of the remaining trip is

$$r_2 = \sqrt{r_{2x}^2 + r_{2y}^2} = \sqrt{(50.0\,\text{km})^2 + (50.0\,\text{km})^2} = 103\,\text{km}\,.$$

The tangent of the angle with the positive $x$ direction is $\tan\phi = r_{2y}/r_{2x} = (90.0\,\text{km})/(50.0\,\text{km}) = 1.80$. The angle is either $60.9°$ or $180° + 60.9° = 241°$. Since the sailor must sail northwest to reach his destination the correct angle is $241°$. This is equivalent to $60.9°$ north of west.

<u>**71**</u>

According to the problem statement $\vec{A} + \vec{B} = 6.0\,\hat{\imath} + 1.0\,\hat{\jmath}$ and $\vec{A} - \vec{B} = -4.0\,\hat{\imath} + 7.0\,\hat{\jmath}$. Add these to obtain $2\vec{A} = 2.0\,\hat{\imath} + 8.0\,\hat{\jmath}$ and then $\vec{A} = 1.0\,\hat{\imath} + 4.0\,\hat{\jmath}$. The magnitude of $\vec{A}$ is

$$A = \sqrt{A_x^2 + A_y^2} = \sqrt{(1.0)^2 + (4.0)^2} = 4.1\,.$$

# Chapter 4

7

The average velocity is the total displacement divided by the time interval. The total displacement $\vec{r}$ is the sum of three displacements, each calculated as the product of a velocity and a time interval. The first has a magnitude of (60.0 km/h)(40.0 min)/(60.0 min/h) = 40.0 km. Its direction is east. If we take the $x$ axis to be toward the east and the $y$ axis to be toward the north, then this displacement is $\vec{r}_1 = (40.0\,\text{km})\,\hat{\text{i}}$.

The second displacement has a magnitude of (60.0 km/h)(20.0 min)/(60.0 min/h) = 20.0 km. Its direction is 50.0° east of north, so it may be written

$$\vec{r}_2 = (20.0\,\text{km})\sin 50.0°\,\hat{\text{i}} + (20.0\,\text{km})\cos 50.0°\,\hat{\text{j}} = (15.3\,\text{km})\,\hat{\text{i}} + (12.9\,\text{km})\,\hat{\text{j}}\,.$$

The third displacement has a magnitude of (60.0 km/h)(50.0 min)/(60.0 min/h) = 50.0 km. Its direction is west, so the displacement may be written $\vec{r}_3 = (-50\,\text{km})\,\hat{\text{i}}$. The total displacement is

$$\begin{aligned}\vec{r} = \vec{r}_1 + \vec{r}_2 + \vec{r}_3 &= (40.0\,\text{km})\,\hat{\text{i}} + (15.3\,\text{km})\,\hat{\text{i}} + (12.9\,\text{km})\,\hat{\text{j}} - (50\,\text{km})\,\hat{\text{i}} \\ &= (5.3\,\text{km})\,\hat{\text{i}} + (12.9\,\text{km})\,\hat{\text{j}}\,.\end{aligned}$$

The total time for the trip is 40 min + 20 min + 50 min = 110 min = 1.83 h. Divide $\vec{r}$ by this interval to obtain an average velocity of $\vec{v}_{\text{avg}} = (2.9\,\text{km/h})\,\hat{\text{i}} + (7.05\,\text{km/h})\,\hat{\text{j}}$. The magnitude of the average velocity is

$$|\vec{v}_{\text{avg}}| = \sqrt{(2.9\,\text{km/h})^2 + (7.05\,\text{km})^2} = 7.6\,\text{km/s}$$

and the angle $\phi$ it makes with the positive $x$ axis satisfies

$$\tan\phi = \frac{7.05\,\text{km/h}}{2.9\,\text{km/h}} = 2.43\,.$$

The angle is $\phi = 68°$.

11

(a) The velocity is the derivative of the position vector with respect to time:

$$\vec{v} = \frac{\text{d}}{\text{d}t}\left(\hat{\text{i}} + 4t^2\,\hat{\text{j}} + t\,\hat{\text{k}}\right) = 8t\,\hat{\text{j}} + \hat{\text{k}}$$

in meters per second for $t$ in seconds.

(b) The acceleration is the derivative of the velocity with respect to time:

$$\vec{a} = \frac{\text{d}}{\text{d}t}\left(8t\,\hat{\text{j}} + \hat{\text{k}}\right) = 8\,\hat{\text{j}}$$

in meters per second squared.

**17**

(a) The velocity of the particle at any time $t$ is given by $\vec{v} = \vec{v}_0 + \vec{a}t$, where $\vec{v}_0$ is the initial velocity and $\vec{a}$ is the acceleration. The $x$ component is $v_x = v_{0x} + a_x t = 3.00\,\mathrm{m/s} - (1.00\,\mathrm{m/s^2})t$ and the $y$ component is $v_y = v_{0y} + a_y t = -(0.500\,\mathrm{m/s^2})t$. When the particle reaches its maximum $x$ coordinate $v_x = 0$. This means $3.00\,\mathrm{m/s} - (1.00\,\mathrm{m/s^2})t = 0$ or $t = 3.00\,\mathrm{s}$. The $y$ component of the velocity at this time is $v_y = (-0.500\,\mathrm{m/s^2})(3.00\,\mathrm{s}) = -1.50\,\mathrm{m/s}$. Thus $\vec{v} = (-1.50\,\mathrm{m/s})\hat{\jmath}$.

(b) The coordinates of the particle at any time $t$ are $x = v_{0x}t + \frac{1}{2}a_x t^2$ and $y = v_{0y}t + \frac{1}{2}a_y t^2$. At $t = 3.00\,\mathrm{s}$ their values are

$$x = (3.00\,\mathrm{m/s})(3.00\,\mathrm{s}) - \frac{1}{2}(1.00\,\mathrm{m/s^2})(3.00\,\mathrm{s})^2 = 4.50\,\mathrm{m}$$

and

$$y = -\frac{1}{2}(0.500\,\mathrm{m/s^2})(3.00\,\mathrm{s})^2 = -2.25\,\mathrm{m}\,.$$

Thus $\vec{r} = (4.50\,\mathrm{m})\hat{\imath} - (2.25\,\mathrm{m})\hat{\jmath}$.

**29**

(a) Take the $y$ axis to be upward and the $x$ axis to be horizontal. Place the origin at the point where the diver leaves the platform. The components of the diver's initial velocity are $v_{0x} = 3.00\,\mathrm{m/s}$ and $v_{0y} = 0$. At $t = 0.800\,\mathrm{s}$ the horizontal distance of the diver from the platform is $x = v_{0x}t = (2.00\,\mathrm{m/s})(0.800\,\mathrm{s}) = 1.60\,\mathrm{m}$.

(b) The driver's $y$ coordinate is $y = -\frac{1}{2}gt^2 = -\frac{1}{2}(9.8\,\mathrm{m/s^2})(0.800\,\mathrm{s})^2 = -3.13\,\mathrm{m}$. The distance above the water surface is $10.0\,\mathrm{m} - 3.13\,\mathrm{m} = 6.86\,\mathrm{m}$.

(c) The driver strikes the water when $y = -10.0\,\mathrm{m}$. The time he strikes is

$$t = \sqrt{-\frac{2y}{g}} = \sqrt{-\frac{2(-10.0\,\mathrm{m})}{9.8\,\mathrm{m/s^2}}} = 1.43\,\mathrm{s}$$

and the horizontal distance from the platform is $x = v_{0x}t = (2.00\,\mathrm{m/s})(1.43\,\mathrm{s}) = 2.86\,\mathrm{m}$.

**31**

(a) Since the projectile is released its initial velocity is the same as the velocity of the plane at the time of release. Take the $y$ axis to be upward and the $x$ axis to be horizontal. Place the origin at the point of release and take the time to be zero at release. Let $x$ and $y$ ($= -730\,\mathrm{m}$) be the coordinates of the point on the ground where the projectile hits and let $t$ be the time when it hits. Then

$$y = -v_0 t \cos\theta_0 - \frac{1}{2}gt^2\,,$$

where $\theta_0 = 53.0°$. This equation gives

$$v_0 = -\frac{y + \frac{1}{2}gt^2}{t\cos\theta_0} = -\frac{-730\,\mathrm{m} + \frac{1}{2}(9.80\,\mathrm{m/s^2})(5.00\,\mathrm{s})^2}{(5.00\,\mathrm{s})\cos(53.0°)} = 202\,\mathrm{m/s}\,.$$

(b) The horizontal distance traveled is $x = v_0 t \sin\theta_0 = (202\,\mathrm{m/s})(5.00\,\mathrm{s})\sin(53.0°) = 806\,\mathrm{m}$.

(c) and (d) The $x$ component of the velocity is

$$v_x = v_0 \sin\theta_0 = (202\,\text{m/s})\sin(53.0°) = 161\,\text{m/s}$$

and the $y$ component is

$$v_y = -v_0\cos\theta_0 - gt = -(202\,\text{m/s})\cos(53°) - (9.80\,\text{m/s}^2)(5.00\,\text{s}) = -171\,\text{m/s}\,.$$

**39**

Take the $y$ axis to be upward and the $x$ axis to the horizontal. Place the origin at the firing point, let the time be zero at firing, and let $\theta_0$ be the firing angle. If the target is a distance $d$ away, then its coordinates are $x = d$ and $y = 0$. The kinematic equations are $d = v_0 t\cos\theta_0$ and $0 = v_0 t \sin\theta_0 - \frac{1}{2}gt^2$. Eliminate $t$ and solve for $\theta_0$. The first equation gives $t = d/v_0\cos\theta_0$. This expression is substituted into the second equation to obtain $2v_0^2\sin\theta_0\cos\theta_0 - gd = 0$. Use the trigonometric identity $\sin\theta_0\cos\theta_0 = \frac{1}{2}\sin(2\theta_0)$ to obtain $v_0^2\sin(2\theta_0) = gd$ or

$$\sin(2\theta_0) = \frac{gd}{v_0^2} = \frac{(9.8\,\text{m/s}^2)(45.7\,\text{m})}{(460\,\text{m/s})^2} = 2.12\times10^{-3}\,.$$

The firing angle is $\theta_0 = 0.0606°$. If the gun is aimed at a point a distance $\ell$ above the target, then $\tan\theta_0 = \ell/d$ or $\ell = d\tan\theta_0 = (45.7\,\text{m})\tan 0.0606° = 0.0484\,\text{m} = 4.84\,\text{cm}$.

**47**

You want to know how high the ball is from the ground when its horizontal distance from home plate is 97.5 m. To calculate this quantity you need to know the components of the initial velocity of the ball. Use the range information. Put the origin at the point where the ball is hit, take the $y$ axis to be upward and the $x$ axis to be horizontal. If $x$ (= 107 m) and $y$ (= 0) are the coordinates of the ball when it lands, then $x = v_{0x}t$ and $0 = v_{0y}t - \frac{1}{2}gt^2$, where $t$ is the time of flight of the ball. The second equation gives $t = 2v_{0y}/g$ and this is substituted into the first equation. Use $v_{0x} = v_{0y}$, which is true since the initial angle is $\theta_0 = 45°$. The result is $x = 2v_{0y}^2/g$. Thus

$$v_{0y} = \sqrt{\frac{gx}{2}} = \sqrt{\frac{(9.8\,\text{m/s}^2)(107\,\text{m})}{2}} = 22.9\,\text{m/s}\,.$$

Now take $x$ and $y$ to be the coordinates when the ball is at the fence. Again $x = v_{0x}t$ and $y = v_{0y}t - \frac{1}{2}gt^2$. The time to reach the fence is given by $t = x/v_{0x} = (97.5\,\text{m})/(22.9\,\text{m/s}) = 4.26\,\text{s}$. When this is substituted into the second equation the result is

$$y = v_{0y}t - \frac{1}{2}gt^2 = (22.9\,\text{m/s})(4.26\,\text{s}) - \frac{1}{2}(9.8\,\text{m/s}^2)(4.26\,\text{s})^2 = 8.63\,\text{m}\,.$$

Since the ball started 1.22 m above the ground, it is 8.63 m + 1.22 m = 9.85 m above the ground when it gets to the fence and it is 9.85 m − 7.32 m = 2.53 m above the top of the fence. It goes over the fence.

**51**

Take the $y$ axis to be upward and the $x$ axis to be horizontal. Place the origin at the point where the ball is kicked, on the ground, and take the time to be zero at the instant it is kicked. $x$ and $y$ are the coordinates of ball at the goal post. You want to find the kicking angle $\theta_0$ so that $y = 3.44\,\text{m}$ when $x = 50\,\text{m}$. Write the kinematic equations for projectile motion: $x = v_0 t \cos\theta_0$ and $y = v_0 t \sin\theta_0 - \frac{1}{2}gt^2$. The first equation gives $t = x/v_0\cos\theta_0$ and when this is substituted into the second the result is

$$y = x\tan\theta_0 - \frac{gx^2}{2v_0^2\cos^2\theta_0}\,.$$

You may solve this by trial and error: systematically try values of $\theta_0$ until you find the two that satisfy the equation. A little manipulation, however, will give you an algebraic solution. Use the trigonometric identity $1/\cos^2\theta_0 = 1 + \tan^2\theta_0$ to obtain

$$\frac{1}{2}\frac{gx^2}{v_0^2}\tan^2\theta_0 - x\tan\theta_0 + y + \frac{1}{2}\frac{gx^2}{v_0^2} = 0\,.$$

This is a quadratic equation for $\tan\theta_0$. To simplify writing the solution, let $c = \frac{1}{2}gx^2/v_0^2 = \frac{1}{2}(9.80\,\text{m/s}^2)(50\,\text{m})^2/(25\,\text{m/s})^2 = 19.6\,\text{m}$. Then the quadratic equation becomes $c\tan^2\theta_0 - x\tan\theta_0 + y + c = 0$. It has the solution

$$\begin{aligned}\tan\theta_0 &= \frac{x \pm \sqrt{x^2 + 4(y+c)c}}{2c}\\ &= \frac{50\,\text{m} \pm \sqrt{(50\,\text{m})^2 - 4(3.44\,\text{m} + 19.6\,\text{m})(19.6\,\text{m})}}{2(19.6\,\text{m})}\,.\end{aligned}$$

The two solutions are $\tan\theta_0 = 1.95$ and $\tan\theta_0 = 0.605$. The corresponding angles are $\theta_0 = 63°$ and $\theta_0 = 31°$. If kicked at any angle between these two, the ball will travel above the cross bar on the goal post.

**53**

Let $h$ be the height of a step and $w$ be the width. To hit step $n$, the ball must fall a distance $nh$ and travel horizontally a distance between $(n-1)w$ and $nw$. Take the origin of a coordinate system to be at the point where the ball leaves the top of the stairway. Take the $y$ axis to be positive in the upward direction and the $x$ axis to be horizontal. The coordinates of the ball at time $t$ are given by $x = v_{0x}t$ and $y = -\frac{1}{2}gt^2$. Equate $y$ to $-nh$ and solve for the time to reach the level of step $n$:

$$t = \sqrt{\frac{2nh}{g}}\,.$$

The $x$ coordinate then is

$$x = v_{0x}\sqrt{\frac{2nh}{g}} = (1.52\,\text{m/s})\sqrt{\frac{2n(0.203\,\text{m})}{9.8\,\text{m/s}^2}} = (0.309\,\text{m})\sqrt{n}\,.$$

Try values of $n$ until you find one for which $x/w$ is less than $n$ but greater than $n - 1$. For $n = 1$, $x = 0.309\,\text{m}$ and $x/w = 1.52$. This is greater than $n$. For $n = 2$, $x = 0.437\,\text{m}$ and $x/w = 2.15$. This is also greater than $n$. For $n = 3$, $x = 0.535\,\text{m}$ and $x/w = 2.64$. This is less than $n$ and greater than $n - 1$. The ball hits the third step.

67

To calculate the centripetal acceleration of the stone you need to know its speed while it is being whirled around. This the same as its initial speed when it flies off. Use the kinematic equations of projectile motion to find that speed. Take the $y$ axis to be upward and the $x$ axis to be horizontal. Place the origin at the point where the stone leaves its circular orbit and take the time to be zero when this occurs. Then the coordinates of the stone when it is a projectile are given by $x = v_0 t$ and $y = -\frac{1}{2}gt^2$. It hits the ground when $x = 10\,\text{m}$ and $y = -2.0\,\text{m}$. Note that the initial velocity is horizontal. Solve the second equation for the time: $t = \sqrt{-2y/g}$. Substitute this expression into the first equation and solve for $v_0$:

$$v_0 = x\sqrt{-\frac{g}{2y}} = (10\,\text{m})\sqrt{-\frac{9.8\,\text{m/s}^2}{2(-2.0\,\text{m})}} = 15.7\,\text{m/s}\,.$$

The magnitude of the centripetal acceleration is $a = v^2/r = (15.7\,\text{m/s})^2/(1.5\,\text{m}) = 160\,\text{m/s}^2$.

73

(a) Take the positive $x$ direction to be to the east and the positive $y$ direction to be to the north. The velocity of ship A is given by

$$\begin{aligned}\vec{v}_A &= -(v_A \sin 45^\circ)\,\hat{\imath} + (v_A \cos 45^\circ)\,\hat{\jmath} = -[(24\,\text{knots})\sin 45^\circ)]\,\hat{\imath} + [(24\,\text{knots})\cos 45^\circ]\,\hat{\jmath} \\ &= -(17.0\,\text{knots})\,\hat{\imath} + (17.0\,\text{knots})\,\hat{\jmath}\end{aligned}$$

and the velocity of ship B is given by

$$\begin{aligned}\vec{v}_B &= -(v_B \sin 40^\circ)\,\hat{\imath} - (v_B \cos 40^\circ)\,\hat{\jmath} = -[(28\,\text{knots})\sin 40^\circ]\,\hat{\imath} - [(28\,\text{knots})\cos 40^\circ]\,\hat{\jmath} \\ &= -(18.0\,\text{knots})\,\hat{\imath} - (21.4\,\text{knots})\,\hat{\jmath}\,.\end{aligned}$$

The velocity of ship A relative to ship B is

$$\begin{aligned}\vec{v}_{AB} &= \vec{v}_A - \vec{v}_B = [(-17.0\,\text{knots}) - (-18\,\text{knots})]\,\hat{\imath} + [(17.0\,\text{knots}) - (-21.4\,\text{knots})]\,\hat{\jmath} \\ &= (1.0\,\text{knots})\,\hat{\imath} + (38.4\,\text{knots})\,\hat{\jmath}\,.\end{aligned}$$

The magnitude is

$$v_{AB} = \sqrt{v_{AB\,x}^2 + v_{AB\,y}^2} = \sqrt{(1.0\,\text{knots})^2 + (38.4\,\text{knots})^2} = 38.4\,\text{knots}\,.$$

(b)The angle $\theta$ that $\vec{v}_{AB}$ makes with the positive $x$ axis is

$$\theta = \tan^{-1}\frac{v_{AB\,y}}{v_{AB\,x}} = \tan^{-1}\frac{38.4\,\text{knots}}{1.0\,\text{knots}} = 88.5^\circ\,.$$

This direction is $1.5^\circ$ east of north.

(c) The time $t$ for the separation to become $d$ is given by $t = d/v_{AB}$. Since a knot is a nautical mile, $t = (160\,\text{nautical miles})/(38.4\,\text{knots}) = 4.2\,\text{h}$.

(d) Ship B will be 1.5° west of south, relative to ship A.

**75**

Relative to the car the velocity of the snowflakes has a vertical component of 8.0 m/s and a horizontal component of 50 km/h = 13.9 m/s. The angle $\theta$ from the vertical is given by $\tan\theta = v_h/v_v = (13.9\,\text{m/s})/(8.0\,\text{m/s}) = 1.74$. The angle is 60°.

**77**

Since the raindrops fall vertically relative to the train, the horizontal component of the velocity of a raindrop is $v_h = 30\,\text{m/s}$, the same as the speed of the train. If $v_v$ is the vertical component of the velocity and $\theta$ is the angle between the direction of motion and the vertical, then $\tan\theta = v_h/v_v$. Thus $v_v = v_h/\tan\theta = (30\,\text{m/s})/\tan 70° = 10.9\,\text{m/s}$. The speed of a raindrop is $v = \sqrt{v_h^2 + v_v^2} = \sqrt{(30\,\text{m/s})^2 + (10.9\,\text{m/s})^2} = 32\,\text{m/s}$.

**91**

(a) Take the positive $y$ axis to be downward and place the origin at the firing point. Then the $y$ coordinate of the bullet is given by $y = \frac{1}{2}gt^2$. If $t$ is the time of flight and $y$ is the distance the bullet hits below the target, then

$$t = \sqrt{\frac{2y}{g}} = \sqrt{\frac{2(0.019\,\text{m})}{9.8\,\text{m/s}^2}} = 6.3 \times 10^{-2}\,\text{s}\,.$$

(b) The muzzle velocity is the initial velocity of the bullet. It is horizontal. If $x$ is the horizontal distance to the target, then $x = v_0 t$ and

$$v_0 = \frac{x}{t} = \frac{30\,\text{m}}{6.3 \times 10^{-2}\,\text{s}} = 4.8 \times 10^2\,\text{m/s}\,.$$

**107**

(a) Use $y = v_{0y}t - \frac{1}{2}gt^2$ and $v_y = v_{0y} - gt$, where the origin is at the point where the ball is hit, the positive $y$ direction is upward, and $v_{0y}$ is the vertical component of the initial velocity. At the highest point $v_y = 0$, so $v_{0y} = gt$ and $y = \frac{1}{2}gt^2 = \frac{1}{2}(9.8\,\text{m/s}^2)(3.0\,\text{s})^2 = 44\,\text{m}$.

(b) Set the time to zero when the ball is at its highest point. The vertical component of the initial velocity is then zero and the ball's initial $y$ coordinate is 44 m. The ball reaches the fence at time $t = 2.5\,\text{s}$. Then its height above the ground is $y = y_0 - \frac{1}{2}gt^2 = 44\,\text{m} - \frac{1}{2}(9.8\,\text{m/s}^2)(2.5\,\text{s})^2 = 13\,\text{m}$.

(c) Since the ball takes 5.5 s to travel the horizontal distance of 97.5 m to the fence, the horizontal component of the initial velocity is $v_{0x} = (97.5\,\text{m})/(5.5\,\text{s}) = 17.7\,\text{m/s}$. Since the ball took 3.0 s to rise from the ground to its highest point it must take the same time, 3.0 s to fall from the highest point to the ground. Thus it hits the ground 0.50 s after clearing the fence. The point where it hits is $(17.7\,\text{m/s})(0.50\,\text{s}) = 8.9\,\text{m}$.

**111**

(a) The position vector of the particle is given by $\vec{r} = \vec{v}_0 t + \frac{1}{2}\vec{a}t^2$, where $\vec{v}_0$ is the velocity at time $t = 0$ and $\vec{a}$ is the acceleration. The $x$ component of this equation is $x = v_{0x}t + \frac{1}{2}a_x t^2$. Since $v_{0x} = 0$ this becomes $x = \frac{1}{2}a_x t^2$. The solution for $t$ is $t = \sqrt{2x/a_x} = \sqrt{2(29\,\text{m})/(4.0\,\text{m/s}^2)} = 3.81\,\text{s}$. The $y$ coordinate then is $y = v_{0y}t + \frac{1}{2}a_y t^2 = (8.0\,\text{m/s})(3.81\,\text{s}) + \frac{1}{2}(2.0\,\text{m/s}^2)(3.81\,\text{s})^2 = 45\,\text{m}$.

(b) The $x$ component of the velocity is $v_x = v_{0x} + a_x t = (4.0\,\text{m/s}^2)(3.81\,\text{s}) = 15.2\,\text{m/s}$ and the $y$ component is $v_y = v_{0y} + \frac{1}{2}a_y t = 8.0\,\text{m/s} + (2.0\,\text{m/s}^2)(3.81\,\text{s}) = 15.6\,\text{m/s}$. The speed is $v = \sqrt{v_x^2 + v_y^2} = \sqrt{(15.2\,\text{m/s})^2 + (15.6\,\text{m/s})^2} = 22\,\text{m/s}$.

**121**

(a) and (b) Take the $x$ axis to be from west to east and the $y$ axis to be from south to north. Sum the two displacements from A to the resting place. The first is $\Delta\vec{r}_1 = (75\,\text{km})(\hat{\imath}\cos 37° + \hat{\jmath}\sin 37°) = (60\,\text{km})\hat{\imath} + (45\,\text{km})\hat{\jmath}$ and the second is $\Delta\vec{r}_2 = -(65\,\text{km})\hat{\jmath}$. The sum is $\Delta r = (60\,\text{km})\hat{\imath} - (20\,\text{km})\hat{\jmath}$. The magnitude of the total displacement is $\Delta r = \sqrt{(60\,\text{km})^2 + (20\,\text{km})^2} = 63\,\text{km}$ and the tangent of the angle it makes with the east is $\tan\theta = (-20\,\text{km})/(60\,\text{km}) = -0.33$. The angle is 18° south of east.

(c) and (d) The total time for the trip and rest is $50\,\text{h} + 35\,\text{h} + 5.0\,\text{h} = 90\,\text{h}$, so the magnitude of the average velocity is $(63\,\text{km})/(90\,\text{h}) = 0.70\,\text{km/h}$. The average velocity is in the same direction as the displacement, 18° south of east.

(e) The average speed is the distance traveled divided by the elapsed time. The distance is $75\,\text{km} + 65\,\text{km} = 140\,\text{km}$, so the average speed is $(140\,\text{km})/(90\,\text{h}) = 1.5\,\text{km/h}$.

(f) and (g) The camel has $120\,\text{h} - 90\,\text{h} = 30\,\text{h}$ to get from the resting place to B. If $\Delta\vec{r}_B$ is the displacement of B from A and $\Delta\vec{r}_{\text{rest}}$ is the displacement of the resting place from A, the displacement of the camel during this time is $\Delta\vec{r}_B - \Delta\vec{r}_{\text{rest}} = (90\,\text{km})\hat{\imath} - (60\,\text{km})\hat{\imath} - (-20\,\text{km})\hat{\jmath} = (30\,\text{km})\hat{\imath} + (20\,\text{km})\hat{\jmath}$. The magnitude of the displacement is $\sqrt{(30\,\text{km})^2 + (20\,\text{km})^2} = 36\,\text{km}$ and the magnitude of the average velocity must be $(36\,\text{km})/(30\,\text{h}) = 1.2\,\text{km/h}$. The angle $\phi$ that the average velocity must make with the east is given by $\tan\phi = (20\,\text{km})/(30\,\text{km}) = 0.66$ and the angle is 34° north of east.

# Chapter 5

## 5

Label the two forces $\vec{F}_1$ and $\vec{F}_2$. According to Newton's second law, $\vec{F}_1 + \vec{F}_2 = m\vec{a}$, so $\vec{F}_2 = m\vec{a} - \vec{F}_1$. In unit vector notation $\vec{F}_1 = (20.0\,\mathrm{N})\,\hat{\mathrm{i}}$ and

$$\vec{a} = -(12\,\mathrm{m/s^2})(\sin 30°)\,\hat{\mathrm{i}} - (12\,\mathrm{m/s^2})(\cos 30°)\,\hat{\mathrm{j}} = -(6.0\,\mathrm{m/s^2})\,\hat{\mathrm{i}} - (10.4\,\mathrm{m/s^2})\,\hat{\mathrm{j}}\,.$$

Thus

$$\vec{F}_2 = (2.0\,\mathrm{kg})(-6.0\,\mathrm{m/s^2})\,\hat{\mathrm{i}} + (2.0\,\mathrm{kg})(-10.4\,\mathrm{m/s^2})\,\hat{\mathrm{j}} - (20.0\,\mathrm{N})\,\hat{\mathrm{i}} = (-32\,\mathrm{N})\,\hat{\mathrm{i}} - (21\,\mathrm{N})\,\hat{\mathrm{j}}\,.$$

(b) and (c) The magnitude of $\vec{F}_2$ is $F_2 = \sqrt{F_{2x}^2 + F_{2y}^2} = \sqrt{(-32\,\mathrm{N})^2 + (-21\,\mathrm{N})^2} = 38\,\mathrm{N}$. The angle that $\vec{F}_2$ makes with the positive $x$ axis is given by $\tan\theta = F_{2y}/F_{2x} = (21\,\mathrm{N})/(32\,\mathrm{N}) = 0.656$. The angle is either $33°$ or $33° + 180° = 213°$. Since both the $x$ and $y$ components are negative the correct result is $213°$. You could also take the angle to be $180° - 213° = -147°$.

## 13

In all three cases the scale is not accelerating, which means that the two cords exert forces of equal magnitude on it. The scale reads the magnitude of either of these forces. In each case the magnitude of the tension force of the cord attached to the salami must be the same as the magnitude of the weight of the salami. You know this because the salami is not accelerating. Thus the scale reading is $mg$, where $m$ is the mass of the salami. Its value is $(11.0\,\mathrm{kg})(9.8\,\mathrm{m/s^2}) = 108\,\mathrm{N}$.

## 19

(a) The free-body diagram is shown in Fig. 5–16 of the text. Since the acceleration of the block is zero, the components of the Newton's second law equation yield $T - mg\sin\theta = 0$ and $F_N - mg\cos\theta = 0$. Solve the first equation for the tension force of the string: $T = mg\sin\theta = (8.5\,\mathrm{kg})(9.8\,\mathrm{m/s^2})\sin 30° = 42\,\mathrm{N}$.

(b) Solve the second equation for $F_N$: $F_N = mg\cos\theta = (8.5\,\mathrm{kg})(9.8\,\mathrm{m/s^2})\cos 30° = 72\,\mathrm{N}$.

(c) When the string is cut it no longer exerts a force on the block and the block accelerates. The $x$ component of the second law becomes $-mg\sin\theta = ma$, so $a = -g\sin\theta = -(9.8\,\mathrm{m/s^2})\sin 30° = -4.9\,\mathrm{m/s^2}$. The negative sign indicates the acceleration is down the plane.

## 25

According to Newton's second law $F = ma$, where $F$ is the magnitude of the force, $a$ is the magnitude of the acceleration, and $m$ is the mass. The acceleration can be found using the equations for constant-acceleration motion. Solve $v = v_0 + at$ for $a$: $a = v/t$. The final velocity is $v = (1600\,\text{km/h})(1000\,\text{m/km})/(3600\,\text{s/h}) = 444\,\text{m/s}$, so $a = (444\,\text{m/s})/(1.8\,\text{s}) = 247\,\text{m/s}^2$ and the magnitude of the force is $F = (500\,\text{kg})(247\,\text{m/s}^2) = 1.2 \times 10^5\,\text{N}$.

**29**

The acceleration of the electron is vertical and for all practical purposes the only force acting on it is the electric force. The force of gravity is much smaller. Take the $x$ axis to be in the direction of the initial velocity and the $y$ axis to be in the direction of the electrical force. Place the origin at the initial position of the electron. Since the force and acceleration are constant the appropriate equations are $x = v_0 t$ and $y = \frac{1}{2}at^2 = \frac{1}{2}(F/m)t^2$, where $F = ma$ was used to substitute for the acceleration $a$. The time taken by the electron to travel a distance $x$ (= 30 mm) horizontally is $t = x/v_0$ and its deflection in the direction of the force is

$$y = \frac{1}{2}\frac{F}{m}\left(\frac{x}{v_0}\right)^2 = \frac{1}{2}\left(\frac{4.5 \times 10^{-16}\,\text{N}}{9.11 \times 10^{-31}\,\text{kg}}\right)\left(\frac{30 \times 10^{-3}\,\text{m}}{1.2 \times 10^7\,\text{m/s}}\right)^2 = 1.5 \times 10^{-3}\,\text{m}\,.$$

**35**

The free-body diagram is shown at the right. $\vec{F}_N$ is the normal force of the plane on the block and $m\vec{g}$ is the force of gravity on the block. Take the positive $x$ axis to be down the plane, in the direction of the acceleration, and the positive $y$ axis to be in the direction of the normal force. The $x$ component of Newton's second law is then $mg \sin\theta = ma$, so the acceleration is $a = g \sin\theta$.

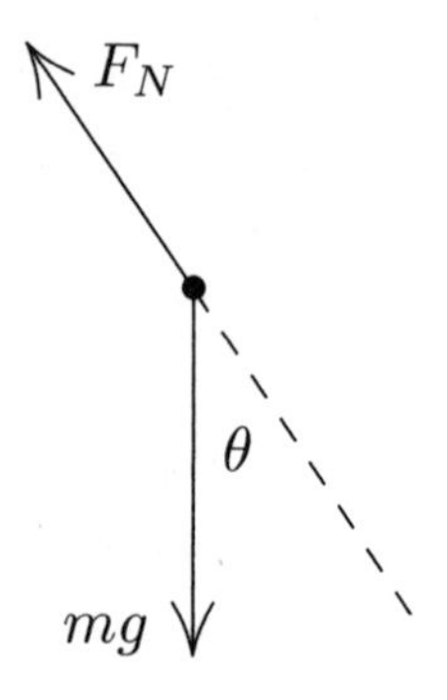


(a) Place the origin at the bottom of the plane. The equations for motion along the $x$ axis are $x = v_0 t + \frac{1}{2}at^2$ and $v = v_0 + at$. The block stops when $v = 0$.

According to the second equation, this is at the time $t = -v_0/a$. The coordinate when it stops is

$$x = v_0\left(\frac{-v_0}{a}\right) + \frac{1}{2}a\left(\frac{-v_0}{a}\right)^2 = -\frac{1}{2}\frac{v_0^2}{a} = -\frac{1}{2}\frac{v_0^2}{g\sin\theta}$$
$$= -\frac{1}{2}\left[\frac{(-3.50\,\text{m/s})^2}{(9.8\,\text{m/s}^2)\sin 32.0^\circ}\right] = -1.18\,\text{m}\,.$$

(b) The time is

$$t = -\frac{v_0}{a} = -\frac{v_0}{g\sin\theta} = -\frac{-3.50\,\text{m/s}}{(9.8\,\text{m/s}^2)\sin 32.0^\circ} = 0.674\,\text{s}\,.$$

(c) Now set $x = 0$ and solve $x = v_0 t + \frac{1}{2}at^2$ for $t$. The result is

$$t = -\frac{2v_0}{a} = -\frac{2v_0}{g\sin\theta} = -\frac{2(-3.50\,\text{m/s})}{(9.8\,\text{m/s}^2)\sin 32.0^\circ} = 1.35\,\text{s}\,.$$

The velocity is

$$v = v_0 + at = v_0 + gt\sin\theta = -3.50\,\text{m/s} + (9.8\,\text{m/s}^2)(1.35\,\text{s})\sin 32^\circ = 3.50\,\text{m/s}\,,$$

as expected since there is no friction. The velocity is down the plane.

**45**

The free-body diagrams for the links are drawn below. The force arrrows are not to scale.

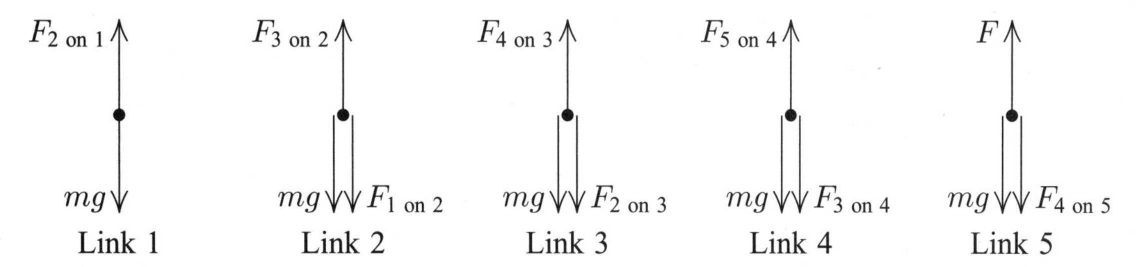


(a) The links are numbered from bottom to top. The forces on the bottom link are the force of gravity $m\vec{g}$, downward, and the force $\vec{F}_{2\text{ on }1}$ of link 2, upward. Take the positive direction to be upward. Then Newton's second law for this link is $F_{2\text{ on }1} - mg = ma$. Thus

$$F_{2\text{ on }1} = m(a+g) = (0.100\,\text{kg})(2.50\,\text{m/s}^2 + 9.8\,\text{m/s}^2) = 1.23\,\text{N}\,.$$

(b) The forces on the second link are the force of gravity $m\vec{g}$, downward, the force $\vec{F}_{1\text{ on }2}$ of link 1, downward, and the force $\vec{F}_{3\text{ on }2}$ of link 3, upward. According to Newton's third law $\vec{F}_{1\text{ on }2}$ has the same magnitude as $\vec{F}_{2\text{ on }1}$. Newton's second law for the second link is $F_{3\text{ on }2} - F_{1\text{ on }2} - mg = ma$, so

$$F_{3\text{ on }2} = m(a+g) + F_{1\text{ on }2} = (0.100\,\text{kg})(2.50\,\text{m/s}^2 + 9.8\,\text{m/s}^2) + 1.23\,\text{N} = 2.46\,\text{N}\,,$$

where Newton's third law was used to substitute the value of $F_{2\text{ on }1}$ for $F_{1\text{ on }2}$.

(c) The forces on the third link are the force of gravity $m\vec{g}$, downward, the force $\vec{F}_{2\text{ on }3}$ of link 2, downward, and the force $\vec{F}_{4\text{ on }3}$ of link 4, upward. Newton's second law for this link is $F_{4\text{ on }3} - F_{2\text{ on }3} - mg = ma$, so

$$F_{4\text{ on }3} = m(a+g) + F_{2\text{ on }3} = (0.100\,\text{N})(2.50\,\text{m/s}^2 + 9.8\,\text{m/s}^2) + 2.46\,\text{N} = 3.69\,\text{N}\,,$$

where Newton's third law was used to substitute the value of $F_{3\text{ on }2}$ for $F_{2\text{ on }3}$.

(d) The forces on the fourth link are the force of gravity $m\vec{g}$, downward, the force $\vec{F}_{3\text{ on }4}$ of link 3, downward, and the force $\vec{F}_{5\text{ on }4}$ of link 5, upward. Newton's second law for this link is $F_{5\text{ on }4} - F_{3\text{ on }4} - mg = ma$, so

$$F_{5\text{ on }4} = m(a+g) + F_{3\text{ on }4} = (0.100\,\text{kg})(2.50\,\text{m/s}^2 + 9.8\,\text{m/s}^2) + 3.69\,\text{N} = 4.92\,\text{N}\,,$$

where Newton's third law was used to substitute the value of $F_{4\text{ on }3}$ for $F_{3\text{ on }4}$.

(e) The forces on the top link are the force of gravity $m\vec{g}$, downward, the force $\vec{F}_{4\text{ on }5}$ of link 4, downward, and the applied force $\vec{F}$, upward. Newton's second law for the top link is $F - F_{4\text{ on }5} - mg = ma$, so

$$F = m(a+g) + F_{4\text{ on }5} = (0.100\,\text{kg})(2.50\,\text{m/s}^2 + 9.8\,\text{m/s}^2) + 4.92\,\text{N} = 6.15\,\text{N}\,,$$

where Newton's third law as used to substitute the value of $F_{5\text{ on }4}$ for $F_{4\text{ on }5}$.

(f) Each link has the same mass and the same acceleration, so the same net force acts on each of them: $F_{\text{net}} = ma = (0.100\,\text{kg})(2.50\,\text{m/s}^2) = 0.25\,\text{N}$.

## 53

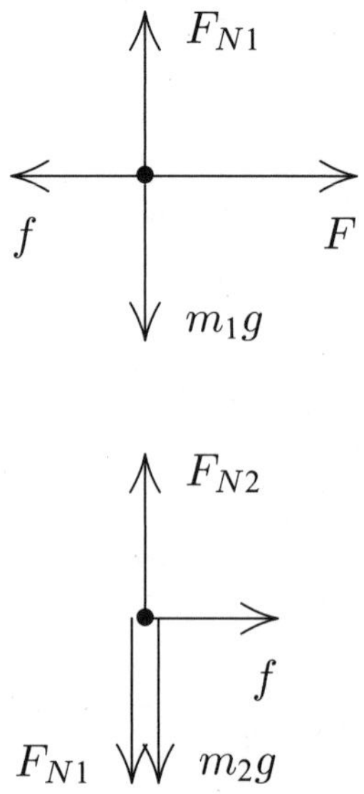

(a) The free-body diagrams are shown to the right. $\vec{F}$ is the applied force and $\vec{f}$ is the force of block 1 on block 2. Note that $\vec{F}$ is applied only to block 1 and that block 2 exerts the force $-\vec{f}$ on block 1. Newton's third law has thereby been taken into account.

$F_{N1}$ $f$ $F$ $m_1g$

$F_{N2}$ $f$ $F_{N1}$ $m_2g$

Newton's second law for block 1 is $F - f = m_1a$, where $a$ is the acceleration. The second law for block 2 is $f = m_2a$. Since the blocks move together they have the same acceleration and the same symbol is used in both equations. Use the second equation to obtain an expression for $a$: $a = f/m_2$. Substitute into the first equation to get $F - f = m_1f/m_2$. Solve for $f$:

$$f = \frac{Fm_2}{m_1 + m_2} = \frac{(3.2\,\text{N})(1.2\,\text{kg})}{2.3\,\text{kg} + 1.2\,\text{kg}} = 1.1\,\text{N}\,.$$

(b) If $\vec{F}$ is applied to block 2 instead of block 1, the force of contact is

$$f = \frac{Fm_1}{m_1 + m_2} = \frac{(3.2\,\text{N})(2.3\,\text{kg})}{2.3\,\text{kg} + 1.2\,\text{kg}} = 2.1\,\text{N}\,.$$

(c) The acceleration of the blocks is the same in the two cases. Since the contact force $f$ is the only horizontal force on one of the blocks it must be just right to give that block the same acceleration as the block to which $\vec{F}$ is applied. In the second case the contact force accelerates a more massive block than in the first, so it must be larger.

## 57

(a) Take the positive direction to be upward for both the monkey and the package. Suppose the monkey pulls downward on the rope with a force of magnitude $F$. According to Newton's third law, the rope pulls upward on the monkey with a force of the same magnitude, so Newton's second law for the monkey is $F - m_mg = m_ma_m$, where $m_m$ is the mass of the monkey and $a_m$ is its acceleration. Since the rope is massless $F$ is the tension in the rope. The rope pulls upward on the package with a force of magnitude $F$, so Newton's second law for the package is $F + F_N - m_pg = m_pa_p$, where $m_p$ is the mass of the package, $a_p$ is its acceleration, and $F_N$ is the normal force of the ground on it.

Now suppose $F$ is the minimum force required to lift the package. Then $F_N = 0$ and $a_p = 0$. According to the second law equation for the package, this means $F = m_pg$. Substitute $m_pg$ for $F$ in the second law equation for the monkey, then solve for $a_m$. You should obtain

$$a_m = \frac{F - m_mg}{m_m} = \frac{(m_p - m_m)g}{m_m} = \frac{(15\,\text{kg} - 10\,\text{kg})(9.8\,\text{m/s}^2)}{10\,\text{kg}} = 4.9\,\text{m/s}^2\,.$$

(b) Newton's second law equations are $F - m_p g = m_p a_p$ for the package and $F - m_m g = m_m a_m$ for the monkey. If the acceleration of the package is downward, then the acceleration of the monkey is upward, so $a_m = -a_p$. Solve the first equation for $F$: $F = m_p(g + a_p) = m_p(g - a_m)$. Substitute the result into the second equation and solve for $a_m$:

$$a_m = \frac{(m_p - m_m)g}{m_p + m_m} = \frac{(15\,\text{kg} - 10\,\text{kg})(9.8\,\text{m/s}^2)}{15\,\text{kg} + 10\,\text{kg}} = 2.0\,\text{m/s}^2 .$$

(c) The result is positive, indicating that the acceleration of the monkey is upward.

(d) Solve the second law equation for the package to obtain

$$F = m_p(g - a_m) = (15\,\text{kg})(9.8\,\text{m/s}^2 - 2.0\,\text{m/s}^2) = 120\,\text{N} .$$

## 61

The forces on the balloon are the force of gravity $m\vec{g}$, down, and the force of the air $\vec{F}_a$, up. Take the positive direction to be up. When the mass is $M$ (before the ballast is thrown out) the acceleration is downward and Newton's second law is $F_a - Mg = -Ma$. After the ballast is thrown out the mass is $M - m$, where $m$ is the mass of the ballast, and the acceleration is upward. Newton's second law is $F_a - (M - m)g = (M - m)a$. The first equation gives $F_a = M(g - a)$ and the second gives $M(g - a) - (M - m)g = (M - m)a$. Solve for $m$: $m = 2Ma/(g + a)$.

## 73

Take the $x$ axis to be horizontal and positive in the direction that the crate slides. Then $F\cos\theta - f = ma_x$, where $m$ is the mass of the crate and $a_x$ is the $x$ component of its acceleration (the only nonvanishing component). In part (a) the acceleration is

$$a_x = \frac{F\cos\theta - f}{m} = \frac{(450\,\text{N})\cos 38° - 125\,\text{N}}{310\,\text{kg}} = 0.74\,\text{m/s}^2 .$$

In part (b) $m = W/g = (310\,\text{N})/(9.8\,\text{m/s}^2) = 31.6\,\text{kg}$ and

$$a_x = \frac{F\cos\theta - f}{m} = \frac{(450\,\text{N})\cos 38° - 125\,\text{N}}{36.1\,\text{kg}} = 7.3\,\text{m/s}^2 .$$

## 79

Let $F$ be the magnitude of the force, $a_1$ ($= 12.0\,\text{m/s}^2$) be the acceleration of object 1, and $a_2$ ($= 3.30\,\text{m/s}^2$) be the acceleration of object 2. According to Newton's second law the masses are $m_1 = F/a_1$ and $m_2 = F/a_2$.

(a) The acceleration of an object of mass $m_2 - m_1$ is

$$a = \frac{F}{m_2 - m_1} = \frac{F}{(F/a_2) - (F/a_1)} = \frac{a_1 a_2}{a_1 - a_2} = \frac{(12.0\,\text{m/s}^2)(3.30\,\text{m/s}^2)}{12.0\,\text{m/s}^2 - 3.30\,\text{m/s}^2} = 4.6\,\text{m/s}^2 .$$

(b) The acceleration of an object of mass $m_1 + m_2$ is

$$a = \frac{F}{m_2 + m_1} = \frac{F}{(F/a_2) + (F/a_1)} = \frac{a_1 a_2}{a_1 + a_2} = \frac{(12.0\,\text{m/s}^2)(3.30\,\text{m/s}^2)}{12.0\,\text{m/s}^2 + 3.30\,\text{m/s}^2} = 2.6\,\text{m/s}^2\,.$$

**91**

(a) Both pieces are stationary, so you know that the net force on each of them is zero. The forces on the bottom piece are the downward force of gravity, with magnitude $m_2 g$, and the upward tension force of the bottom cord, with magnitude $T_b$. Since the net force is zero,

$$T_b = m_2 g = (4.5\,\text{kg})(9.8\,\text{m/s}^2) = 44\,\text{N}\,.$$

(b) The forces on the top piece are the downward force of gravity, with magnitude $m_1 g$, the downward tension force of the bottom cord, with magnitude $T_b$, and the upward force tension of the top cord, with magnitude $T_t$. Since the net force is zero,

$$T_t = T_b + m_1 g = 44\,\text{N} + (3.5\,\text{kg})(9.8\,\text{m/s}^2) = 78\,\text{N}\,.$$

(c) The forces on the bottom piece are the downward force of gravity, with magnitude $m_5 g$, and the upward tension force of the middle cord, with magnitude $T_m$. Since the net force is zero,

$$T_m = m_5 g = (5.5\,\text{kg})(9.8\,\text{m/s}^2) = 54\,\text{N}\,.$$

(d) The forces on the top piece are the downward force of gravity, with magnitude $m_3 g$, the upward tension force of the top cord, with magnitude $T_t$ (= 199 N), and the downward tension force of the middle cord, with magnitude $T_m$. Since the net force is zero,

$$T_m = T_t - m_3 g = 199\,\text{N} - (4.8\,\text{kg})(9.8\,\text{m/s}^2) = 152\,\text{N}\,.$$

**95**

(a) According to Newton's second law the magnitude of the net force on the rider is $F = ma = (60.0\,\text{kg})(3.0\,\text{m/s}^2) = 1.80 \times 10^2\,\text{N}$.

(b) Take the net force to be the vector sum of the force of the motorcycle and the force of Earth: $\vec{F}_{\text{net}} = \vec{F}_m + \vec{F}_E$. Thus $\vec{F}_m = \vec{F}_{\text{net}} - \vec{F}_E$. Now the net force is parallel to the ramp and therefore makes the angle $\theta$ (= 10°) with the horizontal, so $\vec{F}_{\text{net}} = (F\cos\theta)\hat{\text{i}} + (F\sin\theta)\hat{\text{j}}$, where the $x$ axis is taken to be horizontal and the $y$ axis is taken to be vertical. The force of Earth is $\vec{F}_E = -mg\hat{\text{j}}$, so $\vec{F}_m = (F\cos\theta)\hat{\text{i}} + (F\sin\theta + mg)\hat{\text{j}}$. Thus

$$F_{mx} = F\cos\theta = (1.80 \times 10^2\,\text{N})\cos 10° = 1.77 \times 10^2\,\text{N}$$

and

$$F_{my} = (1.80 \times 10^2)\sin 10° + (60.0\,\text{kg})(9.8\,\text{m/s}^2) = 6.19 \times 10^2\,\text{N}\,.$$

The magnitude of the force of the motorcycle is

$$F_m = \sqrt{F_{mx}^2 + F_{my}^2} = \sqrt{(1.77 \times 10^2\,\text{N})^2 + (6.19 \times 10^2\,\text{N})^2} = 6.44 \times 10^2\,\text{N}\,.$$

## 99

The free-body diagrams for the two boxes are shown below.

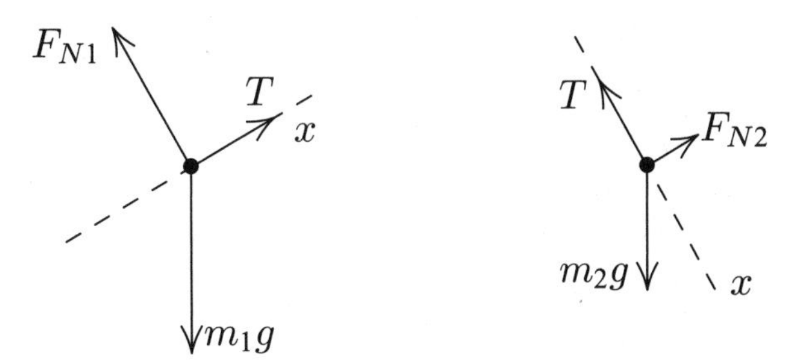


Here $T$ is the tension in the cord, $F_{N1}$ is the normal force of the left incline on box 1, and $F_{N2}$ is the normal force of the right incline on box 2. Different coordinate system are used for the two boxes but the positive $x$ direction are chosen so that the accelerations of the boxes have the same sign. The $x$ component of Newton's second law for box 1 gives $T - m_1 g \sin\theta_1 = m_1 a$ and the $x$ component of the law for box 2 gives $m_2 g \sin\theta_2 - T = m_2 g$. These equations are solved simultaneously for $T$. The result is

$$T = \frac{m_1 m_2 g}{m_1 + m_2}(\sin\theta_1 + \sin\theta_2) = \frac{(3.0\,\text{kg})(2.0\,\text{kg})(9.8\,\text{m/s}^2)}{3.0\,\text{kg} + 2.0\,\text{kg}}(\sin 30^\circ + \sin 60^\circ) = 16\,\text{N}\,.$$

## 101

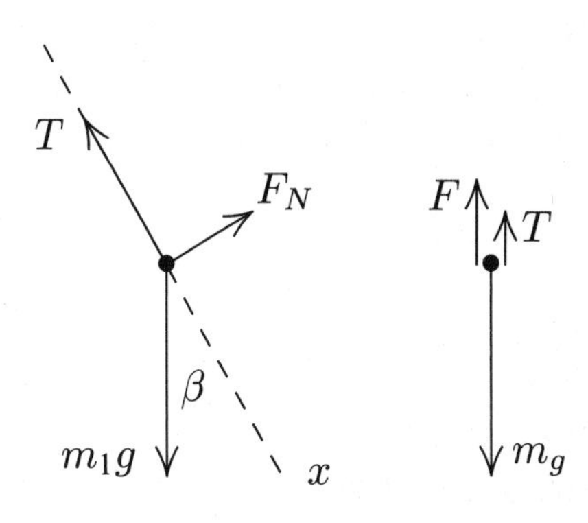


Free-body diagrams for the two tins are shown on the right. $T$ is the tension in the cord and $F_N$ is the normal force of the incline on tin 1. The positive $x$ direction for tin 1 is chosen to be down the incline and the positive $x$ direction for tin 2 is chosen to be downward. The sign of the accelerations of the two tins are both then positive. Newton's second law for tin 1 gives $T + m_1 g \sin\beta = ma$ and for tin 2 gives $m_2 g - T - F = m_2 a$. The second equation is solved for $T$, with the result $T = m_2(g - a) - F = (2.0\,\text{kg})(9.8\,\text{m/s}^2 - 5.5\,\text{m/s}^2) - 6.0\,\text{N} = 2.6\,\text{N}$. The first Newton's law equation is solved for $\sin\beta$, with the result

$$\sin\beta = \frac{m_1 a - T}{m_1 g} = \frac{(1.0\,\text{kg})(5.5\,\text{m/s}^2) - 2.6\,\text{N}}{(1.0\,\text{kg})(9.8\,\text{m/s})} = 0.296\,.$$

The angle is 17°.

# Chapter 6

**1**

(a) The free-body diagram for the bureau is shown on the right. $\vec{F}$ is the applied force, $\vec{f}$ is the force of friction, $\vec{F}_N$ is the normal force of the floor, and $m\vec{g}$ is the force of gravity. Take the $x$ axis to be horizontal and the $y$ axis to be vertical. Assume the bureau does not move and write the Newton's second law equations. The $x$ component is $F - f = 0$ and the $y$ component is $F_N - mg = 0$. The force of friction is then equal in magnitude to the applied force: $f = F$. The normal force is equal in magnitude to the force of gravity: $F_N = mg$. As $F$ increases, $f$ increases until $f = \mu_s F_N$. Then the bureau starts to move. The minimum force that must be applied to start the bureau moving is

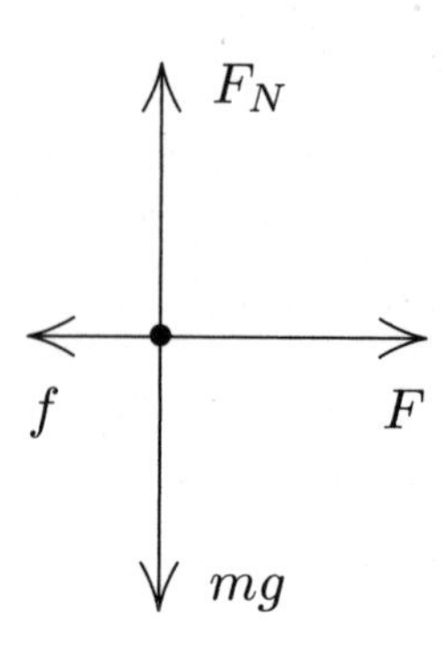


$$F = \mu_s F_N = \mu_s mg = (0.45)(45\,\text{kg})(9.8\,\text{m/s}^2) = 2.0 \times 10^2\,\text{N}.$$

(b) The equation for $F$ is the same but the mass is now $45\,\text{kg} - 17\,\text{kg} = 28\,\text{kg}$. Thus

$$F = \mu_s mg = (0.45)(28\,\text{kg})(9.8\,\text{m/s}^2) = 1.2 \times 10^2\,\text{N}.$$

**3**

(a) The free-body diagram for the crate is shown on the right. $\vec{F}$ is the force of the person on the crate, $\vec{f}$ is the force of friction, $\vec{F}_N$ is the normal force of the floor, and $m\vec{g}$ is the force of gravity. The magnitude of the force of friction is given by $f = \mu_k F_N$, where $\mu_k$ is the coefficient of kinetic friction. The vertical component of Newton's second law is used to find the normal force. Since the vertical component of the acceleration is zero, $F_N - mg = 0$ and $F_N = mg$. Thus

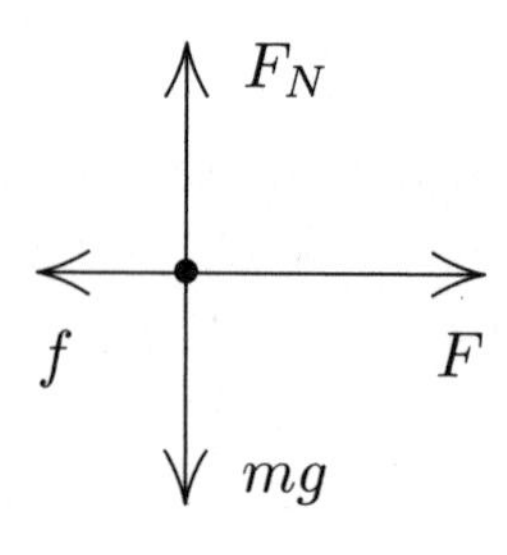


$$f = \mu_k F_N = \mu_k mg = (0.35)(55\,\text{kg})(9.8\,\text{m/s}^2) = 1.9 \times 10^2\,\text{N}.$$

(b) Use the horizontal component of Newton's second law to find the acceleration. Since $F - f = ma$,

$$a = \frac{(F - f)}{m} = \frac{(220\,\text{N} - 189\,\text{N})}{55\,\text{kg}} = 0.56\,\text{m/s}^2.$$

**13**

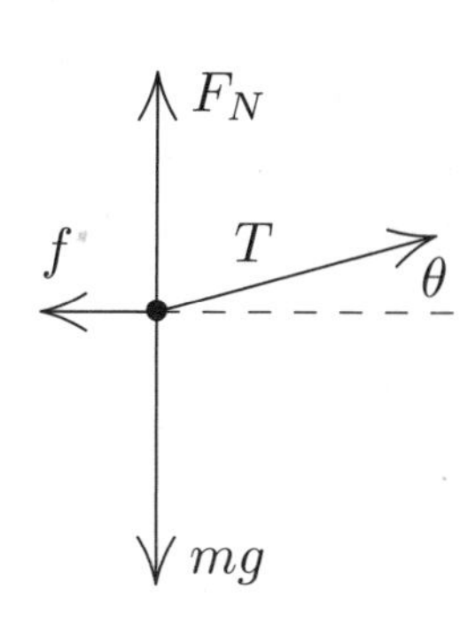


(a) The free-body diagram for the crate is shown on the right. $\vec{T}$ is the tension force of the rope on the crate, $\vec{F}_N$ is the normal force of the floor on the crate, $m\vec{g}$ is the force of gravity, and $\vec{f}$ is the force of friction. Take the $x$ axis to be horizontal on the right and the $y$ axis to be vertically upward. Assume the crate is motionless. The $x$ component of Newton's second law is then $T\cos\theta - f = 0$ and the $y$ component is $T\sin\theta + F_N - mg = 0$, where $\theta$ (= 15°) is the angle between the rope and the horizontal. The first equation gives $f = T\cos\theta$ and the second gives $F_N = mg - T\sin\theta$. If the crate is to remain at rest, $f$ must be less than $\mu_s F_N$, or $T\cos\theta < \mu_s(mg - T\sin\theta)$. When the tension force is sufficient to just start the crate moving $T\cos\theta = \mu_s(mg - T\sin\theta)$. Solve for $T$:

$$T = \frac{\mu_s mg}{\cos\theta + \mu_s \sin\theta} = \frac{(0.50)(68\,\text{kg})(9.8\,\text{m/s}^2)}{\cos 15° + 0.50\sin 15°} = 3.0 \times 10^2\,\text{N}\,.$$

(b) The second law equations for the moving crate are $T\cos\theta - f = ma$ and $F_N + T\sin\theta - mg = 0$. Now $f = \mu_k F_N$. The second equation gives $F_N = mg - T\sin\theta$, as before, so $f = \mu_k(mg - T\sin\theta)$. This expression is substituted for $f$ in the first equation to obtain $T\cos\theta - \mu_k(mg - T\sin\theta) = ma$, so the acceleration is

$$a = \frac{T(\cos\theta + \mu_k \sin\theta)}{m} - \mu_k g\,.$$

Its numerical value is

$$a = \frac{(304\,\text{N})(\cos 15° + 0.35\sin 15°)}{68\,\text{kg}} - (0.35)(9.8\,\text{m/s}^2) = 1.3\,\text{m/s}^2\,.$$

**23**

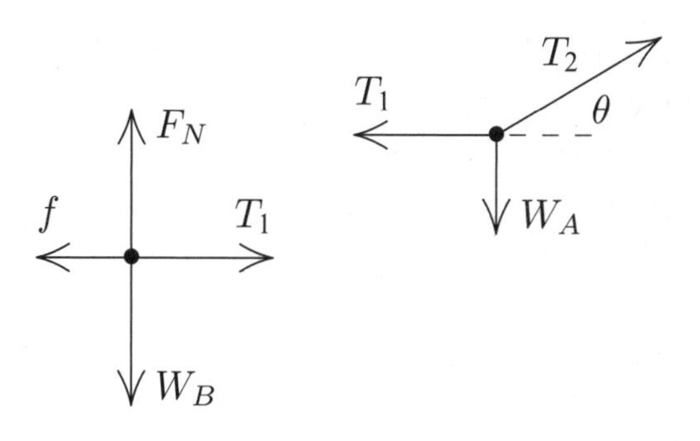


The free-body diagrams for block $B$ and for the knot just above block $A$ are shown on the right. $T_1$ is the magnitude of the tension force of the rope pulling on block $B$, $T_2$ is the magnitude of the tension force of the other rope, $f$ is the magnitude of the force of friction exerted by the horizontal surface on block $B$, $F_N$ is the magnitude of the normal force exerted by the surface on block $B$, $W_A$ is the weight of block $A$, and $W_B$ is the weight of block $B$. $\theta$ (= 30°) is the angle between the second rope and the horizontal.

For each object take the $x$ axis to be horizontal and the $y$ axis to be vertical. The $x$ component of Newton's second law for block $B$ is then $T_1 - f = 0$ and the $y$ component is $F_N - W_B = 0$. The $x$ component of Newton's second law for the knot is $T_2\cos\theta - T_1 = 0$ and the $y$ component is $T_2\sin\theta - W_A = 0$. Eliminate the tension forces and find expressions for $f$ and $F_N$ in terms of $W_A$ and $W_B$, then select $W_A$ so $f = \mu_s F_N$. The second Newton's law equation gives $F_N = W_B$ immediately. The third gives $T_2 = T_1/\cos\theta$. Substitute this expression into the fourth equation to obtain $T_1 = W_A/\tan\theta$. Substitute $W_A/\tan\theta$ for $T_1$ in the first equation to obtain $f = W_A/\tan\theta$. For the blocks to remain stationary $f$ must be less than $\mu_s F_N$ or $W_A/\tan\theta < \mu_s W_B$. The greatest that $W_A$ can be is the value for which $W_A/\tan\theta = \mu_s W_B$. Solve for $W_A$:

$$W_A = \mu_s W_B \tan\theta = (0.25)(711\,\text{N})\tan 30° = 1.0 \times 10^2\,\text{N}\,.$$

<u>27</u>

(a) The free-body diagrams for the two blocks are shown on the right. $T$ is the magnitude of the tension force of the string, $F_{NA}$ is the magnitude of the normal force on block $A$, $F_{NB}$ is the magnitude of the normal force on block $B$, $f_A$ is the magnitude of the friction force on block $A$, $f_B$ is the magnitude of the friction force on block $B$, $m_A$ is the mass of block $A$, and $m_B$ is the mass of block $B$. $\theta$ is the angle of the incline (30°). We have assumed that the incline goes down from right to left and that block $A$ is leading. It is the 3.6-N block.

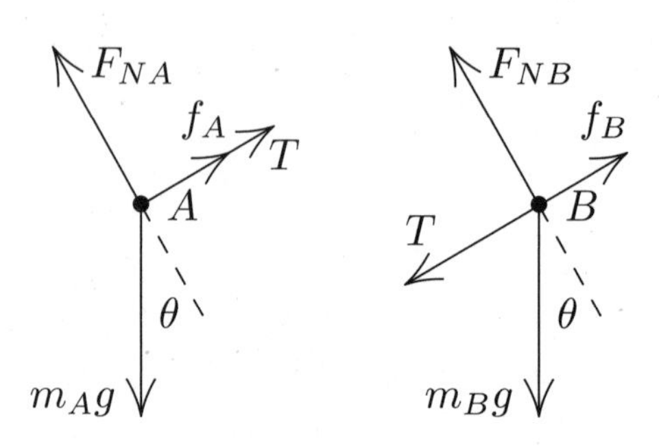


For each block take the $x$ axis to be down the plane and the $y$ axis to be in the direction of the normal force. For block $A$ the $x$ component of Newton's second law is

$$m_A g \sin\theta - f_A - T = m_A a_A$$

and the $y$ component is

$$F_{NA} - m_A g \cos\theta = 0\,.$$

Here $a_A$ is the acceleration of the block. The magnitude of the frictional force is

$$f_A = \mu_{kA} F_{NA} = \mu_{kA} m_A g \cos\theta\,,$$

where $F_{NA} = m_A g \cos\theta$, from the second equation, is substituted. $\mu_{kA}$ is the coefficient of kinetic friction for block $A$. When the expression for $f_A$ is substituted into the first equation the result is

$$m_A g \sin\theta - \mu_{kA} m_A g \cos\theta - T = m_A a_A\,.$$

The same analysis applied to block $B$ leads to

$$m_B g \sin\theta - \mu_{kB} m_B g \cos\theta + T = m_B a_B\,.$$

We must first find out if the rope is taut or slack. Assume the blocks are not joined by a rope and calculate the acceleration of each. If the acceleration of $A$ is greater than the acceleration of $B$, then the rope is taut when it is attached. If the acceleration of $B$ is greater than the acceleration of $A$, then even when the rope is attached $B$ gains speed at a greater rate than $A$ and the rope is slack.

Set $T = 0$ in the equation you derived above and solve for $a_A$ and $a_B$. The results are

$$a_A = g(\sin\theta - \mu_{kA}\cos\theta) = (9.8\,\text{m/s}^2)(\sin 30° - 0.10\cos 30°) = 4.05\,\text{m/s}^2$$

and

$$a_B = g(\sin\theta - \mu_{kB}\cos\theta) = (9.8\,\text{m/s}^2)(\sin 30° - 0.20\cos 30°) = 3.20\,\text{m/s}^2\,.$$

We have learned that when the blocks are joined, the rope is taut, the tension force is not zero, and the two blocks have the same acceleration.

Now go back to $m_A g\sin\theta - \mu_{kA}m_A g\cos\theta - T = m_A a$ and $m_B g\sin\theta - \mu_{kB}m_B g\cos\theta + T = m_B a$, where $a$ has been substituted for both $a_A$ and $a_B$. Solve the first expression for $T$, substitute the result into the second, and solve for $a$. The result is

$$\begin{aligned} a &= g\sin\theta - \frac{\mu_{kA}m_A + \mu_{kB}m_B}{m_A + m_B}g\cos\theta \\ &= (9.8\,\text{m/s}^2)\sin 30° - \left[\frac{(0.10)(3.6\,\text{N}) + (0.20)(7.2\,\text{N})}{3.6\,\text{N} + 7.2\,\text{N}}\right](9.8\,\text{m/s}^2)\cos 30° \\ &= 3.5\,\text{m/s}^2\,. \end{aligned}$$

Strictly speaking, values of the masses rather than weights should be substituted, but the factor $g$ cancels from the numerator and denominator.

(b) Use $m_A g\sin\theta - \mu_{kA}m_A g\cos\theta - T = m_A a$ to find the tension force of the rope:

$$\begin{aligned} T &= m_A g\sin\theta - \mu_{kA}m_A g\cos\theta - m_A a \\ &= (3.6\,\text{N})\sin 30° - (0.10)(3.6\,\text{N})\cos 30° - (3.6\,\text{N}/9.8\,\text{m/s}^2)(3.49\,\text{m/s}^2) = 0.21\,\text{N}\,. \end{aligned}$$

**35**

Let the magnitude of the frictional force be $\alpha v$, where $\alpha = 70\,\text{N}\cdot\text{s/m}$. Take the direction of the boat's motion to be positive. Newton's second law is then $-\alpha v = m\,dv/dt$. Thus

$$\int_{v_0}^{v}\frac{dv}{v} = -\frac{\alpha}{m}\int_0^t dt\,,$$

where $v_0$ is the velocity at time zero and $v$ is the velocity at time $t$. The integrals can be evaluated, with the result

$$\ln\frac{v}{v_0} = -\frac{\alpha t}{m}\,.$$

Take $v = v_0/2$ and solve for $t$:

$$t = \frac{m}{\alpha}\ln 2 = \frac{1000\,\text{kg}}{70\,\text{N}\cdot\text{s/m}}\ln 2 = 9.9\,\text{s}\,.$$

**49**

(a) At the highest point the seat pushes up on the student with a force of magnitude $F_N$ (= 556 N). Earth pulls down with a force of magnitude $W$ (= 667 N). The seat is pushing up with a force that is smaller than the student's weight in magnitude. The student feels light at the highest point.

(b) When the student is at the highest point, the net force toward the center of the circular orbit is $W - F_N$ and, according to Newton's second law, this must be $mv^2/R$, where $v$ is the speed of the student and $R$ is the radius of the orbit. Thus $mv^2/R = W - F_N = 667\,\text{N} - 556\,\text{N} = 111\,\text{N}$. The force of the seat when the student is at the lowest point is upward, so the net force toward the center of the circle is $F_N - W$ and $F_N - W = mv^2/R$. Solve for $F_N$:

$$F_N = \frac{mv^2}{R} + W = 111\,\text{N} + 667\,\text{N} = 778\,\text{N}\,.$$

(c) At the highest point $W - F_N = mv^2/R$, so $F_N = W - mv^2/R$. If the speed is doubled, $mv^2/R$ increases by a factor of 4, to 444 N. Then $F_N = 667\,\text{N} - 444\,\text{N} = 223\,\text{N}$.

(d) At the lowest point $W + F_N = mv^2/R$, so $F_N = W - mv^2/R$, so $F_N = W - mv^2/R$. Since $mv^2/R$ is still 444 N, $F_N = 667\,\text{N} + 444\,\text{N} = 1.11 \times 10^3\,\text{N}$.

**53**

The free-body diagram for the plane is shown on the right. $F$ is the magnitude of the lift on the wings and $m$ is the mass of the plane. Since the wings are tilted by 40° to the horizontal and the lift force is perpendicular to the wings, the angle $\theta$ is 50°. The center of the circular orbit is to the right of the plane, the dashed line along $x$ being a portion of the radius. Take the $x$ axis to be to the right and the $y$ axis to be upward. Then the $x$ component of Newton's second law is $F\cos\theta = mv^2/R$ and the $y$ component is $F\sin\theta - mg = 0$, where $R$ is the radius of the orbit. The first equation gives $F = mv^2/R\cos\theta$ and when this is substituted into the second, $(mv^2/R)\tan\theta = mg$ results. Solve for $R$:

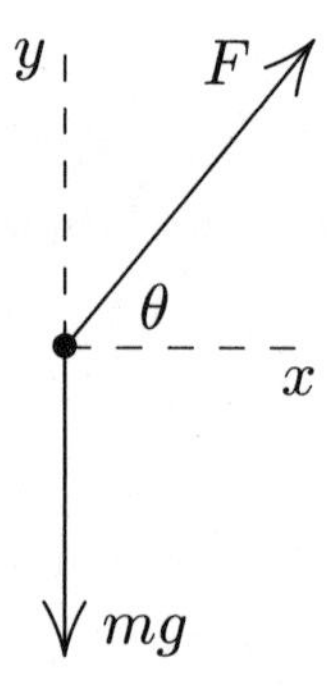


$$R = \frac{v^2}{g}\tan\theta\,.$$

The speed of the plane is $v = 480\,\text{km/h} = 133\,\text{m/s}$, so

$$R = \frac{(133\,\text{m/s})^2}{9.8\,\text{m/s}^2}\tan 50° = 2.2 \times 10^3\,\text{m}\,.$$

**59**

(a) The free-body diagram for the ball is shown on the right. $\vec{T}_u$ is the tension force of the upper string, $\vec{T}_\ell$ is the tension force of the lower string, and $m$ is the mass of the ball. Note that the tension force of the upper string is greater than the tension force of the lower string. It must balance the downward pull of gravity and the force of the lower string.

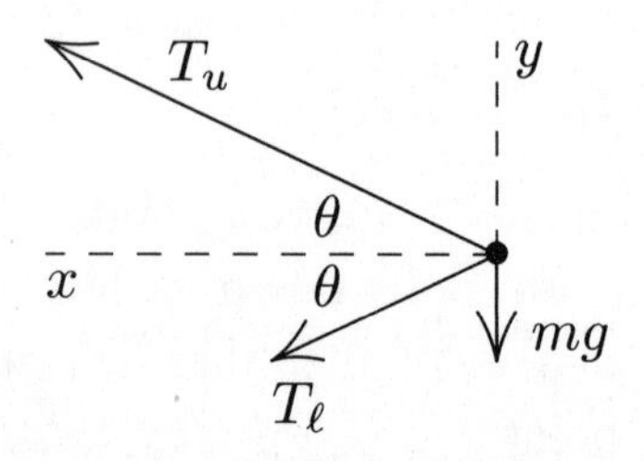

Take the $x$ axis to be to the left, toward the center of the circular orbit, and the $y$ axis to be upward. Since the magnitude of the acceleration is $a = v^2/R$, the $x$ component of Newton's second law is

$$T_u \cos\theta + T_\ell \cos\theta = \frac{mv^2}{R},$$

where $v$ is the speed of the ball and $R$ is the radius of its orbit. The $y$ component is

$$T_u \sin\theta - T_\ell \sin\theta - mg = 0.$$

The second equation gives the tension force of the lower string: $T_\ell = T_u - mg/\sin\theta$. Since the triangle is equilateral $\theta = 30°$. Thus

$$T_\ell = 35\,\text{N} - \frac{(1.34\,\text{kg})(9.8\,\text{m/s}^2)}{\sin 30°} = 8.74\,\text{N}.$$

(b) The net force is radially inward and has magnitude $F_{\text{net, str}} = (T_u + T_\ell)\cos\theta = (35\,\text{N} + 8.74\,\text{N})\cos 30° = 37.9\,\text{N}$.

(c) Use $F_{\text{net, str}} = mv^2/R$. The radius of the orbit is $[(1.70\,\text{m})/2)]\tan 30° = 1.47\,\text{m}$. Thus

$$v = \sqrt{\frac{RF_{\text{net, str}}}{m}} = \sqrt{\frac{(1.47\,\text{m})(37.9\,\text{N})}{1.34\,\text{kg}}} = 6.45\,\text{m/s}.$$

**65**

The first sentence of the problem statement tells us that the maximum force of static friction between the two block is $f_{s,\,\text{max}} = 12\,\text{N}$.

When the force $\vec{F}$ is applied the only horizontal force on the upper block is the frictional force of the lower block, which has magnitude $f$ and is in the forward direction. According to Newton's third law the upper block exerts a force of magnitude $f$ on the lower block and this force is in the rearward direction. The net force on the lower block is $F - f$.

Since the blocks move together their accelerations are the same. Newton's second law for the upper block gives $f = m_t a$ and the second law for the lower block gives $F - f = m_b a$, where $a$ is the common acceleration. The first equation gives $a = f/m_t$. Use this to substitute for $a$ in the second equation and obtain $F - f = (m_b/m_t)f$. Thus

$$F = \left(1 + \frac{m_b}{m_t}\right) f.$$

If $f$ has its maximum value then $F$ has its maximum value, so the maximum force that can be applied with the block moving together is

$$F = \left(1 + \frac{5.0\,\text{kg}}{4.0\,\text{kg}}\right)(12\,\text{N}) = 27\,\text{N}.$$

The acceleration is then

$$a = \frac{f}{m_t} = \frac{12\,\text{N}}{4.0\,\text{N}} = 3.0\,\text{m/s}^2.$$

**77**

(a) The force of friction is the only horizontal force on the bicycle and provides the centripetal force need for the bicycle to round the circle. The magnitude of this force is $f = mv^2/r$, where $m$ is the mass of the bicycle and rider together, $v$ is the speed of the bicycle, and $r$ is the radius of the circle. Thus

$$f = \frac{(85.0\,\text{kg})(9.00\,\text{m/s})^2}{25.0\,\text{m}} = 275\,\text{N}\,.$$

(b) In addition to the frictional force the road also pushes up with a normal force that is equal in magnitude to the weight of the bicycle and rider together. The magnitude of this force is $F_N = mg = (85.0\,\text{kg})(9.8\,\text{m/s}^2) = 833\,\text{N}$. The frictional and normal forces are perpendicular to each other, so the magnitude of the net force of the road on the bicycle is $F_{\text{net}} = \sqrt{f^2 + F_N^2} = \sqrt{(275\,\text{N})^2 + (833\,\text{N})^2} = 877\,\text{N}$.

**81**

The free-body diagrams are shown on the right. $T$ is the tension in the cord, $\vec{F}_{NA}$ is the normal force of the incline on block A, $\vec{F}_{NB}$ is the normal force of the platform on block B, $\theta$ is the angle that the incline makes with the horizontal (which is also the angle between the normal force and the vertical), and $\vec{f}$ is the frictional force of the platform on block B. The $x$ axis for each block is also shown.

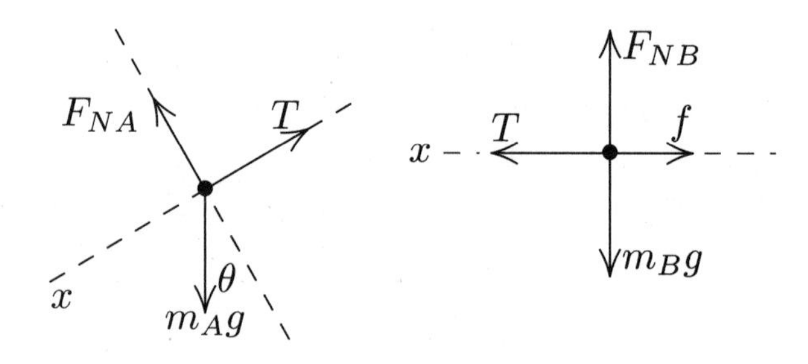


The $x$ component of Newton's second law for block A gives $mg\sin\theta - T = m_A a$, the $x$ component of the second law for block B is $T - f = m_B a$, and the $y$ component gives $F_{NB} - m_B g = 0$. Note that the blocks have the same acceleration.

The magnitude of the frictional force is $\mu_k F_{NB} = \mu_k m_B g$, where $m_B g$ was substituted for $F_{NB}$, and the $x$ component for B becomes $T - \mu_k m_B g = m_B a$. The equations $mg\sin\theta - T = m_A a$ and $T - \mu_k m_B g = m_B a$ are solved simultaneously for $T$ and $a$. The results are

$$T = \frac{m_A m_B(\sin\theta + \mu_k)}{m_A + m_B} = \frac{(4.0\,\text{kg})(2.0\,\text{kg})(\sin 30° + 0.50)}{4.0\,\text{kg} + 2.0\,\text{kg}} = 13\,\text{N}$$

and

$$a = \frac{M_A\sin\theta - \mu_k m_B}{m_A + m_B} g = \frac{(4.0\,\text{kg})\sin 30° - (0.50)(2.0\,\text{kg})}{4.0\,\text{kg} + 2.0\,\text{kg}} = 1.6\,\text{m/s}^2\,.$$

**85**

(a) If $v$ is the speed of the car, $m$ is its mass, and $r$ is the radius of the curve, then the magnitude of the frictional force on the tires of the car must be $f = mv^2/r$ or else the car does not negotiate the curve. Since $m = W/g$, where $W$ is the weight of the car,

$$f = \frac{Wv^2}{gr} = \frac{(10.7 \times 10^3\,\text{N})(13.4\,\text{m/s})^2}{(9.8\,\text{m/s}^2)(61.0\,\text{m})} = 3.21 \times 10^3\,\text{N}\,.$$

(b) The normal force of the road on the car is $F_N = W$ and the maximum possible force of static friction is $f_{s,\,\max} = \mu_s F_N = \mu_s W = (0.350)(10.7 \times 10^3\,\text{N}) = 3.75 \times 10^3\,\text{N}$. Since the frictional force that is required is less than the maximum possible, the car successfully rounds the curve.

**91**

Let $F$ be the magnitude of the applied force and $f$ be the magnitude of the frictional force. Assume the cabinet does not move. Then its acceleration is zero and, according to Newton's second law, $F = f$. The normal force is $F_N = W$, where $W$ is the weight of the cabinet. The maximum force of static friction is $F_{s,\,\max} = \mu_s F_N = \mu_s W = (0.68)(556\,\text{N}) = 378\,\text{N}$. Thus, if $F$ is less than 378 N the cabinet does not move and the frictional force is $f = F$. If $F$ is greater than 378 N, then the cabinet does move and the frictional force is $f = \mu_k F_N = \mu_k W = (0.56)(556\,\text{N}) = 311\,\text{N}$.

(a) The cabinet does not move and $f = 222\,\text{N}$.

(b) The cabinet does not move and $f = 334\,\text{N}$.

(c) The cabinet does move and $f = 311\,\text{N}$.

(d) The cabinet does move and $f = 311\,\text{N}$.

(e) The cabinet moves in attempts (c) and (d).

**99**

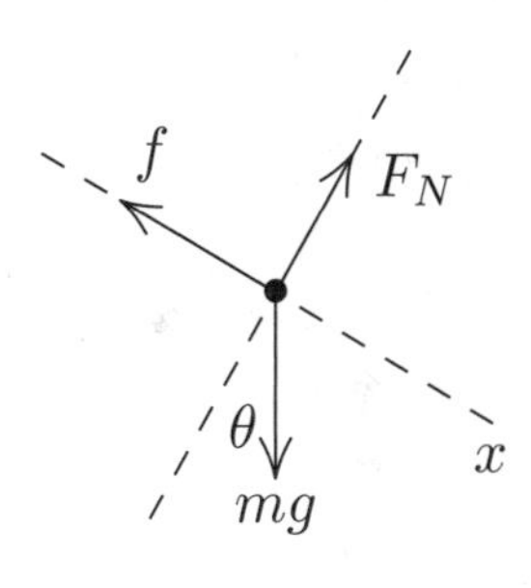


(a) The free-body diagram for the block is shown on the right. The magnitude of the frictional force is denoted by $f$, the magnitude of the normal force is denoted by $F_N$, and the angle between the incline and the horizontal is denoted by $\theta$. Since the block is sliding down the incline the frictional force is up the incline. The positive $x$ direction is taken to be down the incline. For the block when it is sliding with constant velocity the $x$ component of Newton's second law gives $mg\sin\theta - f = 0$ and the $y$ component gives $mg\cos\theta - F_N = 0$. The second equation gives $F_N = mg\cos\theta$, so the magnitude of the frictional force is $f = \mu_k F_N = \mu_k mg\cos\theta$. Use this to substitute for $f$ in the $x$-component equation and obtain $mg\sin\theta - mu_k mg\cos\theta = 0$. Thus the coefficient of kinetic friction is $\mu_k = \tan\theta$.

When the block is sliding up the incline the frictional force has the same magnitude but is directed down the plane. The $x$ component of the second law equation becomes $mg\sin\theta + \mu_k mg\cos\theta = ma$, where $a$ is the acceleration of the block. Thus $a = (\sin\theta + \mu_k\cos\theta)g = 2g\sin\theta$, where $\tan\theta$ was substituted for $\mu_k$ and $\tan\theta = \sin\theta/\cos\theta$ was used.

If $d$ is the displacement of an object with constant acceleration $a$, $v_0$ is its initial speed and $v$ is its final speed, then $v^2 - v_0^2 = 2ad$. Set $v$ equal to zero and $a$ equal to $2g\sin\theta$ and obtain $d = -v_0^2/2a = -v_0^2/4g\sin\theta$. The negative sign indicates that the displacement is up the plane.

(b) Since the coefficient of static friction is greater than the coefficient of kinetic friction the maximum possible static frictional force is greater than the actual frictional force and the block remains at rest once it stops.

**105**

The box is subjected to two horizontal forces: the applied force of the worker, with magnitude $F$, and the frictional force, with magnitude $f$. Newton's second law gives $F - f = ma$, where $m$ is the mass of the box and $a$ is the magnitude of its acceleration. The magnitude of the frictional force is $f = \mu_k F_N$, where $\mu_k$ is the coefficient of kinetic friction and $F_N$ is the magnitude of the normal force of the floor. In this case $F_N = mg$ and $f = \mu_k mg$. The second law equation becomes $F - \mu_k mg = ma$, so $\mu_k = (F - ma)/mg$.

Let $v$ be the final speed of the box and $d$ be the distance it moves. Then $v^2 = 2ad$ and $a = v^2/2d = (1.0\,\text{m/s})^2/2(1.4\,\text{m}) = 0.357\,\text{m/s}^2$. The coefficient of kinetic friction is

$$\mu_k = \frac{F - ma}{mg} = \frac{(85\,\text{N}) - (40\,\text{kg})(0.357\,\text{m/s}^2)}{(40\,\text{kg})(9.8\,\text{m/s}^2)} = 0.18\,.$$

# Chapter 7

**3**

(a) Use Eq. 2–16: $v^2 = v_0^2 + 2ax$, where $v_0$ is the initial velocity, $v$ is the final velocity, $x$ is the displacement, and $a$ is the acceleration. This equation yields

$$v = \sqrt{v_0^2 + 2ax} = \sqrt{(2.4 \times 10^7\,\text{m/s})^2 + 2(3.6 \times 10^{15}\,\text{m/s}^2)(0.035\,\text{m})} = 2.9 \times 10^7\,\text{m/s}\,.$$

(b) The initial kinetic energy is

$$K_i = \tfrac{1}{2}mv_0^2 = \tfrac{1}{2}(1.67 \times 10^{-27}\,\text{kg})(2.4 \times 10^7\,\text{m/s})^2 = 4.8 \times 10^{-13}\,\text{J}\,.$$

The final kinetic energy is

$$K_f = \tfrac{1}{2}mv^2 = \tfrac{1}{2}(1.67 \times 10^{-27}\,\text{kg})(2.9 \times 10^7\,\text{m/s})^2 = 6.9 \times 10^{-13}\,\text{J}\,.$$

The change in kinetic energy is $\Delta K = 6.9 \times 10^{-13}\,\text{J} - 4.8 \times 10^{-13}\,\text{J} = 2.1 \times 10^{-13}\,\text{J}$.

**17**

(a) Let $F$ be the magnitude of the force exerted by the cable on the astronaut. The force of the cable is upward and the force of gravity is $mg$ is downward. Furthermore, the acceleration of the astronaut is $g/10$, upward. According to Newton's second law, $F - mg = mg/10$, so $F = 11mg/10$. Since the force $\vec{F}$ and the displacement $\vec{d}$ are in the same direction the work done by $\vec{F}$ is

$$W_F = Fd = \frac{11mgd}{10} = \frac{11(72\,\text{kg})(9.8\,\text{m/s}^2)(15\,\text{m})}{10} = 1.16 \times 10^4\,\text{J}\,.$$

(b) The force of gravity has magnitude $mg$ and is opposite in direction to the displacement. Since $\cos 180° = -1$, it does work

$$W_g = -mgd = -(72\,\text{kg})(9.8\,\text{m/s}^2)(15\,\text{m}) = -1.06 \times 10^4\,\text{J}\,.$$

(c) The net work done is $W = 1.16 \times 10^4\,\text{J} - 1.06 \times 10^4\,\text{J} = 1.1 \times 10^3\,\text{J}$. Since the astronaut started from rest the work-kinetic energy theorem tells us that this must be her final kinetic energy.

(d) Since $K = \frac{1}{2}mv^2$ her final speed is

$$v = \sqrt{\frac{2K}{m}} = \sqrt{\frac{2(1.1 \times 10^3\,\text{J})}{72\,\text{kg}}} = 5.3\,\text{m/s}\,.$$

**19**

(a) Let $F$ be the magnitude of the force of the cord on the block. This force is upward, while the force of gravity, with magnitude $Mg$, is downward. The acceleration is $g/4$, down. Take the downward direction to be positive. Then Newton's second law is $Mg - F = Mg/4$, so $F = 3Mg/4$. The force is directed opposite to the displacement, so the work it does is $W_F = -Fd = -3Mgd/4$.

(b) The force of gravity is in the same direction as the displacement, so it does work $W_g = Mgd$.

(c) The net work done on the block is $W_T = -3Mgd/4 + Mgd = Mgd/4$. Since the block starts from rest this is its kinetic energy $K$ after it is lowered a distance $d$.

(d) Since $K = \frac{1}{2}Mv^2$, where $v$ is the speed,

$$v = \sqrt{\frac{2K}{M}} = \sqrt{\frac{gd}{2}}$$

after the block is lowered a distance $d$. The result found in (c) was used.

**29**

(a) As the body moves along the $x$ axis from $x_i = 3.0\,\mathrm{m}$ to $x_f = 4.0\,\mathrm{m}$ the work done by the force is

$$W = \int_{x_i}^{x_f} F_x\,dx = \int_{x_i}^{x_f} -6x\,dx = -3x^2\Big|_{x_i}^{x_f} = -3(x_f^2 - x_i^2)$$
$$= -3\left[(4.0)^2 - (3.0)^2\right] = -21\,\mathrm{J}\,.$$

According to the work-kinetic energy theorem, this is the change in the kinetic energy:

$$W = \Delta K = \tfrac{1}{2}m(v_f^2 - v_i^2)\,,$$

where $v_i$ is the initial velocity (at $x_i$) and $v_f$ is the final velocity (at $x_f$). The theorem yields

$$v_f = \sqrt{\frac{2W}{m} + v_i^2} = \sqrt{\frac{2(-21\,\mathrm{J})}{2.0\,\mathrm{kg}} + (8.0\,\mathrm{m/s})^2} = 6.6\,\mathrm{m/s}\,.$$

(b) The velocity of the particle is $v_f = 5.0\,\mathrm{m/s}$ when it is at $x = x_f$. Solve the work-kinetic energy theorem for $x_f$. The net work done on the particle is $W = -3(x_f^2 - x_i^2)$, so the work-kinetic energy theorem yields $-3(x_f^2 - x_i^2) = \frac{1}{2}m(v_f^2 - v_i^2)$. Thus

$$x_f = \sqrt{-\frac{m}{6}(v_f^2 - v_i^2) + x_i^2} = \sqrt{-\frac{2.0\,\mathrm{kg}}{6\,\mathrm{N/m}}\left[(5.0\,\mathrm{m/s})^2 - (8.0\,\mathrm{m/s})^2\right] + (3.0\,\mathrm{m})^2} = 4.7\,\mathrm{m}\,.$$

**35**

(a) The graph shows $F$ as a function of $x$ if $x_0$ is positive. The work is negative as the object moves from $x = 0$ to $x = x_0$ and positive as it moves from $x = x_0$ to $x = 2x_0$. Since the area of a triangle is $\frac{1}{2}$(base)(altitude), the work done from $x = 0$ to $x = x_0$ is $-\frac{1}{2}(x_0)(F_0)$ and the work done from $x = x_0$ to $x = 2x_0$ is $\frac{1}{2}(2x_0 - x_0)(F_0) = \frac{1}{2}(x_0)(F_0)$. The net work is the sum, which is zero.

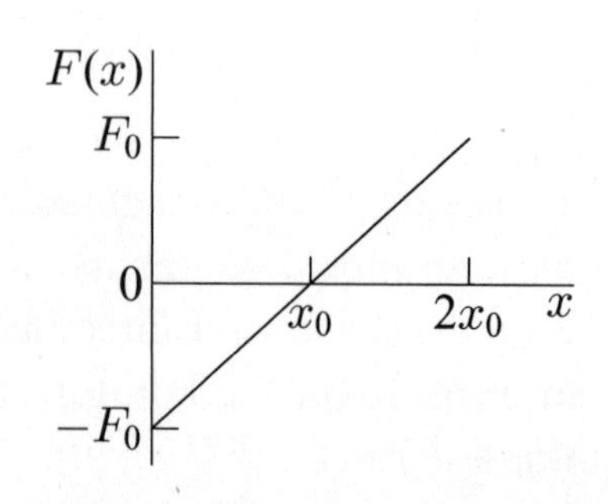

(b) The integral for the work is

$$W = \int_0^{2x_0} F_0 \left( \frac{x}{x_0} - 1 \right) dx = F_0 \left( \frac{x^2}{2x_0} - x \right) \Bigg|_0^{2x_0} = 0\,.$$

**43**

The power associated with force $\vec{F}$ is given by $P = \vec{F} \cdot \vec{v}$, where $\vec{v}$ is the velocity of the object on which the force acts. Let $\phi$ (= 37°) be the angle between the force and the horizontal. Then $P = \vec{F} \cdot \vec{v} = Fv \cos\phi = (122\,\text{N})(5.0\,\text{m/s}) \cos 37° = 4.9 \times 10^2\,\text{W}$.

**45**

(a) The power is given by $P = Fv$ and the work done by $\vec{F}$ from time $t_1$ to time $t_2$ is given by

$$W = \int_{t_1}^{t_2} P\,\text{d}t = \int_{t_1}^{t_2} Fv\,\text{d}t\,.$$

Since $\vec{F}$ is the net force, the magnitude of the acceleration is $a = F/m$ and, since the initial velocity is $v_0 = 0$, the velocity as a function of time is given by $v = v_0 + at = (F/m)t$. Thus

$$W = \int_{t_1}^{t_2} (F^2/m)t\,\text{d}t = \frac{1}{2}(F^2/m)(t_2^2 - t_1^2)\,.$$

For $t_1 = 0$ and $t_2 = 1.0\,\text{s}$,

$$W = \frac{1}{2} \left[ \frac{(5.0\,\text{N})^2}{15\,\text{kg}} \right] (1.0\,\text{s})^2 = 0.83\,\text{J}\,.$$

(b) For $t_1 = 1.0\,\text{s}$ and $t_2 = 2.0\,\text{s}$,

$$W = \frac{1}{2} \left[ \frac{(5.0\,\text{N})^2}{15\,\text{kg}} \right] \left[ (2.0\,\text{s})^2 - (1.0\,\text{s})^2 \right] = 2.5\,\text{J}\,.$$

(c) For $t_1 = 2.0\,\text{s}$ and $t_2 = 3.0\,\text{s}$,

$$W = \frac{1}{2} \left[ \frac{(5.0\,\text{N})^2}{15\,\text{kg}} \right] \left[ (3.0\,\text{s})^2 - (2.0\,\text{s})^2 \right] = 4.2\,\text{J}\,.$$

(d) Substitute $v = (F/m)t$ into $P = Fv$ to obtain $P = F^2 t/m$ for the power at any time $t$. At the end of the third second $P = (5.0\,\text{N})^2(3.0\,\text{s})/15\,\text{kg} = 5.0\,\text{W}$.

**47**

The net work $W_{\text{net}}$ is the sum of the work $W_e$ done by gravity on the elevator, the work $W_c$ done by gravity on the counterweight, and the work $W_s$ done by the motor on the system: $W_{\text{net}} = W_e + W_c + W_s$. Since the elevator moves at constant velocity, its kinetic energy does not change and according to the work-kinetic energy theorem the net work done is zero. This means $W_e + W_c + W_s = 0$. The elevator moves upward through 54 m, so the work done by gravity on it is $W_e = -m_e g d = -(1200\,\text{kg})(9.8\,\text{m/s}^2)(54\,\text{m}) = -6.35 \times 10^5\,\text{J}$. The counterweight moves downward the same distance, so the work done by gravity on it is $W_c = m_c g d = (950\,\text{kg})(9.8\,\text{m/s}^2)(54\,\text{m}) = 5.03 \times 10^5\,\text{J}$. Since $W_T = 0$, the work done by the motor on the system is $W_s = -W_e - W_c = 6.35 \times 10^5\,\text{J} - 5.03 \times 10^5\,\text{J} = 1.32 \times 10^5\,\text{J}$. This work is done in a time interval of $\Delta t = 3.0\,\text{min} = 180\,\text{s}$, so the power supplied by the motor to lift the elevator is

$$P = \frac{W_s}{\Delta t} = \frac{1.32 \times 10^5\,\text{J}}{180\,\text{s}} = 7.35 \times 10^2\,\text{W}\,.$$

**63**

(a) Take the positive $x$ direction to be in the direction of travel of the cart. In time $\Delta t$ the cart moves a distance $\Delta x = v\,\Delta t$, where $v$ is its speed. The work done is $W = F_x\,\Delta x = (Fv\cos\theta)\,\Delta t$, where $\vec{F}$ is the force of the horse and $\theta$ is the angle it makes with the horizontal. Now $6.0\,\text{mi/h} = (6.0\,\text{mi/h})(1.467\,(\text{ft/s})/(\text{mi/h}) = 8.8\,\text{ft/s}$ and $10\,\text{min} = (10\,\text{min})(60\,\text{s/min}) = 600\,\text{s}$, so $F = [(40\,\text{lb})(8.8\,\text{ft/s})\cos 30°](600\,\text{s}) = 1.8 \times 10^5\,\text{ft}\cdot\text{lb}$.

(b) The average power is $P = F_x v = Fv\cos\theta = (40\,\text{lb})(8.8\,\text{ft/s})\cos 30° = 3.0 \times 10^2\,\text{ft}\cdot\text{lb/s}$. Since $1\,\text{ft}\cdot\text{lb/s} = 1.818 \times 10^{-3}\,\text{hp}$, the power is $P = (3.0 \times 10^2\,\text{ft}\cdot\text{lb/s})(1.818 \times 10^{-3}\,\text{hp/ft}\cdot\text{lb/s}) = 0.55\,\text{hp}$.

**69**

(a) The applied force $\vec{F}$ is in the direction of the displacement $\vec{d}$, so the work done by the force is $W_F = Fd = (209\,\text{N})(1.50\,\text{m}) = 314\,\text{J}$.

(b) The crate rises is distance $\Delta y = d\sin\theta$, where $\theta$ is the angle that the incline makes with the horizontal. The work done by the gravitational force of Earth is $W_g = -mgd\sin\theta = -(25.0\,\text{kg})(9.8\,\text{m/s}^2)(1.50\,\text{m})\sin 25.0° = -155\,\text{J}$.

(c) The normal force of the incline on the crate is perpendicular to the displacement of the crate, so this force does no work.

(d) The net work done on the crate is $W_{\text{net}} = W_F + W_g = 314\,\text{J} - 155\,\text{J} = 159\,\text{J}$.

**71**

Let $W_1$ (= 110 N) be the first weight hung on the scale and $x$ be the elongation of the spring with this weight on it. Let $W_2$ (= 240 N) be the second weight hung on the scale and $x_2$ be the elongation of the spring when this weight is hung on it. In each case the spring pulls upward with a force that is equal to the weight hung on it, so according to Hooke's law $W_1 = kx_1$ and $W_2 = kx_2$, where $k$ is the spring constant. Now $x_1$ and $x_2$ are not the readings on the scales but $x_2 - x_1$ is the difference of the scale readings. Subtract the two Hooke's law equations to obtain $W_2 - W_1 = k(x_2 - x_1)$. Thus

$$k = \frac{W_2 - W_1}{x_2 - x_1} = \frac{240\,\text{N} - 110\,\text{N}}{60 \times 10^{-3}\,\text{m} - 40 \times 10^{-3}\,\text{m}} = 6.5 \times 10^3\,\text{N/m}\,.$$

When $W_1$ is hung on the spring the elongation is $x_1 = W_1/k = (110\,\text{N})/(6.5 \times 10^3\,\text{N/m}) = 1.7 \times 10^{-2}\,\text{m} = 17\,\text{mm}$. The reading on the scale is $40\,\text{mm} - 17\,\text{mm} = 23\,\text{mm}$.

(b) When the third weight is hung from the spring the elongation of the spring is $x = 30\,\text{mm} - 23\,\text{mm} = 7.0\,\text{mm}$. The weight is $W = kx = (6.5 \times 10^3\,\text{N/m})(7.0 \times 10{-}3\,\text{m}) = 45\,\text{N}$.

**73**

The elevator is moving upward with constant velocity, so the force $F$ that is moving it must be equal in magnitude to the total weight of the elevator and load. That is, $F = W_{\text{total}} = M_{\text{total}}g$, where $M_{\text{total}}$ is the total mass. The power required is $P = Fv$, where $v$ is the speed of the elevator. Thus $P = M_{\text{total}}gv = (4500\,\text{kg} + 1800\,\text{kg})(9.8\,\text{m/s}^2)(3.80\,\text{m/s}) = 2.35 \times 10^5\,\text{W}$.

**77**

(a)) Since the wind is steady the acceleration of the lunchbox is constant and $x = v_0 t + \frac{1}{2}at^2$, where $v_0$ is the velocity at time zero and $a$ is the acceleration. According to the graph the coordinate is about 0.40 m at time $t = 0.50\,\text{s}$, so $v_0 = (0.40\,\text{m})/(0.50\,\text{s}) = 0.80\,\text{m/s}$. The kinetic energy at $t = 0$ is $K_0 = \frac{1}{2}mv_0^2 = \frac{1}{2}(2\,\text{kg})(0.80\,\text{m/s}^2) = 0.64\,\text{J}$.

(b) At $t = 5.0\,\text{s}$ the velocity is zero, so the kinetic energy is zero.

(c) According to the work-kinetic energy theorem the work done by the wind force is the change in the kinetic energy, which is $-0.64\,\text{J}$.

# Chapter 8

3

(a) The force of gravity is constant, so the work it does is given by $W = \vec{F} \cdot \vec{d}$, where $\vec{F}$ is the force and $\vec{d}$ is the displacement. The force is vertically downward and has magnitude $mg$, where $m$ is the mass of the flake, so this reduces to $W = mgh$, where $h$ is the height from which the flake falls. This is equal to the radius $r$ of the bowl. Thus

$$W = mgr = (2.00 \times 10^{-3}\,\text{kg})(9.8\,\text{m/s}^2)(22.0 \times 10^{-2}\,\text{m}) = 4.31 \times 10^{-3}\,\text{J}\,.$$

(b) The force of gravity is conservative, so the change in gravitational potential energy of the flake-Earth system is the negative of the work done: $\Delta U = -W = -4.31 \times 10^{-3}$ J.

(c) The potential energy when the flake is at the top is greater than when it is at the bottom by $|\Delta U|$. If $U = 0$ at the bottom, then $U = +4.31 \times 10^{-3}$ J at the top.

(d) If $U = 0$ at the top, then $U = -4.31 \times 10^{-3}$ J at the bottom.

(e) All the answers are proportional to the mass of the flake. If the mass is doubled, all answers are doubled.

5

The potential energy stored by the spring is given by $U = \frac{1}{2}kx^2$, where $k$ is the spring constant and $x$ is the displacement of the end of the spring from its position when the spring is in equilibrium. Thus

$$k = \frac{2U}{x^2} = \frac{2(25\,\text{J})}{(0.075\,\text{m})^2} = 8.9 \times 10^3\,\text{N/m}\,.$$

9

(a) Neglect any work done by the force of friction and by air resistance. Then the only force that does work is the force of gravity, a conservative force. Let $K_i$ be the kinetic energy of the truck at the bottom of the ramp and let $K_f$ be its kinetic energy at the top. Let $U_i$ be the gravitational potential energy of the truck-Earth system when the truck is at the bottom and let $U_f$ be the gravitational potential energy when it is at the top. Then $K_f + U_f = K_i + U_i$. If the potential energy is taken to be zero when the truck is at the bottom, then $U_f = mgh$, where $h$ is the final height of the truck above its initial position. $K_i = \frac{1}{2}mv^2$, where $v$ is the initial speed of the truck, and $K_f = 0$ since the truck comes to rest. Thus $mgh = \frac{1}{2}mv^2$ and $h = v^2/2g$. Substitute $v = 130\,\text{km/h} = 36.1\,\text{m/s}$ to obtain

$$h = \frac{(36.1\,\text{m/s})^2}{2(9.8\,\text{m/s}^2)} = 66.5\,\text{m}\,.$$

If $L$ is the length of the ramp, then $L \sin 15° = 66.5$ m or $L = (66.5\,\text{m})/\sin 15° = 257$ m.

The truck is not a particle-like object since its wheels turn and the cylinders of its motor move. However, if there is no frictional force between the tires and the roadway, these moving parts have no influence on the rate with which the truck slows. If there is friction, then when the driver takes his foot off the gas pedal the tires exert a forward frictional force on the road and the road exerts a backward frictional force of the same magnitude on the truck. This, along with air resistance, helps slow the truck. The frictional force is greater if the driver shifts to a lower gear.

(b) The answers do not depend on the mass of the truck. They remain the same if the mass is reduced.

(c) If the speed is decreased $h$ and $L$ both decrease. In fact, $h$ is proportional to the square of the speed. If $v$ is half its former value, then $h$ is one-fourth its former value.

**11**

(a) The only force that does work as the flake falls is the force of gravity and it is a conservative force. If $K_i$ is the kinetic energy of the flake at the edge of the bowl, $K_f$ is its kinetic energy at the bottom, $U_i$ is the gravitational potential energy of the flake-Earth system with the flake at the top, and $U_f$ is the gravitational potential energy with it at the bottom, then $K_f + U_f = K_i + U_i$. Take the potential energy to be zero at the bottom of the bowl. Then the potential energy at the top is $U_i = mgr$, where $r$ is the radius of the bowl and $m$ is the mass of the flake. $K_i = 0$ since the flake starts from rest. Since the problem asks for the speed at the bottom, write $\frac{1}{2}mv^2$ for $K_f$. The energy conservation equation becomes $mgr = \frac{1}{2}mv^2$, so

$$v = \sqrt{2gr} = \sqrt{2(9.8\,\text{m/s}^2)(0.220\,\text{m})} = 2.08\,\text{m/s}\,.$$

(b) Note that the expression for the speed ($v = \sqrt{2gr}$) does not contain the mass of the flake. The speed would be the same, 2.08 m/s, regardless of the mass of the flake.

(c) The final kinetic energy is given by $K_f = K_i + U_i - U_f$. Since $K_i$ is greater than before, $K_f$ is greater. This means the final speed of the flake is greater.

**15**

(a) Take the gravitational potential energy of the marble-Earth system to be zero at the position of the marble when the spring is compressed. The gravitational potential energy when the marble is at the top of its flight is then $U_g = mgh$, where $h$ is the height of the highest point. This is $h = 20\,\text{m}$. Thus

$$U_g = (5.0 \times 10^{-3}\,\text{kg})(9.8\,\text{m/s}^2)(20\,\text{m}) = 0.98\,\text{J}\,.$$

(b) Before firing the marble is at rest and is again at rest at the top of its trajectory. Both the force of the spring and the force of gravity, the only two forces acting, are conservative. Conservation of mechanical energy is expressed as $\Delta U_g + \Delta U_s = 0$, where $U_g$ is the gravitational potential energy and $U_s$ is the spring potential energy. This means $\Delta U_s = -\Delta U_g = -0.98\,\text{J}$.

(c) Take the spring potential energy to be zero when the spring has its equilibrium length. Then its initial potential energy is $U_s = 0.98\,\text{J}$. This must be $\frac{1}{2}kx^2$, where $k$ is the spring constant and

$x$ is the initial compression. Solve for $k$:

$$k = \frac{2U_s}{x^2} = \frac{2(0.98\,\text{J})}{(0.080\,\text{m})^2} = 3.1 \times 10^2\,\text{N/m}\,.$$

**31**

Information given in the second sentence allows us to compute the spring constant. Solve $F = kx$ for $k$:

$$k = \frac{F}{x} = \frac{270\,\text{N}}{0.02\,\text{m}} = 1.35 \times 10^4\,\text{N/m}\,.$$

(a) Now consider the block sliding down the incline. If it starts from rest at a height $h$ above the point where it momentarily comes to rest, its initial kinetic energy is zero and the initial gravitational potential energy of the block-Earth system is $mgh$, where $m$ is the mass of the block. We have taken the zero of gravitational potential energy to be at the point where the block comes to rest. We also take the initial potential energy stored in the spring to be zero. Suppose the block compresses the spring a distance $x$ before coming momentarily to rest. Then the final kinetic energy is zero, the final gravitational potential energy is zero, and the final spring potential energy is $\frac{1}{2}kx^2$. The incline is frictionless and the normal force it exerts on the block does no work, so mechanical energy is conserved. This means $mgh = \frac{1}{2}kx^2$, so

$$h = \frac{kx^2}{2mg} = \frac{(1.35 \times 10^4\,\text{N/m})(0.055\,\text{m})^2}{2(12\,\text{kg})(9.8\,\text{m/s}^2)} = 0.174\,\text{m}\,.$$

If the block traveled down a length of incline equal to $\ell$, then $\ell \sin 30° = h$, so $\ell = h/\sin 30° = (0.174\,\text{m})/\sin 30° = 0.35\,\text{m}$.

(b) Just before it touches the spring it is 0.055 m away from the place where it comes to rest and so is a vertical distance $h' = (0.055\,\text{m}) \sin 30° = 0.0275\,\text{m}$ above its final position. The gravitational potential energy is then

$$mgh' = (12\,\text{kg})(9.8\,\text{m/s}^2)(0.0275\,\text{m}) = 3.23\,\text{J}\,.$$

On the other hand, its initial potential energy is

$$mgh = (12\,\text{kg})(9.8\,\text{m/s}^2)(0.174\,\text{m}) = 20.5\,\text{J}\,.$$

The difference is its final kinetic energy: $K_f = 20.5\,\text{J} - 3.23\,\text{J} = 17.2\,\text{J}$. Its final speed is

$$v = \sqrt{\frac{2K_f}{m}} = \sqrt{\frac{2(17.2\,\text{J})}{12\,\text{kg}}} = 1.7\,\text{m/s}\,.$$

**45**

(a) The force exerted by the rope is constant, so the work it does is $W = \vec{F} \cdot \vec{d}$, where $\vec{F}$ is the force and $\vec{d}$ is the displacement. Thus

$$W = Fd \cos\theta = (7.68\,\text{N})(4.06\,\text{m}) \cos 15.0° = 30.1\,\text{J}\,.$$

(b) The increase in thermal energy is $\Delta E_{\text{th}} = fd = (7.42\,\text{N})(4.06\,\text{m}) = 30.1\,\text{J}$.

(c) We can use Newton's second law of motion to obtain the frictional and normal forces, then use $\mu_k = f/N$ to obtain the coefficient of friction. Place the $x$ axis along the path of the block and the $y$ axis normal to the floor. The $x$ component of Newton's second law is $F\cos\theta - f = 0$ and the $y$ component is $F_N + F\sin\theta - mg = 0$, where $m$ is the mass of the block, $F_N$ is the normal force of the floor, $F$ is the force exerted by the rope, and $\theta$ is the angle between that force and the horizontal. The first equation gives

$$f = F\cos\theta = (7.68\,\text{N})\cos 15.0^\circ = 7.42\,\text{N}$$

and the second gives

$$F_N = mg - F\sin\theta = (3.57\,\text{kg})(9.8\,\text{m/s}^2) - (7.68\,\text{N})\sin 15.0^\circ = 33.0\,\text{N}\,.$$

Thus $\mu_k = f/F_N = (7.42\,\text{N})/(33.0\,\text{N}) = 0.225$.

**47**

(a) Take the initial gravitational potential energy to be $U_i = 0$. Then the final gravitational potential energy is $U_f = -mgL$, where $L$ is the length of the tree. The change is $U_f - U_i = -mgL = -(25\,\text{kg})(9.8\,\text{m/s}^2)(12\,\text{m}) = -2.9 \times 10^3\,\text{J}$.

(b) The kinetic energy is $K = \frac{1}{2}mv^2 = \frac{1}{2}(25\,\text{kg})(5.6\,\text{m/s})^2 = 3.9 \times 10^2\,\text{J}$.

(c) The changes in the mechanical and thermal energies must sum to zero. Since the change in thermal energy is $\Delta E_{\text{th}} = fL$, where $f$ is the magnitude of the average frictional force,

$$f = -\frac{\Delta K + \Delta U}{L} = -\frac{3.9 \times 10^2\,\text{J} - 2.9 \times 10^3\,\text{J}}{12\,\text{m}} = 2.1 \times 10^2\,\text{N}\,.$$

**69**

The change in the potential energy of the block-Earth system as the block goes from A to B is the same for the two cases and, since mechanical energy is conserved, the change in the kinetic energy of the block is the same. The change in the kinetic energy is

$$\Delta K = \frac{1}{2}m(v_B^2 - v_A^2) = \frac{1}{2}m[(2.60\,\text{m/s})^2 - (2.00\,\text{m/s})^2] = (1.38\,\text{J/kg})m\,.$$

For the second trial

$$K_B = K_A + \Delta K = \frac{1}{2}m(4.00\,\text{m/s})^2 + (1.38\,\text{J/kg})m = (9.38\,\text{J/kg})m$$

and the speed at B is

$$v_B = \sqrt{2K_B/m} = \sqrt{2(9.38\,\text{J/kg})} = 4.33\,\text{m/s}\,.$$

**75**

Since the blocks start from rest and the mechanical energy of the system consisting of the blocks and Earth is conserved, the final kinetic energy is the negative of the change in potential energy.

If block B falls a distance $d$, block A moves a distance $d$ along the incline and rises a vertical distance $d \sin\theta$, where $\theta$ is the angle of the incline. Thus

$$K = -\Delta U = -\left[(-m_B g d) + (m_A g d \sin\theta)\right] = gd\left[m_B - m_A \sin\theta\right]$$
$$= (9.8\,\mathrm{m/s^2})(0.25\,\mathrm{m})\left[2.0\,\mathrm{kg} - (1.0\,\mathrm{kg})\sin 30°\right] = 3.7\,\mathrm{J}\,.$$

**83**

(a) Use conservation of mechanical energy. Let $K_i$ be the initial kinetic energy, $K_f$ be the final kinetic energy, and $\ell$ be the compression of the spring. Then the change in kinetic energy is $\Delta K = K_f - K_i$. The block travels the distance $d+\ell$ along the incline and the vertical component of its displacement has magnitude $(d+\ell)\sin\theta$, where $\theta$ is the angle of the incline. Thus the change in the potential energy is $\Delta U = mg(d+\ell)\sin\theta + \frac{1}{2}k\ell^2$, where $k$ is the spring constant. Since mechanical energy is conserved the final kinetic energy of the block is

$$K_f = K_i - mg(d+\ell)\sin\theta - \frac{1}{2}k\ell^2$$
$$= 16\,\mathrm{J} - (1.0\,\mathrm{kg})(9.8\,\mathrm{m/s^2})(0.60\,\mathrm{m} + 0.20\,\mathrm{m})\sin 40° - \frac{1}{2}(200\,\mathrm{N/m})(0.20\,\mathrm{m})^2 = 7.0\,\mathrm{J}\,.$$

(b) Now $K_f = 0$ and $K_i$ is the unknown. Conservation of mechanical energy gives

$$K_i = mg(d+\ell)\sin\theta + \frac{1}{2}k\ell^2$$
$$= (1.0\,\mathrm{kg})(9.8\,\mathrm{m/s^2})(0.60\,\mathrm{m} + 0.40\,\mathrm{m})\sin 40° + \frac{1}{2}(200\,\mathrm{N/m})(0.40\,\mathrm{m})^2 = 22\,\mathrm{J}\,.$$

**87**

Neither the kinetic energy or the potential energy changes, so conservation of energy tells us that the change in the total thermal energy is equal to the work done by the applied force: $\Delta E_{\mathrm{th}} = W$. If $F$ is the magnitude of the applied force and $d$ is the distance the cube travels, then $W = Fd$. The thermal energies of the cube and the floor both change, so $\Delta E_{\mathrm{th}} = \Delta E_{\mathrm{th,\,cube}} + \Delta E_{\mathrm{th,\,floor}}$ and

$$\Delta E_{\mathrm{th,\,floor}} = Fd - \Delta E_{\mathrm{th,\,cube}} = (15\,\mathrm{N})(3.0\,\mathrm{m}) - 20\,\mathrm{J} = 25\,\mathrm{J}\,.$$

**109**

(a) Take the potential energy of the ball-Earth system to be zero when the ball is at the bottom of its swing. Then the initial potential energy is $2mgL$, where $m$ is the mass of the ball and $L$ is length of the rod. The initial kinetic energy is zero since the ball is at rest. Write $\frac{1}{2}mv^2$, where $v$ is the speed of the ball, for the final kinetic energy, at the bottom of the swing. Since mechanical energy is conserved $2mgL = \frac{1}{2}mv^2$ and

$$v = 2\sqrt{gL} = 2\sqrt{(9.8\,\mathrm{m/s^2})(0.62\,\mathrm{m})} = 4.9\,\mathrm{m/s}\,.$$

(b) At the bottom of the swing the force of gravity is downward and the tension force of the rod is upward. If $T$ is the magnitude of the tension force, Newton's second law is $T - mg = mv^2/L$, so

$$T = mg + mv^2/L = mg + 4mg = 5mg = 5(0.092\,\text{kg})(9.8\,\text{m/s}^2) = 4.5\,\text{N}.$$

(c) The diagram on the right is the free-body diagram for the ball when the tension force of the rod has the same magnitude as the force of gravity. We wish to solve for $\theta$. The component of the force of gravity along the radial direction is $mg\cos\theta$ and is outward. The net inward force is $T - mg\cos\theta$ and, according to Newton's second law this must equal $mv^2/L$, where $v$ is the speed of the ball. Thus $T = mv^2/L + mg\cos\theta$.

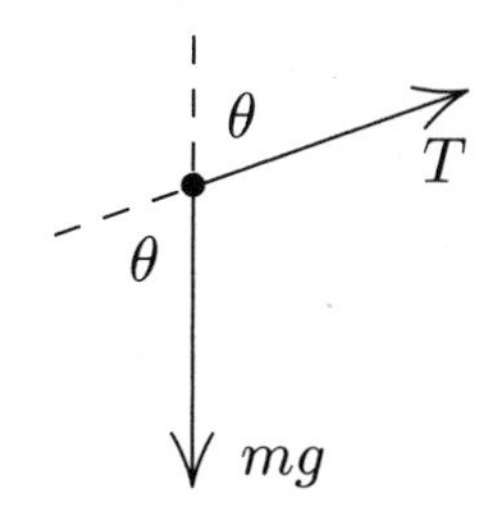


We now need to find the speed of the ball in terms of $\theta$. Take the potential energy to be zero when the rod is horizontal. Since it starts from rest its kinetic energy is also zero. As can be seen on the diagram on the right, when the rod makes the angle $\theta$ with the vertical, the ball has dropped through a vertical distance $L\cos\theta$. The potential energy is then $-mgL\cos\theta$. Write the kinetic energy as $\frac{1}{2}mv^2$ and the conservation of energy equation as $0 = -mgL\cos\theta + \frac{1}{2}mv^2$. Thus $mv^2 = 2mgL\cos\theta$. Substitute this expression into the equation developed above for $T$: $T = 2mg\cos\theta + mg\cos\theta = 3mg\cos\theta$. According to the condition of the problem, this must be equal to $mg$, so $3mg\cos\theta = mg$, or $\cos\theta = 1/3$. This means $\theta = 71°$.

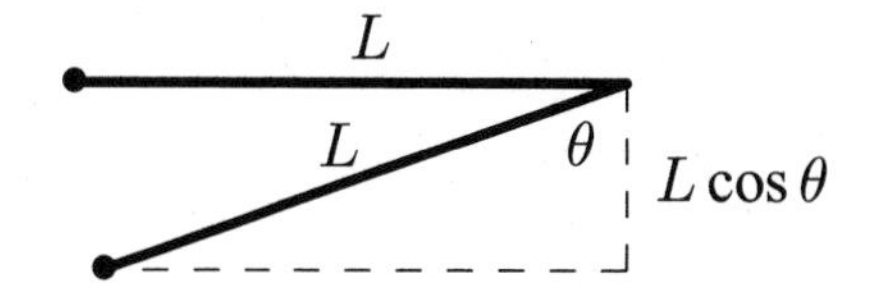


(d) Notice that the mass of the ball cancels from the equation for $\cos\theta$, so $\theta$ does not depend on the mass. The answer to (c) remains the same.

**111**

(a) At the top of its flight the velocity of the ball has only a horizontal component and this is the same as the horizontal component of the initial velocity. Let $v_0$ be the initial speed and $\theta_0$ be the angle with which the ball is thrown. Then the kinetic energy at the top of the flight is

$$K = \frac{1}{2}m(v_0\cos\theta_0)^2 = \frac{1}{2}(50\times 10^{-3}\,\text{kg})[(8.0\,\text{m/s})\cos 30°]^2 = 1.2\,\text{J}.$$

(b) Use conservation of mechanical energy. When the ball goes from the window to the point 3.0 m below, the potential energy changes by $\Delta U = -mgd$, where $d$ (= 3.0 m) is the distance from the window to the lower point. Since mechanical energy is conserved the change in the kinetic energy is $+mgd$. If $v_0$ is the initial speed and $v$ is the speed at the lower point, $\frac{1}{2}mv^2 = \frac{1}{2}mv_0^2 + mgd$ and

$$v = \sqrt{v_0^2 + 2gd} = \sqrt{(8.0\,\text{m/s})^2 + 2(9.8\,\text{m/s}^2)(3.0\,\text{m}} = 11\,\text{m/s}.$$

(c) and (d) Notice that the mass cancels from the conservation of energy equation and none of the quantities in that equation depend on the initial angle, so the answer to part (b) does not depend on the mass or the angle.

**119**

(a) After the cue loses contact with the disk all the kinetic energy of the disk is converted to thermal energy. Thus the increase in thermal energy of the disk and court is $\Delta E_{th} = \frac{1}{2}mv^2 = \frac{1}{2}(0.42\,\text{kg})(4.2\,\text{m/s})^2 = 3.7\,\text{J}$.

(b) The change in thermal energy is $\Delta E_{th} = fd$, where $f$ is the force of friction of the court on the disk and $d$ is the distance the disk travels. Thus $f = \Delta E_{th}/d = (3.7\,\text{J})/12\,\text{m}) = 0.31\,\text{N}$. Over the entire 14 m the increase in thermal energy is $\Delta E_{th} = (0.31\,\text{N})(14\,\text{N}) = 4.3\,\text{J}$.

(c) All of the energy transferred from the cue to the disk ends up as thermal energy so the work done by the cue is 4.3 J.

**121**

(a) and (b) The force is the negative of the slope of the curve. Take the potential energy to be $-2.8\,\text{J}$ when the particle is at $x = 1.0\,\text{m}$ and $-17.5\,\text{J}$ when the particle is at $x = 4.0\,\text{m}$. Then

$$F = -\frac{(-17.3\,\text{J}) - (-2.8\,\text{J})}{4.0\,\text{m} - 1.0\,\text{m}} = 4.8\,\text{N}\,.$$

The magnitude is 4.8 N and it is in the positive $x$ direction.

(c) and (d) When the particle is at $x = 2.0\,\text{m}$ the potential energy is about $U = 7.7\,\text{J}$ and the kinetic energy is $K = \frac{1}{2}mv^2 = \frac{1}{2}(2.0\,\text{kg})(-1.5\,\text{m/s})^2 = 2.2\,\text{J}$. The mechanical energy is $E_{mec} = K + U = 2.2\,\text{J} - 7.7\,\text{J} = -5.5\,\text{J}$. According to the graph, the potential energy is $-5.5\,\text{J}$ when the particle is at $x = 1.5\,\text{m}$ and $x = 13.5\,\text{m}$. The particle moves between these two coordinates.

(e) When the particle is at $x = 7.0\,\text{m}$ the potential energy is about $U = -17.5\,\text{J}$, so the kinetic energy is $K = E_{mec} - U = (-5.5\,\text{J}) - (-17.5\,\text{J}) = 12\,\text{J}$. The speed of the particle is

$$v = \sqrt{\frac{2K}{m}} = \sqrt{\frac{2(12\,\text{J})}{2.0\,\text{kg})}} = 3.5\,\text{m/s}\,.$$

**123**

(a) Let $d$ be the distance the car travels. Its vertical position lowers by $d \sin\theta$, where $\theta$ is the angle of the incline, so the potential energy changes by $-\Delta U = mgd \sin\theta$, where $m$ is the mass of the car. The kinetic energy changes by $\Delta K = \frac{1}{2}m(v_f^2 - v_i^2)$, where $v_i$ is the initial speed and $v_f$ is the final speed. The change in the mechanical energy is $\Delta E_{mec} = \Delta K + \Delta U = -mgd \sin\theta + \frac{1}{2}m(v_f^2 - v_i^2)$. Convert the given speeds to meters per second. They are $v_i = 8.33\,\text{m/s}$ and $v_f = 11.1\,\text{m/s}$, so

$$\begin{aligned}\Delta E_{mec} &= -(1500\,\text{kg})(9.8\,\text{m/s}^2)(50\,\text{m})\sin 5.0^\circ + \frac{1}{2}(1500\,\text{kg})\left[(11.1\,\text{m/s})^2 - (8.33\,\text{m/s})^2\right] \\ &= -2.4 \times 10^4\,\text{J}\,.\end{aligned}$$

The mechanical energy decreases by $2.4 \times 10^4$ J.

(b) The change in mechanical energy is given by $-fd$, where $f$ is the magnitude of the frictional force. Thus $f = -\Delta E_{\text{mec}}/d = -(-2.4 \times 10^4\,\text{J})/(50\,\text{m}) = 4.7 \times 10^2\,\text{N}$.

**127**

(a) The potential energy does not change, so the change in the mechanical energy is equal to the change in the kinetic energy. Let $m$ be the mass of the block, $v_0$ be its speed at the beginning of the acceleration period and $v$ be its speed at the end. Then $\Delta E_{\text{mec}} = \frac{1}{2}m\left[v^2 - v_0^2\right] = \frac{1}{2}(15\,\text{kg})\left[(30\,\text{m/s})^2 - (10\,\text{m/s})^2\right] = 6.0 \times 10^3\,\text{J}$.

(b) The average rate with which energy is transferred to the block is the total energy transferred divided by the time for the transferral. Since the acceleration $a$ is constant the time is given by $\Delta t = (\Delta v)/a = (30\,\text{m/s} - 10\,\text{m/s})/(2.0\,\text{m/s}^2 = 10\,\text{s}$ and the average rate of transferral is $\Delta E_{\text{mec}}/\Delta t = (6.0 \times 10^3\,\text{J})/(10\,\text{s}) = 6.0 \times 10^2\,\text{W}$.

(c) and (d) If the accelerating force has magnitude $F$, then the instantaneous rate of energy transfer is given by $P = Fv = mav$, where Newton's second law was used to substitute $ma$ for $F$. For $v = 10\,\text{m/s}$, $P = (15\,\text{kg})(2.0\,\text{m/s}^2)(10\,\text{m/s}) = 3.0 \times 10^2\,\text{W}$ and for $v = 30\,\text{m/s}$, $P = (15\,\text{kg})(2.0\,\text{m/s}^2)(30\,\text{m/s}) = 9.0 \times 10^2\,\text{W}$

**131**

The kinetic energy gained per unit time is equal to the potential energy lost per unit time. If mass $\Delta m$ of water passes over the falls the gain in kinetic energy is $\Delta K = \Delta m\,gh$, where $h$ is the height of the falls. The rate of production of electrical energy is

$$P = \frac{3}{4}\frac{\Delta m\,gh}{\Delta t} = \frac{3}{4}(1200\,\text{m}^3/\text{s})(1000\,\text{k/m}^3)(9.8\,\text{m/s}^2)(100\,\text{m}) = 8.8 \times 10^8\,\text{W}\,.$$

**133**

(a) When the ball is at D the potential energy is $mgL$ greater than when it is at A and the kinetic energy is $\frac{1}{2}mv_0^2$ less. Since mechanical energy is conserved, $-\frac{1}{2}mv_0^2 + mgL = 0$ and $v_0 = \sqrt{2gL}$.

(b) Let $T$ be the tension in the rod when the ball is at B and let $v$ be the speed of the ball then. Newton's second law gives $T - mg = mv^2/L$. Use conservation of mechanical energy to find $v^2$. When the ball is at B the potential energy is $mgL$ less than when it is at A and the kinetic energy is greater by $\frac{1}{2}m(v^2 - v_0^2)$, so $\frac{1}{2}m(v^2 - v_0^2) - mgL = 0$ and $v^2 = v_0^2 + 2gL = 4gL$, where $2gL$ was substituted for $v_0^2$. Thus $T = mg + mv^2/L = mg + 4mg = 5mg$.

(c) When the ball is at C the potential energy is the same as when it is at A and the kinetic energy is $\frac{1}{2}mv_0^2$ less. All of the kinetic energy is converted to thermal energy. The decrease in mechanical energy is $\frac{1}{2}mv_0^2 = \frac{1}{2}m2gL = mgL$, where $2gL$ was substituted for $v_0^2$.

(d) When the ball has settled at B the potential energy is $mgL$ less than when it started at A and the kinetic energy is $\frac{1}{2}mv_0^2 = mgL$ less. The mechanical energy has decreased by $2mgL$.

# Chapter 9

**15**

You need to find the coordinates of the point where the shell explodes and the velocity of the fragment that does not fall straight down. These become the initial conditions for a projectile motion problem to determine where it lands.

Consider first the motion of the shell from firing to the time of the explosion. Place the origin at the firing point, take the $x$ axis to be horizontal, and take the $y$ axis to be vertically upward. The $y$ component of the velocity is given by $v = v_{0y} - gt$ and this is zero at time $t = v_{0y}/g = (v_0/g)\sin\theta_0$, where $v_0$ is the initial speed and $\theta_0$ is the firing angle. The coordinates of the highest point on the trajectory are

$$x = v_{0x}t = v_0 t\cos\theta_0 = \frac{v_0^2}{g}\sin\theta_0\cos\theta_0 = \frac{(20\,\text{m/s})^2}{9.8\,\text{m/s}^2}\sin 60^\circ\cos 60^\circ = 17.7\,\text{m}$$

and

$$y = v_{0y}t - \frac{1}{2}gt^2 = \frac{1}{2}\frac{v_0^2}{g}\sin^2\theta_0 = \frac{1}{2}\frac{(20\,\text{m/s})^2}{9.8\,\text{m/s}^2}\sin^2 60^\circ = 15.3\,\text{m}\,.$$

Since no horizontal forces act, the horizontal component of the velocity of the center of mass is constant. At the highest point the velocity of the shell is $v_0\cos\theta_0$, in the positive $x$ direction. This is the velocity of the center of mass. Let $M$ be the mass of the shell and let $V_0$ be the velocity of the fragment that does not fall straight down. Then the velocity of the center of mass is given by $MV_0/2M = V_0/2$, since the masses of the fragments are the same. Since the velocity of the center of mass is constant, $v_0\cos\theta_0 = V_0/2$. This means

$$V_0 = 2v_0\cos\theta_0 = 2(20\,\text{m/s})\cos 60^\circ = 20\,\text{m/s}\,.$$

Now consider a projectile launched horizontally at time $t = 0$ with a speed of $20\,\text{m/s}$ from the point with coordinates $x_0 = 17.7\,\text{m}$, $y_0 = 15.3\,\text{m}$. Its $y$ coordinate is given by $y = y_0 - \frac{1}{2}gt^2$, and when it lands this is zero. The time of landing is $t = \sqrt{2y_0/g}$ and the $x$ coordinate of the landing point is

$$x = x_0 + V_0 t = x_0 + V_0\sqrt{\frac{2y_0}{g}} = 17.7\,\text{m} + (20\,\text{m/s})\sqrt{\frac{2(15.3\,\text{m})}{9.8\,\text{m/s}^2}} = 53\,\text{m}\,.$$

**23**

(a) Take the initial direction of motion to be positive and let $J$ be the magnitude of the impulse, $m$ be the mass of the ball, $v_i$ be the initial velocity of the ball, and $v_f$ be the final velocity of the ball. The impulse is in the negative $x$ direction and the impulse-momentum theorem yields $-J = mv_f - mv_i$. Solve for $v_f$ to obtain

$$v_f = \frac{mv_i - J}{m} = \frac{(0.40\,\text{kg})(14\,\text{m/s}) - 32.4\,\text{N}\cdot\text{s}}{0.40\,\text{kg}} = -67\,\text{m/s}\,.$$

The final speed of the ball is $67\,\text{m/s}$.

(b) The negative sign indicates that the direction of the velocity is opposite to the initial direction of travel. That is, it is in the negative $x$ direction.

(c) The magnitude of the average force is $F_{avg} = J/\Delta t = (32.4\,\text{N}\cdot\text{s})/(27 \times 10^{-3}\,\text{s} = 1.2 \times 10^3\,\text{N}$.

(d) The impulse is in the negative $x$ direction, the same as the force.

**35**

(a) Take the force to be in the positive direction, at least for earlier times. Then the impulse is

$$J = \int_0^{3.0\times10^{-3}} F\,\text{d}t = \int_0^{3.0\times10^{-3}} \left[(6.0 \times 10^6)t - (2.0 \times 10^9)t^2\right]\,\text{d}t$$

$$= \left[\frac{1}{2}(6.0 \times 10^6)t^2 - \frac{1}{3}(2.0 \times 10^9)t^3\right]_0^{3.0\times10^{-3}} = 9.0\,\text{N}\cdot\text{s}\,.$$

The impulse is in the positive direction.

(b) Since $J = F_{avg}\,\Delta t$, where $F_{avg}$ is the average force and $\Delta t$ is the duration of the kick,

$$F_{avg} = \frac{J}{\Delta t} = \frac{9.0\,\text{N}\cdot\text{s}}{3.0 \times 10^{-3}\,\text{s}} = 3.0 \times 10^3\,\text{N}\,.$$

(c) To find time at which the maximum force occurs set the derivative of $F$ with respect to time equal to zero and solve for $t$. The result is $t = 1.5 \times 10^{-3}$ s. At that time the force is

$$F_{max} = (6.0 \times 10^6)(1.5 \times 10^{-3}) - (2.0 \times 10^9)(1.5 \times 10^{-3})^2 = 4.5 \times 10^3\,\text{N}\,.$$

(d) During the kick the ball gains momentum equal to the impulse. Since it starts from rest, its momentum just after the player's foot loses contact is $p = J$. Let $m$ be the mass of the ball and $v$ be its speed as it leaves the foot. Then, since $v = p/m$,

$$v = \frac{J}{m} = \frac{9.0\,\text{N}\cdot\text{s}}{0.45\,\text{kg}} = 20\,\text{m/s}\,.$$

**39**

No external forces with horizontal components act on the man-stone system and the vertical forces sum to zero, so the total momentum of the system is conserved. Since the man and the stone are initially at rest the total momentum is zero both before and after the stone is kicked. Let $m_s$ be the mass of the stone and $v_s$ be its velocity after it is kicked; let $m_m$ be the mass of the man and $v_m$ be his velocity after he kicks the stone. Then $m_s v_s + m_m v_m = 0$ and $v_m = -m_s v_s/m_m$. Take the axis to be positive in the direction the stone travels. Then $v_m = -(0.068\,\text{kg})(4.0\,\text{m/s})/(91\,\text{kg}) = -3.0 \times 10^{-3}\,\text{m/s}$. The negative sign indicates that the man moves in the direction opposite to the direction of motion of the stone.

**47**

(a) Let $m$ be the mass and $v_i\,\hat{\text{i}}$ be the velocity of the body before the explosion. Let $m_1$, $m_2$, and $m_3$ be the masses of the fragments. (The mass of the third fragment is 6.00 kg.) Write

$v_1\,\hat{\text{j}}$ for the velocity of fragment 1, $-v_2\,\hat{\text{i}}$ for the velocity of fragment 2, and $v_{3x}\,\hat{\text{i}} + v_{3y}\,\hat{\text{j}}$ for the velocity of fragment 3. Since the original body and two of the fragments all move in the $xy$ plane the third fragment must also move in that plane. Conservation of linear momentum leads to $mv_i\,\hat{\text{i}} = m_1v_1\,\hat{\text{j}} - m_2v_2\,\hat{\text{i}} + m_3v_{3x}\,\hat{\text{i}} + m_3v_{3y}\,\hat{\text{j}}$, or $(mv_i + m_2v_2 - m_3v_{3x})\hat{\text{i}} - (m_1v_1 + m_3v_{3y})\hat{\text{j}} = 0$. The $x$ component of this equation gives

$$v_{3x} = \frac{mv_i + m_2v_2}{m_3} = \frac{(20.0\,\text{kg})(200\,\text{m/s}) + (4.00\,\text{kg})(500\,\text{m/s})}{6.0\,\text{kg}} = 1.00 \times 10^3\,\text{m/s}\,.$$

The $y$ component gives

$$v_{3y} = -\frac{m_1v_1}{m_3} = -\frac{(10.0\,\text{kg})(100\,\text{m/s})}{6.0\,\text{kg}} = -167\,\text{m/s}\,.$$

Thus $\vec{v}_3 = (1.00 \times 10^3\,\text{m/s})\hat{\text{i}} - (167\,\text{m/s})\hat{\text{j}}$. The velocity has a magnitude of $1.01 \times 10^3$ m/s and is 9.48° below the $x$ axis.

(b) The initial kinetic energy is

$$K_i = \frac{1}{2}mv_i^2 = \frac{1}{2}(20.0\,\text{kg})(200\,\text{m/s})^2 = 4.00 \times 10^5\,\text{J}\,.$$

The final kinetic energy is

$$\begin{aligned} K_f &= \frac{1}{2}m_1v_1^2 + \frac{1}{2}m_2v_2^2 + \frac{1}{2}m_3v_3^2 \\ &= \frac{1}{2}\left[(10.0\,\text{kg})(100\,\text{m/s})^2 + (4.00\,\text{kg})(500\,\text{m/s})^2 + (6.00\,\text{kg})(1014\,\text{m/s})^2\right] \\ &= 3.63 \times 10^6\,\text{J}\,. \end{aligned}$$

The energy released in the explosion is $3.63 \times 10^6\,\text{J} - 4.00 \times 10^5\,\text{J} = 3.23 \times 10^6\,\text{J}$.

**61**

(a) Let $m_1$ be the mass of the cart that is originally moving, $v_{1i}$ be its velocity before the collision, and $v_{1f}$ be its velocity after the collision. Let $m_2$ be the mass of the cart that is originally at rest and $v_{2f}$ be its velocity after the collision. Then, according to Eq. 9–67,

$$v_{1f} = \frac{m_1 - m_2}{m_1 + m_2}v_{1i}\,.$$

Solve for $m_2$ to obtain

$$m_2 = \frac{v_{1i} - v_{1f}}{v_{1i} + v_{1f}}m_1 = \left(\frac{1.2\,\text{m/s} - 0.66\,\text{m/s}}{1.2\,\text{m/s} + 0.66\,\text{m/s}}\right)(0.340\,\text{kg}) = 0.099\,\text{kg}\,.$$

(b) The velocity of the second cart is given by Eq. 9–68:

$$v_{2f} = \frac{2m_1}{m_1 + m_2}v_{1i} = \left[\frac{2(0.340\,\text{kg})}{0.340\,\text{kg} + 0.099\,\text{kg}}\right](1.2\,\text{m/s}) = 1.9\,\text{m/s}\,.$$

(c) The speed of the center of mass is

$$v_{com} = \frac{m_1 v_{1i} + m_2 v_{2i}}{m_1 + m_2} = \frac{(0.340\,\text{kg})(1.2\,\text{m/s})}{0.340\,\text{kg} + 0.099\,\text{kg}} = 0.93\,\text{m/s}\,.$$

Values for the initial velocities were used but the same result is obtained if values for the final velocities are used. The acceleration of the center of mass is zero.

**63**

(a) Let $m_1$ be the mass of the body that is originally moving, $v_{1i}$ be its velocity before the collision, and $v_{1f}$ be its velocity after the collision. Let $m_2$ be the mass of the body that is originally at rest and $v_{2f}$ be its velocity after the collision. Then, according to Eq. 9–67,

$$v_{1f} = \frac{m_1 - m_2}{m_1 + m_2} v_{1i}\,.$$

Solve for $m_2$ to obtain

$$m_2 = \frac{v_{1i} - v_{1f}}{v_{1f} + v_{1i}} m_1\,.$$

Substitute $v_{1f} = v_{1i}/4$ to obtain $m_2 = 3m_1/5 = 3(2.0\,\text{kg})/5 = 1.2\,\text{kg}$.

(b) The speed of the center of mass is

$$v_{com} = \frac{m_1 v_{1i} + m_2 v_{2i}}{m_1 + m_2} = \frac{(2.0\,\text{kg})(4.0\,\text{m/s})}{2.0\,\text{kg} + 1.2\,\text{kg}} = 2.5\,\text{m/s}\,.$$

**77**

(a) The thrust of the rocket is given by $T = Rv_{rel}$, where $R$ is the rate of fuel consumption and $v_{rel}$ is the speed of the exhaust gas relative to the rocket. For this problem $R = 480\,\text{kg/s}$ and $v_{rel} = 3.27 \times 10^3\,\text{m/s}$, so $T = (480\,\text{kg/s})(3.27 \times 10^3\,\text{m/s}) = 1.57 \times 10^6\,\text{N}$.

(b) The mass of fuel ejected is given by $M_{fuel} = R\Delta t$, where $\Delta t$ is the time interval of the burn. Thus $M_{fuel} = (480\,\text{kg/s})(250\,\text{s}) = 1.20 \times 10^5\,\text{kg}$. The mass of the rocket after the burn is $M_f = M_i - M_{fuel} = 2.55 \times 10^5\,\text{kg} - 1.20 \times 10^5\,\text{kg} = 1.35 \times 10^5\,\text{kg}$.

(c) Since the initial speed is zero, the final speed is given by Eq. 9–88:

$$v_f = v_{rel} \ln \frac{M_i}{M_f} = (3.27 \times 10^3\,\text{m/s}) l0n \frac{2.55 \times 10^5\,\text{kg}}{1.35 \times 10^5\,\text{kg}} = 2.08 \times 10^3\,\text{m/s}\,.$$

**79**

(a) Take the $x$ axis to be positive to the right in Fig. 9–72 of the text and take the $y$ axis to be perpendicular to that direction. Consider first the slow barge and suppose the mass of coal shoveled in time $\Delta t$ is $\Delta M$. If $\vec{v}_s$ is the velocity of the barge and $\vec{U}$ is the velocity of the coal as it leaves the barge, then the change in the momentum of the coal-barge system during this interval is $\Delta\vec{P} = \Delta M(\vec{U} - \vec{v}_s)$. The momentum of the coal changed from $\vec{v}_s\Delta M$ to $\vec{U}\Delta M$ and the momentum of the barge did not change. The force that must be exerted on the barge to keep its velocity constant is $\vec{F}_s = \Delta\vec{P}/\Delta t = (\Delta M/\Delta t)(\vec{U} - \vec{v}_s)$. Now $\vec{v}_s = v_s\,\hat{\mathrm{i}}$ and if the coal is shoveled perpendicularly to the length of the boat then $\vec{U} = v_s\,\hat{\mathrm{i}} + U_y\,\hat{\mathrm{j}}$. Thus $\vec{F}_s = (\Delta M/\Delta t)U_y\,\hat{\mathrm{j}}$. $U_y$ is the slight transverse speed the coal must be given to get it from one barge to the other. It is not given in the problem statement, so we assume it is so small it may be neglected. The force that must be applied to the slower barge is essentially zero.

Now consider the faster barge, which receives coal with mass $\Delta M$. Initially the coal has velocity $\vec{U}$ but after it comes to rest relative to the barge its velocity is $\vec{v}_f$, the same as the velocity of the barge. The momentum of the coal changes from $\Delta M\vec{U}$ to $\Delta M\vec{v}_f$ and the momentum of the barge does not change. The force that must be applied to the barge is $\vec{F}_f = (\Delta M/\Delta t)(\vec{v}_f - \vec{U})$. Now $\vec{v}_f = v_f\,\hat{\mathrm{i}}$ and $\vec{U} = v_s\,\hat{\mathrm{i}} + U_y\,\hat{\mathrm{j}}$, so the $x$ component of the force is

$$F_{fx} = \frac{\Delta M}{\Delta t}(v_f - v_s) = \left(\frac{1000\,\mathrm{kg}}{60\,\mathrm{s}}\right)(20\,\mathrm{km/h} - 10\,\mathrm{km/h})\left(\frac{1000\,\mathrm{m/km}}{3600\,\mathrm{s/h}}\right) = 46\,\mathrm{N}\,.$$

The rate with which coal is shoveled is converted from kg/min to kg/s in the first factor and the barge speeds are converted from km/h to m/s by the last factor.

(b) The $y$ component of the force that is applied to the faster barge is $F_{fy} = -(\Delta M/\Delta t)U_y$. If $U_y$ is small, $F_{fy}$ is essentially zero.

**91**

(a) If $m$ is the mass of a pellet and $v$ is its velocity as it hits the wall, then its momentum is $p = mv = (2.0 \times 10^{-3}\,\mathrm{kg})(500\,\mathrm{m/s}) = 1.0\,\mathrm{kg\cdot m/s}$, toward the wall.

(b) The kinetic energy of a pellet is $K = \frac{1}{2}mv^2 = \frac{1}{2}(2.0 \times 10^{-3}\,\mathrm{kg})(500\,\mathrm{m/s})^2 = 2.5 \times 10^2\,\mathrm{J}$.

(c) The force on the wall is given by the rate at which momentum is transferred from the pellets to the wall. Since the pellets do not rebound, each pellet that hits transfers momentum $p = 1.00\,\mathrm{kg\cdot m/s}$. If $\Delta N$ pellets hit in time $\Delta t$, then the average rate at which momentum is transferred is

$$F_{\mathrm{avg}} = \frac{p\,\Delta N}{\Delta t} = (1.0\,\mathrm{kg\cdot m/s})(10\,\mathrm{s^{-1}}) = 10\,\mathrm{N}\,.$$

The force on the wall is in the direction of the initial velocity of the pellets.

(d) If $\Delta t$ is the time interval for a pellet to be brought to rest by the wall, then the average force exerted on the wall by a pellet is

$$F_{\mathrm{avg}} = \frac{p}{\Delta t} = \frac{1.0\,\mathrm{kg\cdot m/s}}{0.6 \times 10^{-3}\,\mathrm{s}} = 1.7 \times 10^3\,\mathrm{N}\,.$$

The force is in the direction of the initial velocity of the pellet.

(e) In part (d) the force is averaged over the time a pellet is in contact with the wall, while in part (c) it is averaged over the time for many pellets to hit the wall. Most of this time no pellet is in contact with the wall, so the average force in part (c) is much less than the average force in (d).

**93**

(a) The initial momentum of the car is $\vec{p}_i = m\vec{v}_i = (1400\,\mathrm{kg})(5.3\,\mathrm{m/s})\,\hat{\mathrm{j}} = (7400\,\mathrm{kg\cdot m/s})\,\hat{\mathrm{j}}$ and the final momentum is $\vec{p}_f = (7400\,\mathrm{kg\cdot m/s})\,\hat{\mathrm{i}}$. The impulse on it equals the change in its momentum: $\vec{J} = \vec{p}_f - \vec{p}_i = (7400\,\mathrm{kg\cdot m/s})(\hat{\mathrm{i}} - \hat{\mathrm{j}})$.

(b) The initial momentum of the car is $\vec{p}_i = (7400\,\text{kg}\cdot\text{m/s})\,\hat{\text{i}}$ and the final momentum is $\vec{p}_f = 0$. The impulse acting on it is $\vec{J} = \vec{p}_f - \vec{p}_i = -(7400\,\text{kg}\cdot\text{m/s})\,\hat{\text{i}}$.

(c) The average force on the car is

$$\vec{F}_{\text{avg}} = \frac{\Delta\vec{p}}{\Delta t} = \frac{\vec{J}}{\Delta t} = \frac{(7400\,\text{kg}\cdot\text{m/s})(\hat{\text{i}} - \hat{\text{j}})}{4.6\,\text{s}} = (1600\,\text{N})(\hat{\text{i}} - \hat{\text{j}})$$

and its magnitude is $F_{\text{avg}} = (1600\,\text{N})\sqrt{2} = 2300\,\text{N}$.

(d) The average force is

$$\vec{F}_{\text{avg}} = \frac{\vec{J}}{\Delta t} = \frac{(-7400\,\text{kg}\cdot\text{m/s})\,\hat{\text{i}}}{350\times 10^{-3}\,\text{s}} = (-2.1\times 10^4\,\text{N})\,\hat{\text{i}}$$

and its magnitude is $F_{\text{avg}} = 2.1\times 10^4\,\text{N}$.

(e) The average force is given above in unit vector notation. Its $x$ and $y$ components have equal magnitudes. The $x$ component is positive and the $y$ component is negative, so the force is 45° below the positive $x$ axis.

**97**

Let $m_F$ be the mass of the freight car and $v_F$ be its initial velocity. Let $m_C$ be the mass of the caboose and $v$ be the common final velocity of the two when they are coupled. Conservation of the total momentum of the two-car system leads to $m_F v_F = (m_F + m_C)v$, so $v = v_F m_F/(m_F + m_C)$. The initial kinetic energy of the system is

$$K_i = \frac{1}{2} m_F v_F^2$$

and the final kinetic energy is

$$K_f = \frac{1}{2}(m_F + m_C)v^2 = \frac{1}{2}(m_F + m_C)\frac{m_F^2 v_F^2}{(m_F + m_C)^2} = \frac{1}{2}\frac{m_F^2 v_F^2}{(m_F + m_C)}\,.$$

Since 27% of the original kinetic energy is lost $K_f = 0.73K_i$, or

$$\frac{1}{2}\frac{m_F^2 v_F^2}{(m_F + m_C)} = (0.73)\left(\frac{1}{2} m_F v_F^2\right)\,.$$

Following some obvious cancellations this becomes $m_F/(m_F + m_C) = 0.73$. Solve for $m_C$: $m_C = (0.27/0.73)m_F = 0.37m_F = (0.37)(3.18\times 10^4\,\text{kg}) = 1.18\times 10^4\,\text{kg}$.

**101**

(a) Let $v_{1i}$ be the speed of ball 1 before the collisions and $v_{1f}$ be its speed afterwards. Let $v_{2f}$ be the speed of ball 2 after the collision. Let $m$ be the mass of each ball. Then conservation of momentum leads to the $x$ component equation $mv_{1i} = mv_{1f}]cos\theta_1 + mv_{2f}\cos\theta_2$ and the $y$ component equation $0 = -mv_{1f}\sin\theta_1 + mv_{2f}\sin\theta_2$. The masses cancel from these equations. The $x$ component equation gives $v_{1f}\cos\theta_1 = v_{1i} - v_{2f}\cos\theta_2$ and the $y$ component equation gives

$v_{1f}\sin\theta_1 = v_{2f}\sin\theta_2$. Square these equations and add them, then use $\sin^2\theta_1 + \cos^2\theta_1 = 1$ to obtain

$$v_{1f}^2 = (v_{2f}\sin\theta_2)^2 + (v_{1f} - v_{2f}\cos\theta_2)^2 = [(1.1\,\text{m/s})\sin 60°]^2 + [2.2\,\text{m/s} - (1.1\,\text{m/s})\cos 60°]^2$$
$$= 3.62\,\text{m}^2/\text{s}^2\,.$$

The speed is $v_{1f} = \sqrt{3.62\,\text{m}^2/\text{s}^2} = 1.9\,\text{m/s}$.

(b) Divide $v_{1f}\sin\theta_1 = v_{2f}\sin\theta_2$ by $v_{1f}\cos\theta_1 = v_{1i} - v_{2f}\cos\theta_2$ and use $\tan\theta_1 = (\sin\theta_1)/(\cos\theta_1)$ to obtain

$$\tan\theta_1 = \frac{v_{2f}\sin\theta_2}{v_{1i} - v_{2f}\cos\theta_2} = \frac{(1.1\,\text{m/s})\sin 60°}{(2.2\,\text{m/s}) - (1.1\,\text{m/s})\cos 60°} = 0.577\,.$$

The angle is $\theta_1 = 30°$.

(c) The initial kinetic energy is $\frac{1}{2}mv_{1i}^2 = \frac{1}{2}m(2.2\,\text{m/s})^2 = 2.42m$, in joules if $m$ is in kilograms. The final kinetic energy is $\frac{1}{2}mv_{1f}^2 + \frac{1}{2}v_{2f}^2 = \frac{1}{2}m(1.9\,\text{m/s})^2 + \frac{1}{2}m(1.1\,\text{m/s})^2 = 2.4m$, in joules if $m$ is in kilograms. The collision is elastic (at least to the number of significant digits given in the problem).

**107**

(a) The acceleration of the center of mass of the two-particle system is the net external force on particles of the system divided by the total mass of the system:

$$\vec{a}_{\text{com}} = \frac{\vec{F}_{\text{net}}}{m_1 + m_2} = \frac{[(-4.00\,\text{N})\hat{\imath} + (5.00\,\text{N})\hat{\jmath}] + [(2.00\,\text{N})\hat{\imath} + (-4.00\,\text{N})\hat{\jmath}]}{2.00\times10^{-3}\,\text{kg} + 4.00\times10^{-3}\,\text{kg}}$$
$$= (-3.33\times10^2\,\text{m/s}^2)\hat{\imath} + (1.67\times10^2\,\text{m/s}^2)\hat{\jmath}\,.$$

Since the acceleration is constant and the center of mass is initially at rest, the displacement during the interval is

$$\Delta\vec{r}_{\text{com}} = \frac{1}{2}\vec{a}(\Delta t)^2 = \frac{1}{2}[(-3.33\times10^2\,\text{m/s}^2)\hat{\imath} + (1.67\times10^2\,\text{m/s}^2)\hat{\jmath}](2.00\times10^{-3}\,\text{s})^2$$
$$= (-6.67\times10^{-4}\,\text{m})\hat{\imath} + (3.33\times10^{-4}\,\text{m})\hat{\jmath}\,.$$

The magnitude of the displacement is

$$|\Delta\vec{r}| = \sqrt{(-6.67\times10^{-4}\,\text{m})^2 + (3.33\times10^{-4}\,\text{m})^2} = 7.45\times10^{-4}\,\text{m}\,.$$

(b) If $\theta$ is the angle made by the displacement and the positive $x$ direction then $\tan\theta = (3.33\times10^{-4}\,\text{m})/(-6.67\times10^{-4}\,\text{m}) = -4.99$ and $\theta = -26.5°$ or $153°$. Since the displacement has a negative $x$ component and a positive $y$ component the correct answer is $153°$.

(c) The velocity of the center of mass is

$$\vec{v}_{\text{com}} = \vec{a}_{\text{com}}\,\Delta t = [(-3.33\times10^2\,\text{m/s}^2)\hat{\imath} + (1.67\times10^2\,\text{m/s}^2)\hat{\jmath}](2.00\times10^{-3}\,\text{s})$$
$$= (-0.667\,\text{m/s})\hat{\imath} + (0.333\,\text{m/s})\hat{\jmath}$$

and the kinetic energy of the center of mass is

$$\begin{aligned}K_{com} &= \frac{1}{2}(m_1+m_2)v_{com}^2 = \frac{1}{2}(m_1+m_2)(v_{com\,x}^2+v_{com\,y}^2)\\ &= \frac{1}{2}(2.00\times10^{-3}\,\text{km}+4.00\times10^{-3}\,\text{kg})[(0.667\,\text{m/s})^2+(0.333\,\text{m/s})^2]\\ &= 1.67\times10^{-3}\,\text{J}\,.\end{aligned}$$

**113**

Let $M_s$ be the mass of the sled and $v_0$ be its initial speed. Let $M_w$ be the mass of water scooped up and $v_f$ be the final speed of the sled and the water it contains. Before the water is scooped up the momentum of the sled-water system is $M_s v_0$ and afterwards it is $(M_s+M_w)v_f$. The final speed is

$$v_f = \frac{M_s v_0}{M_s+M_w} = \frac{(2900\,\text{kg})(250\,\text{m/s})}{2900\,\text{kg}+920\,\text{kg}} = 190\,\text{m/s}\,.$$

You should recognize this as a completely inelastic collision between the sled and the water.

**115**

(a) Put the origin at the center of Earth. Then the distance $r_{com}$ of the center of mass of the Earth-Moon system is given by

$$r_{com} = \frac{m_M r_M}{m_M+m_E}\,,$$

where $m_M$ is the mass of the Moon, $m_E$ is the mass of Earth, and $r_M$ is their separation. These values are given in Appendix C. The numerical result is

$$r_{com} = \frac{(7.36\times10^{22}\,\text{kg})(3.82\times10^{8}\,\text{m})}{7.36\times10^{22}\,\text{kg}+5.98\times10^{24}\,\text{kg}} = 4.64\times10^{6}\,\text{m}\,.$$

(b) The radius of Earth is $R_E = 6.37\times10^6$ m, so $r_{com}/R_E = 0.73$ and the distance from the center of earth to the center of mass of the earth-Moon system is 73% of Earth's radius.

**117**

(a) The thrust is $T = Rv_{rel}$, where $R$ is the mass rate of fuel consumption and $v_{rel}$ is the speed of the fuel relative to the rocket. This should be equal to the gravitational force $Mg$, where $M$ is the mass of the rocket (including fuel). Thus $Rv_{rel} = Mg$ and $R = Mg/v_{rel} = (6100\,\text{kg})(9.8\,\text{m/s}^2)/(1200\,\text{m/s}) = 50\,\text{kg/s}$.

(b) Now $Rv_{rel} - Mg = Ma$, where $a$ is the acceleration. This means

$$R = \frac{M(g+a)}{v_{rel}} = \frac{(6100\,\text{kg})(9.8\,\text{m/s}^2+21\,\text{m/s}^2)}{1200\,\text{m/s}} = 1.6\times10^2\,\text{kg/s}\,.$$

**129**

Write Eq. 9–68 in the form $v_{2f} = 2mv_{1i}/(m+M)$, where $m$ is the mass of the incident object and $M$ is the mass of the target. Solve for $M$:

$$M = \frac{m(2v_{1i}-v_{2f})}{v_{2f}} = \frac{(3.0\,\text{kg})[2(8.0\,\text{m/s})-(6.0\,\text{m/s})]}{6.0\,\text{m/s}} = 5.0\,\text{kg}\,.$$

# Chapter 10

**13**

Take the time $t$ to be zero at the start of the interval. Then at the end of the interval $t = 4.0$ s, and the angle of rotation is $\theta = \omega_0 t + \frac{1}{2}\alpha t^2$. Solve for $\omega_0$:

$$\omega_0 = \frac{\theta - \frac{1}{2}\alpha t^2}{t} = \frac{120\,\text{rad} - \frac{1}{2}(3.0\,\text{rad/s}^2)(4.0\,\text{s})^2}{4.0\,\text{s}} = 24\,\text{rad/s}\,.$$

Now use $\omega = \omega_0 + \alpha t$ to find the time when the wheel is at rest ($\omega = 0$):

$$t = -\frac{\omega_0}{\alpha} = -\frac{24\,\text{rad/s}}{3.0\,\text{rad/s}^2} = -8.0\,\text{s}\,.$$

That is, the wheel started from rest 8.0 s before the start of the 4.0 s interval.

**21**

(a) Use 1 rev = $2\pi$ rad and 1 min = 60 s to obtain

$$\omega = \frac{200\,\text{rev}}{1\,\text{min}} = \frac{(200\,\text{rev})(2\pi\,\text{rad/rev})}{(1\,\text{min})(60\,\text{s/min})} = 20.9\,\text{rad/s}\,.$$

(b) The speed of a point on the rim is given by $v = \omega r$, where $r$ is the radius of the flywheel and $\omega$ must be in radians per second. Thus $v = (20.9\,\text{rad/s})(0.60\,\text{m} = 12.5\,\text{m/s}$.

(c) If $\omega$ is the angular velocity at time $t$, $\omega_0$ is the angular velocity at $t = 0$, and $\alpha$ is the angular acceleration, then since the angular acceleration is constant $\omega = \omega_0 + \alpha t$ and

$$\alpha = \frac{\omega - \omega_0}{t} = \frac{(1000\,\text{rev/min}) - (200\,\text{rev/min})}{1.0\,\text{min}} = 800\,\text{rev/min}^2\,.$$

(d) The flywheel turns through the angle $\theta$, which is

$$\theta = \omega_0 t + \frac{1}{2}\alpha t^2 = (200\,\text{rev/min})(1.0\,\text{min}) + \frac{1}{2}(800\,\text{rev/min}^2)(1.0\,\text{min})^2 = 600\,\text{rev}\,.$$

**29**

(a) Earth makes one rotation per day and 1 d is $(24\,\text{h})(3600\,\text{s/h}) = 8.64 \times 10^4$ s, so the angular speed of Earth is $(2\pi\,\text{rad})/(8.64 \times 10^4\,\text{s}) = 7.3 \times 10^{-5}\,\text{rad/s}$.

(b) Use $v = \omega r$, where $r$ is the radius of its orbit. A point on Earth at a latitude of 40° goes around a circle of radius $r = R\cos 40°$, where $R$ is the radius of Earth ($6.37 \times 10^6$ m). Its speed is $v = \omega(R\cos 40°) = (7.27 \times 10^{-5}\,\text{rad/s})(6.37 \times 10^6\,\text{m})(\cos 40°) = 3.6 \times 10^2\,\text{m/s}$.

(c) At the equator (and all other points on Earth) the value of $\omega$ is the same ($7.3 \times 10^{-5}$ rad/s).

(d) The latitude is 0° and the speed is $v = \omega R = (7.3 \times 10^{-5}\,\text{rad/s})(6.37 \times 10^6\,\text{m}) = 4.6 \times 10^2\,\text{m/s}$.

**33**

The kinetic energy is given by $K = \frac{1}{2}I\omega^2$, where $I$ is the rotational inertia and $\omega$ is the angular velocity. Use

$$\omega = \frac{(602\,\text{rev/min})(2\pi\,\text{rad/rev})}{60\,\text{s/min}} = 63.0\,\text{rad/s}\,.$$

Then

$$I = \frac{2K}{\omega^2} = \frac{2(24400\,\text{J})}{(63.0\,\text{rad/s})^2} = 12.3\,\text{kg}\cdot\text{m}^2\,.$$

**35**

Use the parallel axis theorem: $I = I_{\text{com}} + Mh^2$, where $I_{\text{com}}$ is the rotational inertia about a parallel axis through the center of mass, $M$ is the mass, and $h$ is the distance between the two axes. In this case the axis through the center of mass is at the 0.50 m mark, so $h = 0.50\,\text{m} - 0.20\,\text{m} = 0.30\,\text{m}$. Now

$$I_{\text{com}} = \frac{1}{12}M\ell^2 = \frac{1}{12}(0.56\,\text{kg})(1.0\,\text{m})^2 = 4.67 \times 10^{-2}\,\text{kg}\cdot\text{m}^2\,,$$

so

$$I = 4.67 \times 10^{-2}\,\text{kg}\cdot\text{m}^2 + (0.56\,\text{kg})(0.30\,\text{m})^2 = 9.7 \times 10^{-2}\,\text{kg}\cdot\text{m}^2\,.$$

**37**

Since the rotational inertia of a cylinder of mass $M$ and radius $R$ is $I = \frac{1}{2}MR^2$, the kinetic energy of a cylinder when it rotates with angular velocity $\omega$ is

$$K = \frac{1}{2}I\omega^2 = \frac{1}{4}MR^2\omega^2\,.$$

(a) For the first cylinder

$$K = \frac{1}{4}(1.25\,\text{kg})(0.25\,\text{m})^2(235\,\text{rad/s})^2 = 1.1 \times 10^3\,\text{J}\,.$$

(b) For the second

$$K = \frac{1}{4}(1.25\,\text{kg})(0.75\,\text{m})^2(235\,\text{rad/s})^2 = 9.7 \times 10^3\,\text{J}\,.$$

**41**

Use the parallel-axis theorem. According to Table 10–2, the rotational inertia of a uniform slab about an axis through the center and perpendicular to the large faces is given by

$$I_{\text{com}} = \frac{M}{12}(a^2 + b^2)\,.$$

A parallel axis through a corner is a distance $h = \sqrt{(a/2)^2 + (b/2)^2}$ from the center, so

$$I = I_{\text{com}} + Mh^2 = \frac{M}{12}(a^2 + b^2) + \frac{M}{4}(a^2 + b^2) = \frac{M}{3}(a^2 + b^2)$$
$$= \frac{0.172\,\text{kg}}{3}\left[(0.035\,\text{m})^2 + (0.084\,\text{m})^2\right] = 4.7 \times 10^{-4}\,\text{kg}\cdot\text{m}^2 .$$

**45**

Two forces act on the ball, the force of the rod and the force of gravity. No torque about the pivot point is associated with the force of the rod since that force is along the line from the pivot point to the ball. As can be seen from the diagram, the component of the force of gravity that is perpendicular to the rod is $mg\sin\theta$, so if $\ell$ is the length of the rod then the torque associated with this force has magnitude $\tau = mg\ell\sin\theta = (0.75\,\text{kg})(9.8\,\text{m/s}^2)(1.25\,\text{m})\sin 30° = 4.6\,\text{N}\cdot\text{m}$. For the position of the ball shown the torque is counterclockwise.

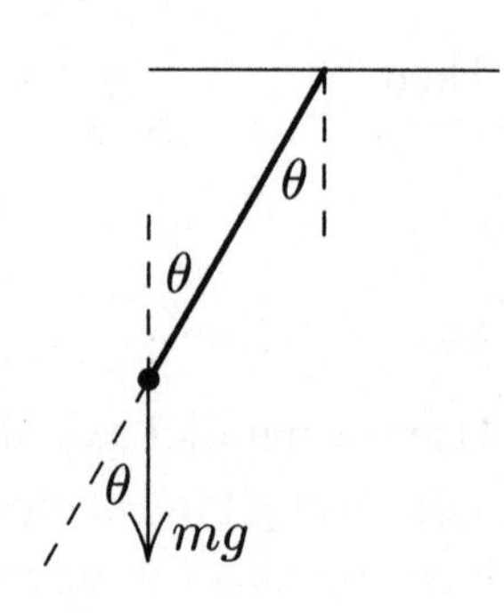


**47**

Take a torque that tends to cause a counterclockwise rotation from rest to be positive and a torque that tends to cause a clockwise rotation from rest to be negative. Thus a positive torque of magnitude $r_1F_1\sin\theta_1$ is associated with $\vec{F}_1$ and a negative torque of magnitude $r_2F_2\sin\theta_2$ is associated with $\vec{F}_2$. Both of these are about $O$. The net torque about $O$ is

$$\tau = r_1F_1\sin\theta_1 - r_2F_2\sin\theta_2$$
$$= (1.30\,\text{m})(4.20\,\text{N})\sin 75.0° - (2.15\,\text{m})(4.90\,\text{N})\sin 60.0° = -3.85\,\text{N}\cdot\text{m} .$$

**49**

(a) Use the kinematic equation $\omega = \omega_0 + \alpha t$, where $\omega_0$ is the initial angular velocity, $\omega$ is the final angular velocity, $\alpha$ is the angular acceleration, and $t$ is the time. This gives

$$\alpha = \frac{\omega - \omega_0}{t} = \frac{6.20\,\text{rad/s}}{220 \times 10^{-3}\,\text{s}} = 28.2\,\text{rad/s}^2 .$$

(b) If $I$ is the rotational inertia of the diver, then according to Newton's second law for rotation, the magnitude of the torque acting on her is $\tau = I\alpha = (12.0\,\text{kg}\cdot\text{m}^2)(28.2\,\text{rad/s}^2) = 3.38 \times 10^2\,\text{N}\cdot\text{m}$.

**63**

Let $\ell$ be the length of the stick. Since its center of mass is $\ell/2$ from either end, its initial potential energy is $\frac{1}{2}mg\ell$, where $m$ is its mass, and its initial kinetic energy is zero. Its final potential energy is zero and its final kinetic energy is $\frac{1}{2}I\omega^2$, where $I$ is its rotational inertia for rotation about an axis through one end and $\omega$ is its angular velocity just before it hits the floor. Conservation of energy yields $\frac{1}{2}mg\ell = \frac{1}{2}I\omega^2$, or $\omega = \sqrt{mg\ell/I}$. The free end of the stick is a distance $\ell$ from the rotation axis, so its speed as it hits the floor is $v = \omega\ell = \sqrt{mg\ell^3/I}$. According to Table 10–2, $I = \frac{1}{3}m\ell^2$, so

$$v = \sqrt{3g\ell} = \sqrt{3(9.8\,\text{m/s}^2)(1.00\,\text{m})} = 5.42\,\text{m/s}\,.$$

<u>69</u>

(a) Choose clockwise rotation of the pulley to be positive and take its angular position $\theta$ to be zero at time $t = 0$. Then the angular position at time $t$ is $\theta = \frac{1}{2}\alpha t^2$, where $\alpha$ is its angular acceleration. Thus $\alpha = 2\theta/t^2 = 2(1.30\,\text{rad})/(91.0 \times 10^{-3}\,\text{s})^2 = 3.14 \times 10^2\,\text{rad/s}^2$.

(b) The string does not slip on the pulley, so the acceleration of either block is $a = r\alpha = (0.0240\,\text{m})(3.14 \times 10^2\,\text{rad/s}^2) = 7.54\,\text{m/s}^2$.

(c) The forces on the hanging block are the tension force of the string and the gravitational force of Earth. Newton's second law for this block gives $mg - T_1 = ma$, so $T_1 = m(g - a) = (6.20\,\text{kg})(9.8\,\text{m/s}^2 - 7.54\,\text{m/s}^2) = 14.0\,\text{N}$.

(d) The net torque on the pulley is $r(T_1 - T_2)$, so $r(T_1 - T_2) = I\alpha$, where $I$ is the rotational inertia of the pulley. Thus

$$T_2 = T_1 - \frac{I\alpha}{r} = 14.0\,\text{N} - \frac{(7.40 \times 10^{-4}\,\text{kg}\cdot\text{m}^2)(3.14 \times 10^2\,\text{rad/s}^2)}{0.024\,\text{m}} = 4.32\,\text{N}\,.$$

<u>79</u>

Use conservation of energy. Take the potential energy to be zero when the rod is horizontal. If $L$ is the length of the rod, the center of mass of the rod is initially a distance $(L\sin\theta)/2$ above the pin and the initial potential energy is $U_i = mgL(\sin\theta)/2$, where $m$ is the mass of the rod. The initial kinetic energy is zero since the rod starts from rest. The final kinetic energy is rotational and is given by $K_f = \frac{1}{2}I\omega_f^2$, where $I$ is the rotational inertia of the rod and $\omega_f$ is its angular speed as it passes the horizontal. The conservation law gives $mg(L\sin\theta)/2 = \frac{1}{2}I\omega_f^2$, so $\omega_f = \sqrt{mg(L\sin\theta)/I}$.

We now need the rotational inertia for rotation about the pin. According to Table 10–2 the rotational inertia of the rod about an axis through the center of mass is $I_{\text{com}} = (1/12)mL^2$. The parallel-axis theorem tells us that the rotational inertia for rotation about the pin is $I = (1/12)mL^2 + m(L/2)^2 = (1/3)mL^2$. The angular speed as the rod passes the horizontal is

$$\omega_f = \sqrt{\frac{3mg(L\sin\theta)}{mL^2}} = \sqrt{\frac{3g\sin\theta}{L}} = \sqrt{\frac{3(9.8\,\text{m/s}^2)\sin 40^\circ}{2.0\,\text{m}}} = 3.1\,\text{rad/s}\,.$$

<u>87</u>

Take the positive direction to be toward the right for the block and take clockwise to be the positive direction of rotation for the wheel. Let $T$ be the tension force of the cord. The horizontal component of Newton's second law for the block gives $P - T = ma$, where $m$ is the mass of the block and $a$ is its acceleration. The torque on the wheel is $Tr$, where $r$ is the radius of the wheel, so Newton's second law for rotation gives $Tr = I\alpha$, where $I$ is the rotational inertia of the wheel and $\alpha$ is its angular acceleration. Since the cord does not slip on the wheel, $a = r\alpha$. When this substitution is made for $a$ is the equation for the block, the result is $P - T = mr\alpha$, so $T = P - mr\alpha$. Use this to substitute for $T$ in the equation for the wheel. The result is $Pr - mr^2\alpha = I\alpha$ and the solution for $\alpha$ is

$$\alpha = \frac{Pr}{mr^2 + I} = \frac{(3.0\,\text{N})(0.20\,\text{m})}{(2.0\,\text{kg})(0.20\,\text{m})^2 + 0.050\,\text{kg}\cdot\text{m}^2} = 4.6\,\text{rad/s}^2\,.$$

**89**

(a) For constant angular acceleration $\omega = \omega_0 + \alpha t$, so $\alpha = (\omega - \omega_0)/t$. Take $\omega = 0$ and to obtain the units requested use $t = (30\,\text{s})/(60\,\text{s/min}) = 0.50\,\text{min}$. Then

$$\alpha = -\frac{33.33\,\text{rev/min}}{0.50\,\text{min}} = -66.7\,\text{rev/min}^2\,.$$

The negative sign indicates that the direction of the angular acceleration is opposite that of the angular velocity.

(b) The angle through which the turntable turns is

$$\begin{aligned}\theta &= \omega_0 t + \frac{1}{2}\alpha t^2 = (33.33\,\text{rev/min})(0.50\,\text{min}) + \frac{1}{2}(-66.7\,\text{rev/min}^2)(0.50\,\text{min})^2\\ &= 8.3\,\text{rev}\,.\end{aligned}$$

**91**

(a) According to Table 10–2, the rotational inertia of a uniform solid cylinder about its central axis is given by $I_C = \frac{1}{2}MR^2$, where $M$ is its mass and $R$ is its radius. For a hoop with mass $M$ and radius $R_H$ Table 10–2 gives $I_H = MR_H^2$ for the rotational inertia. If the two bodies have the same mass, then they will have the same rotational inertia if $R^2/2 = R_H^2$, or $R_H = R/\sqrt{2}$.

(b) You want the rotational inertia to be given by $I = Mk^2$, where $M$ is the mass of the arbitrary body and $k$ is the radius of the equivalent hoop. Thus $k = \sqrt{I/M}$.

**115**

(a) The kinetic energy of the box is given by $K_b = \frac{1}{2}mv^2$, where $m$ is its mass and $v$ is its speed. Thus the speed of the box is $v = \sqrt{2K_b/m}$ and, since the cord does not slip on the wheel, the angular speed of the wheel is $\omega = v/r = \sqrt{2K_b/mr^2}$. The rotational kinetic energy of the wheel is

$$K_w = \frac{1}{2}I\omega^2 = \frac{1}{2}I\frac{2K_b}{mr^2} = \frac{IK_b}{mr^2} = \frac{(0.40\,\text{kg}\cdot\text{m}^2)(6.0\,\text{J})}{(6.0\,\text{kg})(0.20\,\text{m})^2} = 10\,\text{J}\,.$$

(b) Use conservation of energy. As the box falls a distance $h$ the potential energy of the box-wheel-Earth-mount system changes by $\Delta U = -mgh$ and the change in the kinetic energy is $\Delta K = 6.0\,\text{J} + 10\,\text{J} = 16\,\text{J}$. Since $\Delta K + \Delta U = 0$, $h = \Delta K/mg = (16\,\text{J})/(6.0\,\text{kg})(9.8\,\text{m/s}^2) = 0.27\,\text{m}$.

# Chapter 11

**5**

The work required to stop the hoop is the negative of the initial kinetic energy of the hoop. The initial kinetic energy is given by $K = \frac{1}{2}I\omega^2 + \frac{1}{2}mv^2$, where $I$ is its rotational inertia, $m$ is its mass, $\omega$ is its angular speed about its center of mass, and $v$ is the speed of its center of mass. The rotational inertia of the hoop is given by $I = mR^2$, where $R$ is its radius. Since the hoop rolls without sliding the angular speed and the speed of the center of mass are related by $\omega = v/R$. Thus

$$K = \frac{1}{2}mR^2\left(\frac{v^2}{R^2}\right) + \frac{1}{2}mv^2 = mv^2 = (140\,\text{kg})(0.150\,\text{m/s})^2 = 3.15\,\text{J}\,.$$

The work required is $W = -3.15\,\text{J}$.

**17**

(a) An expression for the acceleration is derived in the text and appears as Eq. 11−13:

$$a_{\text{com}} = -\frac{g}{1 + I_{\text{com}}/MR_0^2}\,,$$

where $M$ is the mass of the yo-yo, $I_{\text{com}}$ is its rotational inertia about the center, and $R_0$ is the radius of its axle. The upward direction is taken to be positive. Substitute $I_{\text{com}} = 950\,\text{g}\cdot\text{cm}^2$, $M = 120\,\text{g}$, $R_0 = 0.32\,\text{cm}$, and $g = 980\,\text{cm/s}^2$ to obtain

$$a_{\text{com}} = \frac{980\,\text{cm/s}^2}{1 + (950\,\text{g}\cdot\text{cm}^2)/(120\,\text{g})(0.32\,\text{cm})^2} = 13\,\text{cm/s}^2\,.$$

(b) Solve the kinematic equation $y_{\text{com}} = \frac{1}{2}a_{\text{com}}t^2$ for $t$ and substitute $y_{\text{com}} = 120\,\text{cm}$:

$$t = \sqrt{\frac{2y_{\text{com}}}{a_{\text{com}}}} = \sqrt{\frac{2(120\,\text{cm})}{13\,\text{cm/s}^2}} = 4.4\,\text{s}\,.$$

(c) As it reaches the end of the string its linear speed is $v_{\text{com}} = a_{\text{com}}t = (13\,\text{cm/s}^2)(4.4\,\text{s}) = 55\,\text{cm/s}$.

(d) The translational kinetic energy is $K = \frac{1}{2}mv_{\text{com}}^2 = \frac{1}{2}(0.120\,\text{kg})(0.55\,\text{m/s})^2 = 1.8\times10^{-2}\,\text{J}$.

(e) The angular speed is given by $\omega = v_{\text{com}}/R_0$ and the rotational kinetic energy is $K = \frac{1}{2}I_{\text{com}}\omega^2 = \frac{1}{2}I_{\text{com}}v_{\text{com}}^2/R_0^2 = \frac{1}{2}(9.50\times10^{-5}\,\text{kg}\cdot\text{m}^2)(0.55\,\text{m/s})^2/(3.2\times10^{-3}\,\text{m})^2 = 1.4\,\text{J}$.

(f) The angular speed is $\omega = v_{\text{com}}/R_0 = (0.55\,\text{m/s})/(3.2\times10^{-3}\,\text{m}) = 1.7\times10^2\,\text{rad/s} = 27\,\text{rev/s}$.

**23**

(a) Let $\vec{F} = F_x\hat{\text{i}} + F_y\hat{\text{j}}$ and $\vec{r} = x\hat{\text{i}} + y\hat{\text{j}}$. Then

$$\vec{\tau} = \vec{r}\times\vec{F} = (x\hat{\text{i}} + y\hat{\text{j}})\times(F_x\hat{\text{i}} + F_y\hat{\text{j}}) = (xF_y - yF_x)\hat{\text{k}}\,.$$

The last result can be obtained by multiplying out the quantities in parentheses and using $\hat{\imath} \times \hat{\jmath} = \hat{k}$, $\hat{\jmath} \times \hat{\imath} = -\hat{k}$, $\hat{\imath} \times \hat{\imath} = 0$, and $\hat{\jmath} \times \hat{\jmath} = 0$. Numerically,

$$\vec{\tau} = [(3.0\,\text{m})(6.0\,\text{N}) - (4.0\,\text{m})(-8.0\,\text{N})]\,\hat{k} = (50\,\text{N}\cdot\text{m})\,\hat{k}\,.$$

(b) Use the definition of the vector product: $|\vec{r} \times \vec{F}| = rF\sin\phi$, where $\phi$ is the angle between $\vec{r}$ and $\vec{F}$ when they are drawn with their tails at the same point. Now $r = \sqrt{x^2+y^2} = \sqrt{(3.0\,\text{m})^2+(4.0\,\text{m})^2} = 5.0\,\text{m}$ and $F = \sqrt{F_x^2+F_y^2} = \sqrt{(-8.0\,\text{N})^2+(6.0\,\text{N})^2} = 10\,\text{N}$. Thus $rF = (5.0\,\text{m})(10\,\text{N}) = 50\,\text{N}\cdot\text{m}$, the same as the magnitude of the vector product. This means $\sin\phi = 1$ and $\phi = 90°$.

**29**

(a) Use $\vec{\ell} = m\vec{r} \times \vec{v}$, where $\vec{r}$ is the position vector of the object, $\vec{v}$ is its velocity vector, and $m$ is its mass. The position and velocity vectors have nonvanishing $x$ and $z$ components, so they are written $\vec{r} = x\,\hat{\imath} + z\,\hat{k}$ and $\vec{v} = v_x\,\hat{\imath} + v_z\,\hat{k}$. Evaluate the vector product term by term, making sure to keep the order of the factors intact:

$$\vec{r} \times \vec{v} = (x\,\hat{\imath} + z\,\hat{k}) \times (v_x\,\hat{\imath} + v_z\,\hat{k}) = xv_x\,\hat{\imath}\times\hat{\imath} + xv_z\,\hat{\imath}\times\hat{k} + zv_x\,\hat{k}\times\hat{\imath} + zv_z\,\hat{k}\times\hat{k}\,.$$

Now use $\hat{\imath} \times \hat{\imath} = 0$, $\hat{\imath} \times \hat{k} = -\hat{\jmath}$, $\hat{k} \times \hat{\imath} = +\hat{\jmath}$, and $\hat{k} \times \hat{k} = 0$ to obtain

$$\vec{r} \times \vec{v} = (-xv_z + zv_x)\,\hat{\jmath}\,.$$

Thus

$$\begin{aligned}\vec{\ell} &= m(-xv_z + zv_x)\,\hat{\jmath} \\ &= (0.25\,\text{kg})\left[-(2.0\,\text{m})(5.0\,\text{m/s}) + (-2.0\,\text{m})(-5.0\,\text{m/s})\right]\hat{\jmath} = 0\,.\end{aligned}$$

(b) Use $\vec{\tau} = \vec{r} \times \vec{F}$, with $\vec{F} = F\,\hat{\jmath}$:

$$\begin{aligned}\vec{\tau} &= (x\,\hat{\imath} + z\,\hat{k}) \times (F\,\hat{\jmath}) = xF\,\hat{\imath}\times\hat{\jmath} + zF\,\hat{k}\times\hat{\jmath} = xF\,\hat{k} - zF\,\hat{\imath} \\ &= (2.0\,\text{m})(4.0\,\text{N})\,\hat{\imath} - (-2.0\,\text{m})(4.0\,\text{N})\,\hat{k} = (8.0\,\text{N}\cdot\text{m})\,\hat{\imath} + (8.0\,\text{N}\cdot\text{m})\,\hat{k}\,.\end{aligned}$$

**33**

(a) The angular momentum is given by the vector product $\vec{\ell} = m\vec{r} \times \vec{v}$, where $\vec{r}$ is the position vector of the particle and $\vec{v}$ is its velocity. Since the position and velocity vectors are in the $xy$ plane we may write $\vec{r} = x\,\hat{\imath} + y\,\hat{\jmath}$ and $\vec{v} = v_x\,\hat{\imath} + v_y\,\hat{\jmath}$. Thus

$$\vec{r} \times \vec{v} = (x\,\hat{\imath} + y\,\hat{\jmath}) \times (v_x\,\hat{\imath} + v_y\,\hat{\jmath}) = xv_x\,\hat{\imath}\times\hat{\imath} + xv_y\,\hat{\imath}\times\hat{\jmath} + yv_x\,\hat{\jmath}\times\hat{\imath} + yv_y\,\hat{\jmath}\times\hat{\jmath}\,.$$

Use $\hat{\imath} \times \hat{\imath} = 0$, $\hat{\imath} \times \hat{\jmath} = \hat{k}$, $\hat{\jmath} \times \hat{\imath} = -\hat{k}$, and $\hat{\jmath} \times \hat{\jmath} = 0$ to obtain

$$\vec{r} \times \vec{v} = (xv_y - yv_x)\,\hat{k}\,.$$

Thus

$$\begin{aligned}\vec{\ell} &= m(xv_y - yv_x)\,\hat{k} \\ &= (3.0\,\text{kg})\left[(3.0\,\text{m})(-6.0\,\text{m/s}) - (8.0\,\text{m})(5.0\,\text{m/s})\right]\hat{k} = (-1.7\times 10^2\,\text{kg}\cdot\text{m}^2/\text{s})\,\hat{k}\,.\end{aligned}$$

(b) The torque is given by $\vec{\tau} = \vec{r}\times\vec{F}$. Since the force has only an $x$ component we may write $\vec{F} = F_x\,\hat{\text{i}}$ and

$$\vec{\tau} = (x\,\hat{\text{i}} + y\,\hat{\text{j}})\times(F_x\,\hat{\text{i}}) = -yF_x\,\hat{k} = -(8.0\,\text{m})(-7.0\,\text{N})\,\hat{k} = (56\,\text{N}\cdot\text{m})\,\hat{k}\,.$$

(c) According to Newton's second law for rotation, $\vec{\tau} = d\vec{\ell}/dt$, so the time rate of change of the angular momentum is $56\,\text{kg}\cdot\text{m}^2/\text{s}^2$, in the positive $z$ direction.

**37**

(a) Since $\tau = dL/dt$, the average torque acting during any interval is given by $\tau_{\text{avg}} = (L_f - L_i)/\Delta t$, where $L_i$ is the initial angular momentum, $L_f$ is the final angular momentum, and $\Delta t$ is the time interval. Thus

$$\tau_{\text{avg}} = \frac{0.800\,\text{kg}\cdot\text{m}^2/\text{s} - 3.00\,\text{kg}\cdot\text{m}^2/\text{s}}{1.50\,\text{s}} = -1.47\,\text{N}\cdot\text{m}\,.$$

In this case the negative sign simply indicates that the direction of the torque is opposite the direction of the initial angular momentum, which is taken to be positive.

(b) The angle turned is $\theta = \omega_0 t + \frac{1}{2}\alpha t^2$. If the angular acceleration $\alpha$ is uniform, then so is the torque and $\alpha = \tau/I$. Furthermore, $\omega_0 = L_i/I$, so

$$\theta = \frac{L_i t + \frac{1}{2}\tau t^2}{I} = \frac{(3.00\,\text{kg}\cdot\text{m}^2/\text{s})(1.50\,\text{s}) + \frac{1}{2}(-1.47\,\text{N}\cdot\text{m})(1.50\,\text{s})^2}{0.140\,\text{kg}\cdot\text{m}^2} = 20.3\,\text{rad}\,.$$

(c) The work done on the wheel is

$$W = \tau\theta = (-1.47\,\text{N}\cdot\text{m})(20.3\,\text{rad}) = -29.8\,\text{J}\,.$$

(d) The average power is the work done by the flywheel (the negative of the work done on the flywheel) divided by the time interval:

$$P_{\text{avg}} = -\frac{W}{\Delta t} = -\frac{-29.8\,\text{J}}{1.50\,\text{s}} = 19.9\,\text{W}\,.$$

**43**

(a) No external torques act on the system consisting of the man, bricks, and platform, so the total angular momentum of that system is conserved. Let $I_i$ be the initial rotational inertia of the system and let $I_f$ be the final rotational inertia. If $\omega_i$ is the initial angular velocity and $\omega_f$ is the final angular velocity, then $I_i\omega_i = I_f\omega_f$ and

$$\omega_f = \left(\frac{I_i}{I_f}\right)\omega_i = \left(\frac{6.0\,\text{kg}\cdot\text{m}^2}{2.0\,\text{kg}\cdot\text{m}^2}\right)(1.2\,\text{rev/s}) = 3.6\,\text{rev/s}\,.$$

(b) The initial kinetic energy is $K_i = \frac{1}{2}I_i\omega_i^2$, the final kinetic energy is $K_f = \frac{1}{2}I_f\omega_f^2$, and their ratio is

$$\frac{K_f}{K_i} = \frac{I_f\omega_f^2}{I_i\omega_i^2} = \frac{(2.0\,\text{kg}\cdot\text{m}^2)(3.6\,\text{rev/s})^2}{(6.0\,\text{kg}\cdot\text{m}^2)(1.2\,\text{rev/s})^2} = 3.0\,.$$

(c) The man did work in decreasing the rotational inertia by pulling the bricks closer to his body. This energy came from the man's store of internal energy.

**45**

(a) No external torques act on the system consisting of the two wheels, so its total angular momentum is conserved. Let $I_1$ be the rotational inertia of the wheel that is originally spinning and $I_2$ be the rotational inertia of the wheel that is initially at rest. If $\omega_i$ is the initial angular velocity of the first wheel and $\omega_f$ is the common final angular velocity of each wheel, then $I_1\omega_i = (I_1 + I_2)\omega_f$ and

$$\omega_f = \frac{I_1}{I_1 + I_2}\omega_i\,.$$

Substitute $I_2 = 2I_1$ and $\omega_i = 800\,\text{rev/min}$ to obtain $\omega_f = 267\,\text{rev/min}$.

(b) The initial kinetic energy is $K_i = \frac{1}{2}I_1\omega_i^2$ and the final kinetic energy is $K_f = \frac{1}{2}(I_1 + I_2)\omega_f^2$. The fraction lost is

$$\frac{\Delta K}{K_i} = \frac{K_i - K_f}{K_i} = \frac{I_1\omega_i^2 - (I_1 + I_2)\omega_f^2}{I_1\omega_i^2} = \frac{\omega_i^2 - 3\omega_f^2}{\omega_i^2}$$
$$= \frac{(800\,\text{rev/min})^2 - 3(267\,\text{rev/min})^2}{(800\,\text{rev/min})^2} = 0.667\,.$$

**49**

No external torques act on the system consisting of the train and wheel. The total angular momentum of the system is initially zero and remains zero. Let $I$ ($= MR^2$) be the rotational inertia of the wheel. Its final angular momentum is $L_w = I\omega = MR^2\omega$, where $M$ is the mass of the wheel. The speed of the track is $\omega R$ and the speed of the train is $\omega R - v$. The angular momentum of the train is $L_t = m(\omega R - v)R$, where $m$ is its mass. The direction of rotation of the track is taken to be positive. If the train is moving slowly relative to the track, its velocity and angular momentum are positive; if it is moving fast its velocity and angular momentum are negative. Conservation of angular momentum yields $0 = MR^2\omega + m(\omega R - v)R$. When this equation is solved for $\omega$, the result is

$$\omega = \frac{mvR}{(M + m)R^2} = \frac{mv}{(M + m)R}\,.$$

Substitute $M = 1.1m$, $R = 0.43\,\text{m}$, and $v = 0.15\,\text{m/s}$ to obtain

$$\omega = \frac{m(0.15\,\text{m/s})}{(1.1m + m)(0.43\,\text{m})} = 0.17\,\text{rad/s}\,.$$

**67**

(a) If we consider a short time interval from just before the wad hits to just after it hits and sticks, we may use the principle of conservation of angular momentum. The initial angular momentum is the angular momentum of the falling putty wad. The wad initially moves along a line that is $d/2$ distant from the axis of rotation, where $d$ is the length of the rod. The angular momentum of the wad is $mvd/2$. After the wad sticks, the rod has angular velocity $\omega$ and angular momentum $I\omega$, where $I$ is the rotational inertia of the system consisting of the rod with the two balls and the wad at its end. Conservation of angular momentum yields $mvd/2 = I\omega$. If $M$ is the mass of one of the balls, $I = (2M + m)(d/2)^2$. When $mvd/2 = (2M + m)(d/2)^2\omega$ is solved for $\omega$, the result is

$$\omega = \frac{2mv}{(2M + m)d} = \frac{2(0.0500\,\text{kg})(3.00\,\text{m/s})}{[2(2.00\,\text{kg}) + 0.0500\,\text{kg}](0.500\,\text{m})} = 0.148\,\text{rad/s}.$$

(b) The initial kinetic energy is $K_i = \frac{1}{2}mv^2$, the final kinetic energy is $K_f = \frac{1}{2}I\omega^2$, and their ratio is $K_f/K_i = I\omega^2/mv^2$. When $I = (2M+m)d^2/4$ and $\omega = 2mv/(2M+m)d$ are substituted, this becomes

$$\frac{K_f}{K_i} = \frac{m}{2M + m} = \frac{0.0500\,\text{kg}}{2(2.00\,\text{kg}) + 0.0500\,\text{kg}} = 0.0123.$$

(c) As the rod rotates the sum of the kinetic and potential energies of the Earth-rod-wad system is conserved. If one of the balls is lowered a distance $h$, the other is raised the same distance and the sum of the potential energies of the balls does not change. We need consider only the potential energy of the putty wad. It moves through a 90° arc to reach the lowest point on its path, gaining kinetic energy and losing gravitational potential energy as it goes. It then swings up through an angle $\theta$, losing kinetic energy and gaining potential energy, until it momentarily comes to rest. Take the lowest point on the path to be the zero of potential energy. It starts a distance $d/2$ above this point, so its initial potential energy is $U_i = mgd/2$. If it swings through the angle $\theta$, measured from its lowest point, then its final position is $(d/2)(1 - \cos\theta)$ above the lowest point and its final potential energy is $U_f = mg(d/2)(1 - \cos\theta)$. The initial kinetic energy is the sum of the kinetic energies of the balls and wad: $K_i = \frac{1}{2}I\omega^2 = \frac{1}{2}(2M + m)(d/2)^2\omega^2$. At its final position the rod is instantaneously stopped, so the final kinetic energy is $K_f = 0$. Conservation of energy yields $mgd/2 + \frac{1}{2}(2M + m)(d/2)^2\omega^2 = mg(d/2)(1 - \cos\theta)$. When this equation is solved for $\cos\theta$, the result is

$$\begin{aligned}\cos\theta &= -\frac{1}{2}\left(\frac{2M + m}{mg}\right)\left(\frac{d}{2}\right)\omega^2\\ &= -\frac{1}{2}\left[\frac{2(2.00\,\text{kg}) + 0.0500\,\text{kg}}{(0.0500\,\text{kg})(9.8\,\text{m/s}^2)}\right]\left(\frac{0.500\,\text{m}}{2}\right)(0.148\,\text{rad/s})^2 = -0.0226.\end{aligned}$$

The result for $\theta$ is 91.3°. The total angle of the swing is 90° + 91.3° = 181°.

**73**

(a) and (b) The diagram on the right shows the particles and their lines of motion. The origin is marked $O$ and may be anywhere. The angular momentum of particle 1 has magnitude $\ell_1 = mvr_1 \sin\theta_1 = mv(d+h)$ and it is into the page. The angular momentum of particle 2 has magnitude $\ell_2 = mvr_2 \sin\theta_2 = mvh$ and it is out of the page. The net angular momentum has magnitude $L = mv(d+h) - mvh = mvd = (2.90 \times 10^{-4}\,\text{kg})(5.46\,\text{m/s})(0.0420\,\text{m}) = 6.65\times10^{-5}\,\text{kg·m}^2/\text{s}^2$ and is into the page. This result is independent of the location of the origin.

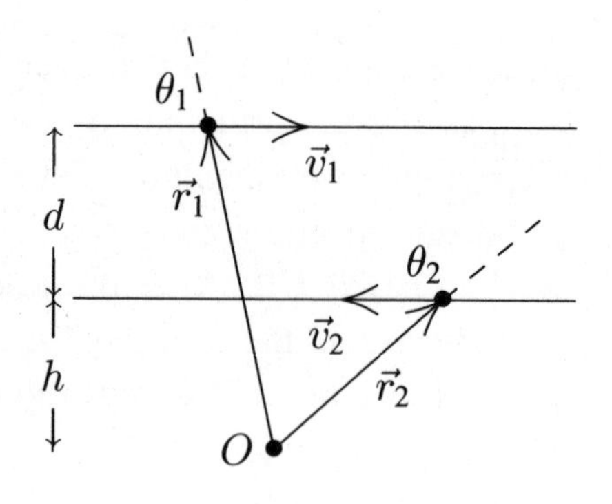


(c) and (d) Suppose particle 2 is traveling to the right. Then $L = mv(d+h)+mvh = mv(d+2h)$. This result depends on $h$, the distance from the origin to one of the lines of motion. If the origin is midway between the lines of motion, then $h = -d/2$ and $L = 0$.

**77**

Use conservation of energy. If the wheel moves a distance $d$ along the incline its center of mass drops a vertical distance $h = d\sin\theta$, where $\theta$ is the angle of the incline. The potential energy of the Earth-wheel system changes by $\Delta U = -mgh = -mgd\sin\theta$, where $m$ is the mass of the wheel. The change in the kinetic energy is $\Delta K = \frac{1}{2}mv_{\text{com}}^2 + \frac{1}{2}I\omega^2$, where $v_{\text{com}}$ is the final speed of the center of mass, $\omega$ is the final angular speed of the wheel, and $I$ is the rotational inertia of the wheel. Since the wheel rolls without sliding $v_{\text{com}} = \omega r$, where $r$ is the radius of the axle. Thus $\Delta K = \frac{1}{2}\omega^2(mr^2 + I)$. Since energy is conserved $mgd\sin\theta = \frac{1}{2}\omega^2(mr^2 + I)$ and

$$\omega^2 = \frac{2mgd\sin\theta}{mr^2 + I} = \frac{2(10.0\,\text{kg})(9.8\,\text{m/s}^2)(2.00\,\text{m})\sin 30^\circ}{(10.0\,\text{kg})(0.200\,)^2 + (0.600\,\text{kg}\cdot\text{m}^2)} = 196\,\text{rad}^2/\text{s}^2\,.$$

(a) The rotational kinetic energy is $K_{\text{rot}} = \frac{1}{2}I\omega^2 = \frac{1}{2}(0.600\,\text{kg}\cdot\text{m}^2)(196\,\text{rad}^2/\text{s}^2) = 58.8\,\text{J}$.

(b) The square of the speed of the center of mass is $v_{\text{com}}^2 = \omega^2 r^2 = (196\,\text{rad}^2/\text{s}^2)(0.200\,\text{m})^2 = 7.84\,\text{m}^2/\text{s}^2$ and the translational kinetic energy is $K_{\text{trans}} = \frac{1}{2}mv_{\text{com}}^2 = \frac{1}{2}(10.0\,\text{kg})(7.84\,\text{m}^2/\text{s}^2) = 39.2\,\text{J}$.

**85**

(a) In terms of the radius of gyration $k$ the rotational inertia of the merry-go-round is $I = Mk^2$ and its value is $(180\,\text{kg})(0.910\,\text{m})^2 = 149\,\text{kg}\cdot\text{m}^2$.

(b) Recall that an object moving along a straight line has angular momentum about any point that is not on the line. Its magnitude is $mvd$, where $m$ is the mass of the object, $v$ is the speed of the object, and $d$ is the distance from the origin to the line of motion. In particular, the angular momentum of the child about the center of the merry-go-round is $L_c = mvR$, where $R$ is the radius of the merry-go-round. Its value is $(44.0\,\text{kg})(3.00\,\text{m/s})(1.20\,\text{m}) = 158\,\text{kg}\cdot\text{m}^2/\text{s}$.

(c) No external torques act on the system consisting of the child and the merry-go-round, so the total angular momentum of the system is conserved. The initial angular momentum is given by $mvR$; the final angular momentum is given by $(I + mR^2)\omega$, where $\omega$ is the final common angular velocity of the merry-go-round and child. Thus $mvR = (I + mR^2)\omega$ and

$$\omega = \frac{mvR}{I + mR^2} = \frac{158\,\mathrm{kg \cdot m^2/s}}{149\,\mathrm{kg \cdot m^2} + (44.0\,\mathrm{kg})(1.20\,\mathrm{m})^2} = 0.744\,\mathrm{rad/s}\,.$$

**87**

The car is moving along the $x$ axis, going in the negative $x$ direction. Let $\vec{r}$ be the vector from the reference point to the particle and $\vec{v}$ be the velocity of the car. Then the angular momentum of the car is given by the vector product $m\vec{r} \times \vec{v}$, which is $-myv_x\,\hat{\mathrm{k}}$ since $\vec{v}$ has only an $x$ component and, in all cases, $\vec{r}$ has only $x$ and $y$ components. According to Newton's second law in angular form the torque is the rate of change of the angular momentum.

(a) and (b) The reference point is at the origin, so $y = 0$ and $\vec{\ell} = 0$. The angular momentum is constant so the torque is also zero.

(c) The reference point is a distance $|\vec{y}| = 5.0\,\mathrm{m}$ from the $x$ axis so the magnitude of the angular momentum is $\ell = (3.0\,\mathrm{kg})(2.0\,\mathrm{m/s^4})t^3(5.0\,\mathrm{m} = (30\,\mathrm{kg \cdot m^2/s^4})t^3$. Since $y$ is positive the angular momentum is in the negative $z$ direction. Thus $\vec{\ell} = -(30\,\mathrm{kg \cdot m^2/s^4})t^3\,\hat{\mathrm{k}}$.

(d) The torque is

$$\vec{\tau} = \frac{d\vec{\ell}}{dt} = \frac{d\ t)}{dt}[-(30\,\mathrm{kg \cdot m^2/s^4})t^3\,\hat{\mathrm{k}}] = -(90\,\mathrm{kg \cdot m^2/s^4})t^2\,\hat{\mathrm{k}}\,.$$

(e) and (f) The reference point is the same distance from the $x$ axis as in parts (c) and (d), so the magnitudes of the angular momentum and torque are the same. Now, however, $y$ is negative, so $\vec{\ell}$ ad $\vec{\tau}$ are in the positive $z$ direction. Thus $\vec{\ell} = +(30\,\mathrm{kg \cdot m^2/s^4})t^3\,\hat{\mathrm{k}}$ and $\vec{\tau} = +(90\,\mathrm{kg \cdot m^2/s^4})t^2\,\hat{\mathrm{k}}$.

**95**

Two horizontal forces act on the cylinder: the applied force $\vec{F}_{\mathrm{app}}$ in the positive $x$ direction and the frictional force $\vec{f}$ along the $x$ axis. Newton's second law for the center of mass is $F_{\mathrm{app}} + f_x = ma_{\mathrm{com},\,x}$, where $m$ is the mass of the cylinder and $a_{\mathrm{com}\,x}$ is the $x$ component of the acceleration of its center of mass. $\vec{F}_{\mathrm{app}}$ acts the top of the cylinder and $\vec{f}$ acts at the bottom, both at the rim, so the magnitude of the net torque on the cylinder is $R(F - f_x)$ and Newton's second law for rotation is $R(F - f_x) = I\alpha$, where $I$ is the rotational inertia of the cylinder and $\alpha$ is its angular acceleration. Since the cylinder rolls without sliding the acceleration of the center of mass and the angular acceleration are related by $a_{\mathrm{com}} = \alpha R$. Solve these equations simultaneously for $a_{\mathrm{com}}$, $\alpha$, and $f_x$. The solutions are

$$a_{\mathrm{com}} = \frac{2R^2 F_{\mathrm{app}}}{mR^2 + I}\,,$$

$$\alpha = \frac{2RF_{\mathrm{app}}}{mR^2 + I}\,,$$

and

$$f_x = \frac{2mR^2 F_{app}}{mR^2 + I} - F.$$

According to Table 10–2 the rotational inertia of the cylinder for rotation about its central axis is given by $I = \frac{1}{2}mR^2$. Substitution of this expression leads to

$$a_{com} = \frac{4F_{app}}{3m} = \frac{4(12\,\mathrm{N})}{3(10\,\mathrm{kg})} = 1.6\,\mathrm{m/s^2},$$

$$\alpha = \frac{4F_{app}}{3mR} = \frac{4(12\,\mathrm{N})}{3(10\,\mathrm{kg})(0.10\,\mathrm{m})} = 16\,\mathrm{rad/s^2},$$

and

$$f_x = \frac{F_{app}}{3} = \frac{12\,\mathrm{N}}{3} = 4.0\,\mathrm{N}.$$

Since $f_x$ is positive, the frictional force is in the positive $x$ direction and can be written $\vec{f} = (4.0\,\mathrm{N})\hat{\mathrm{i}}$.

# Chapter 12

5

Three forces act on the sphere: the tension force $\vec{T}$ of the rope (which is along the rope), the force of the wall $\vec{F}_N$ (which is horizontally away from the wall), and the force of gravity $m\vec{g}$ (which is downward). Since the sphere is in equilibrium they sum to zero. Let $\theta$ be the angle between the rope and the vertical. Then the vertical component of Newton's second is $T\cos\theta - mg = 0$. The horizontal component is $F_N - T\sin\theta = 0$.

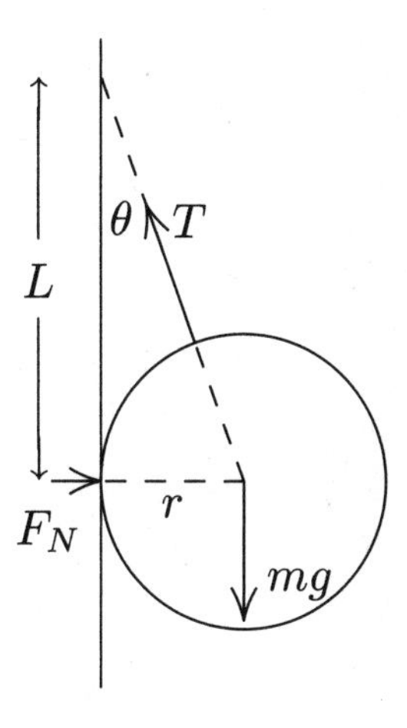


(a) Solve the first equation for $T$: $T = mg/\cos\theta$. Substitute $\cos\theta = L/\sqrt{L^2+r^2}$ to obtain

$$T = \frac{mg\sqrt{L^2+r^2}}{L} = \frac{(0.85\,\text{kg})(9.8\,\text{m/s}^2)\sqrt{(0.080\,\text{m})^2+(0.042\,\text{m})^2}}{0.080\,\text{m}}$$
$$= 9.4\,\text{N}.$$

(b) Solve the second equation for $F_N$: $F_N = T\sin\theta$. Use $\sin\theta = r/\sqrt{L^2+r^2}$ to obtain

$$F_N = \frac{Tr}{\sqrt{L^2+r^2}} = \frac{mg\sqrt{L^2+r^2}}{L}\frac{r}{\sqrt{L^2+r^2}} = \frac{mgr}{L} = \frac{(0.85\,\text{N})(9.8\,\text{m/s}^2)(0.042\,\text{m})}{0.080\,\text{m}} = 4.4\,\text{N}.$$

7

The board is in equilibrium, so the sum of the forces and the sum of the torques on it are each zero. Place the $x$ axis along the diving board. Take the upward direction to be positive. Take the vertical component of the force of the left pedestal to be $F_1$ and suppose this pedestal is at $x = 0$. Take the vertical component of the force of the right pedestal to be $F_2$ and suppose this pedestal is at $x = d$. Let $W$ be the weight of the diver, applied at $x = L$. Set the expression for the sum of the forces equal to zero:

$$F_1 + F_2 - W = 0.$$

Set the expression for the torque about the right pedestal equal to zero:

$$F_1 d + W(L-d) = 0.$$

(a) and (b) The second equation gives

$$F_1 = -\frac{L-d}{d}W = -\left(\frac{3.0\,\text{m}}{1.5\,\text{m}}\right)(580\,\text{N}) = -1.2\times10^3\,\text{N}.$$

The result is negative, indicating that this force is downward.

(c) and (d) The first equation gives

$$F_2 = W - F_1 = 580\,\mathrm{N} + 1.2 \times 10^3\,\mathrm{N} = 1.8 \times 10^3\,\mathrm{N}\,.$$

The result is positive, indicating that this force is upward.

(e) and (f) The force of the diving board on the left pedestal is upward (opposite to the force of the pedestal on the diving board), so this pedestal is being stretched. The force of the diving board on the right pedestal is downward, so this pedestal is being compressed.

**11**

Place the $x$ axis along the meter stick, with the origin at the zero position on the scale. The forces on it are shown on the diagram to the right. The coins are at $x = x_1$ (= 0.120 m) and $m$ is their total mass. The knife edge is at $x = x_2$ (= 0.455 m) and its force is $\vec{F}$. The mass of the meter stick is $M$ and the force of gravity acts at the center of the stick, $x = x_3$ (= 0.500 m).

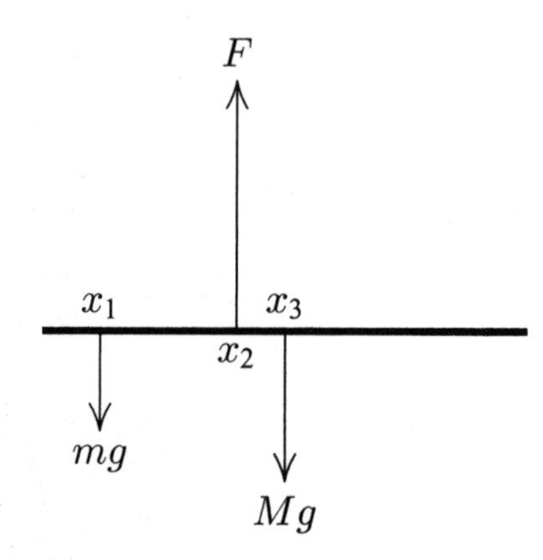


Since the meter stick is in equilibrium the sum of the torques about $x_2$ must vanish: $Mg(x_3 - x_2) - mg(x_2 - x_1) = 0$. Thus,

$$M = \frac{x_2 - x_1}{x_3 - x_2}\,m = \left(\frac{0.455\,\mathrm{m} - 0.120\,\mathrm{m}}{0.500\,\mathrm{m} - 0.455\,\mathrm{m}}\right)(10.0\,\mathrm{g}) = 74.4\,\mathrm{g}\,.$$

**21**

Consider the wheel as it leaves the lower floor. There is no longer a force of the floor on the wheel, and the only forces on it are the force $F$ applied horizontally at the axle, the force of gravity $mg$ vertically downward at the center of the wheel, and the force of the step corner, shown as the two components $f_h$ and $f_v$. If the minimum force is applied the wheel does not accelerate, so both the total force and the total torque on it are zero.

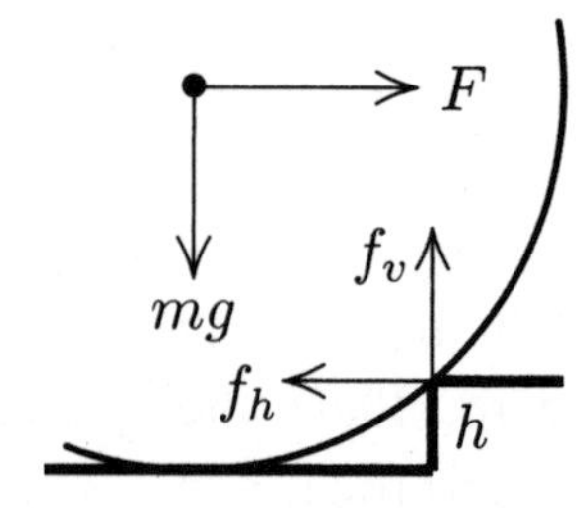


Calculate the torque around the step corner. Look at the second diagram to see that the distance from the line of $F$ to the corner is $r - h$, where $r$ is the radius of the wheel and $h$ is the height of the step. The distance from the line of $mg$ to the corner is $\sqrt{r^2 + (r-h)^2} = \sqrt{2rh - h^2}$. Thus $F(r-h) - mg\sqrt{2rh - h^2} = 0$. The solution for $F$ is

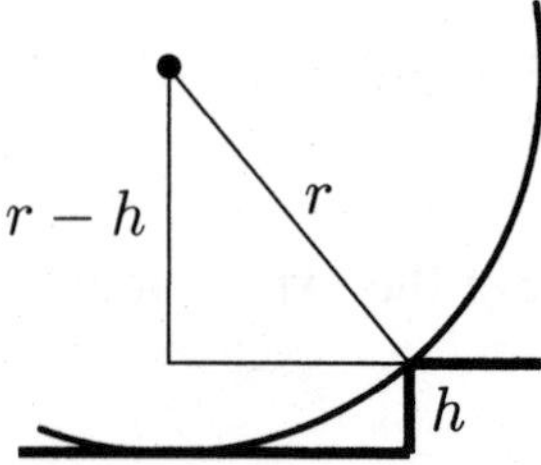


$$F = \frac{\sqrt{2rh - h^2}}{r - h}\,mg$$

$$= \frac{\sqrt{2(0.0600\,\mathrm{m})(0.0300\,\mathrm{m}) - (0.0300\,\mathrm{m})^2}}{0.0600\,\mathrm{m} - 0.0300\,\mathrm{m}}(0.800\,\mathrm{kg})(9.8\,\mathrm{m/s^2}) = 13.6\,\mathrm{N}\,.$$

**33**

(a) Examine the box when it is about to tip. Since it will rotate about the lower right edge, that is where the normal force of the floor is applied. This force is labeled $F_N$ on the diagram to the right. The force of friction is denoted by $f$, the applied force by $F$, and the force of gravity by $W$. Note that the force of gravity is applied at the center of the box. When the minimum force is applied the box does not accelerate, so the sum of the horizontal force components vanishes:

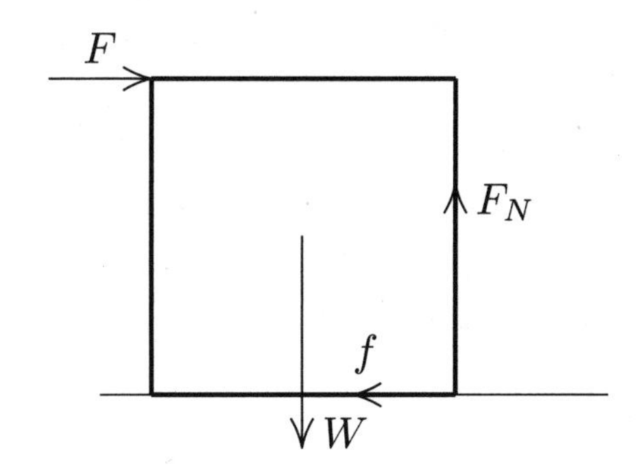


$$F - f = 0\,,$$

the sum of the vertical force components vanishes:

$$F_N - W = 0\,,$$

and the sum of the torques vanishes:

$$FL - \frac{WL}{2} = 0\,.$$

Here $L$ is the length of a side of the box and the origin was chosen to be at the lower right edge. Solve the torque equation for $F$:

$$F = \frac{W}{2} = \frac{890\,\text{N}}{2} = 445\,\text{N}\,.$$

(b) The coefficient of static friction must be large enough that the box does not slip. The box is on the verge of slipping if $\mu_s = f/F_N$. According to the equations of equilibrium $F_N = W = 890\,\text{N}$ and $f = F = 445\,\text{N}$, so $\mu_s = (445\,\text{N})/(890\,\text{N}) = 0.50$.

(c) The box can be rolled with a smaller applied force if the force points upward as well as to the right. Let $\theta$ be the angle the force makes with the horizontal. The torque equation then becomes $FL\cos\theta + FL\sin\theta - WL/2 = 0$, with the solution

$$F = \frac{W}{2(\cos\theta + \sin\theta)}\,.$$

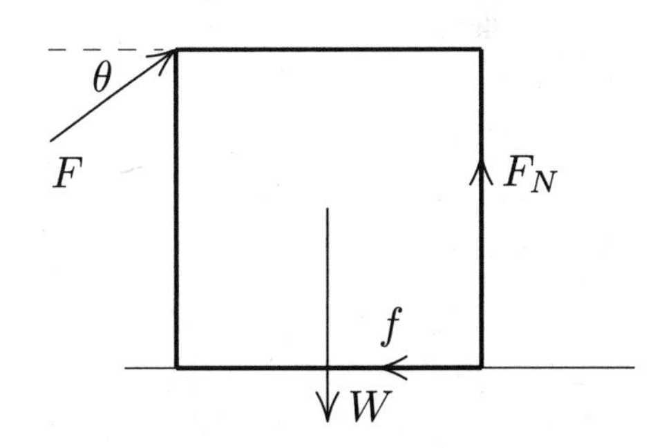


You want $\cos\theta + \sin\theta$ to have the largest possible value. This occurs if $\theta = 45°$, a result you can prove by setting the derivative of $\cos\theta + \sin\theta$ equal to zero and solving for $\theta$. The minimum force needed is

$$F = \frac{W}{4\cos 45°} = \frac{890\,\text{N}}{4\cos 45°} = 315\,\text{N}\,.$$

**43**

(a) The shear stress is given by $F/A$, where $F$ is the magnitude of the force applied parallel to one face of the aluminum rod and $A$ is the cross-sectional area of the rod. In this case $F$ is the weight of the object hung on the end: $F = mg$, where $m$ is the mass of the object. If $r$ is the radius of the rod then $A = \pi r^2$. Thus the shear stress is

$$\frac{F}{A} = \frac{mg}{\pi r^2} = \frac{(1200\,\text{kg})(9.8\,\text{m/s}^2)}{\pi(0.024\,\text{m})^2} = 6.5 \times 10^6\,\text{N/m}^2\,.$$

(b) The shear modulus $G$ is given by

$$G = \frac{F/A}{\Delta x/L}\,,$$

where $L$ is the protrusion of the rod and $\Delta x$ is its vertical deflection at its end. Thus

$$\Delta x = \frac{(F/A)L}{G} = \frac{(6.5 \times 10^6\,\text{N/m}^2)(0.053\,\text{m})}{3.0 \times 10^{10}\,\text{N/m}^2} = 1.1 \times 10^{-5}\,\text{m}\,.$$

**55**

(a) The forces acting on the bucket are the force of gravity, down, and the tension force of cable A, up. Since the bucket is in equilibrium and its weight is $W_B = m_B g = (817\,\text{kg})(9.8\,\text{m/s}^2) = 8.01 \times 10^3\,\text{N}$, the tension force of cable A is $T_A = 8.01 \times 10^3\,\text{N}$.

(b) Use the coordinate axes defined in the diagram. Cable A makes an angle of 66° with the negative $y$ axis, cable B makes an angle of 27° with the positive $y$ axis, and cable C is along the $x$ axis. The $y$ components of the forces must sum to zero since the knot is in equilibrium. This means $T_B \cos 27° - T_A \cos 66° = 0$ and

$$T_B = \frac{\cos 66°}{\cos 27°} T_A = \left(\frac{\cos 66°}{\cos 27°}\right)(8.01 \times 10^3\,\text{N}) = 3.65 \times 10^3\,\text{N}\,.$$

(c) The $x$ components must also sum to zero. This means $T_C + T_B \sin 27° - T_A \sin 66° = 0$ and

$$T_C = T_A \sin 66° - T_B \sin 27° = (8.01 \times 10^3\,\text{N}) \sin 66° - (3.65 \times 10^3\,\text{N}) \sin 27° = 5.66 \times 10^3\,\text{N}\,.$$

**61**

(a) The volume of the slab is $(43\,\text{m})(12\,\text{m})(2.5\,\text{m}) = 1.29 \times 10^3\,\text{m}^3$ and, since the density is $3.2 \times 10^3\,\text{kg/m}^3$, its mass is $m = (1.29 \times 10^3\,\text{m}^3)(3.2 \times 10^3\,\text{kg/m}^3) = 4.13 \times 10^6\,\text{kg}$. The component of the gravitational force parallel to the bedrock surface is $mg \sin\theta = (4.13 \times 10^6\,\text{kg})(9.8\,\text{m/s}^2) \sin 26° = 1.77 \times 10^7\,\text{N}$.

(b) The maximum possible force of static friction is $f_{\text{max}} = \mu_s F_N$, where $\mu_s$ is the coefficient of static friction and $F_N$ is the normal force of the bedrock surface on the slab. Newton's second law (with the acceleration equal to zero) gives the normal force as $mg\cos\theta$, so $f_{\text{max}} = \mu_s mg \cos\theta = (0.39)(4.13 \times 10^6\,\text{kg})(9.8\,\text{m/s}^2) \cos 26° = 1.42 \times 10^7\,\text{N}$.

(c) The bolts must support a total shearing force of $1.77 \times 10^7\,\text{N} - 1.42 \times 10^7\,\text{N} = 3.5 \times 10^6\,\text{N}$. Each bolt can support a shearing force of $(3.6 \times 10^8\,\text{N/m}^2)((6.4 \times 10^{-4}\,\text{m}^2 = 2.3 \times 10^5\,\text{N}$ so the number of bolts required is $(3.5 \times 10^6\,\mathrm{N})/(2.3 \times 10^5\,\mathrm{N}) = 15.2$. Round up to the nearest integer: 16 bolts are required.

## 63

Let $T_A$ be the tension in the horizontal cord, $f$ be the magnitude of the frictional force on block A, $F_N$ be the magnitude of the normal force on that block, and $m_A$ be the mass of that block. Assume block A is stationary. Then Newton's second law for block A gives $T_A - f = 0$ and $F_N - m_A g = 0$. Thus $f = T_A$ and $F_N = m_A g$. If the block does not slip $f$ must be less than $\mu_s F_N$, where $\mu_s$ is the coefficient of static friction between the block and the table surface. This means $T_A$ must be less than $\mu_s m_A g$.

Now consider block B. Let $T_B$ be the tension in the cord attached to it and $M_b$ be its mass. Newton's second gives $T_B - m_B g = 0$, so $T_B = m_B g$.

Next consider the knot where the three cords join and let $T$ be the tension in the third cord. Newton's second law gives $T \sin\theta - T_A = 0$ and $T\cos\theta - T_B = 0$. The second equation gives $T = T_B/\cos\theta$ and since $T_B = m_B g$, this means $T = m_B g/\cos\theta$. Substitution into the first equation gives $T_A = T\sin\theta = m_B g \sin\theta/\cos\theta = m_B g \tan\theta$.

If block B does not slip $m_B g \tan\theta$ must be less than $\mu_s m_A g$. Since $m_B$ is the greatest it can be without block A slipping, $m_B \tan\theta = \mu_s m_A g$ or

$$\mu_s = \frac{m_B}{m_A}\tan\theta = \frac{5.0\,\mathrm{kg}}{10\,\mathrm{kg}}\tan 30^\circ = 0.29\,.$$

## 65

The force diagram for the rod is shown on the right. $\vec{T}$ is the tension force of the rope and $\vec{F}_h$ is the force of the hinge. The angle $\alpha$ is $180^\circ - \theta_1 - \theta_2 = 180^\circ - 60^\circ - \theta_2 = 120^\circ - \theta_2$. The net torque about the hinge is $TL\sin\alpha - mg(L/2)\sin\theta_1$ and this must be zero if the rod is to be in equilibrium. Thus

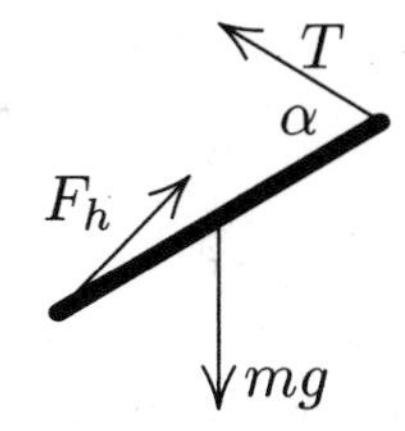


$$\sin\alpha = \frac{mg(L/2)\sin\theta_1}{TL} = \frac{mg(L/2)\sin\theta_1}{(mg/2)L} = \sin\theta_1\,,$$

since $T = mg/2$. This means $\alpha = \theta_1$ and $\theta_2 = 120^\circ - 60^\circ = 60^\circ$.

## 81

The force diagram for the cube is shown on the right. $\vec{F}_N$ is the normal force of the floor on the cube, $\vec{f}$ is the force of friction of the floor, and $m$ is the mass of the cube. Assume the cube is stationary but it is about to tip. Only the lower right edge of the cube exerts a force on the floor and the line of action of the normal force is through the right side of the cube. The horizontal component of Newton's second law for the center of

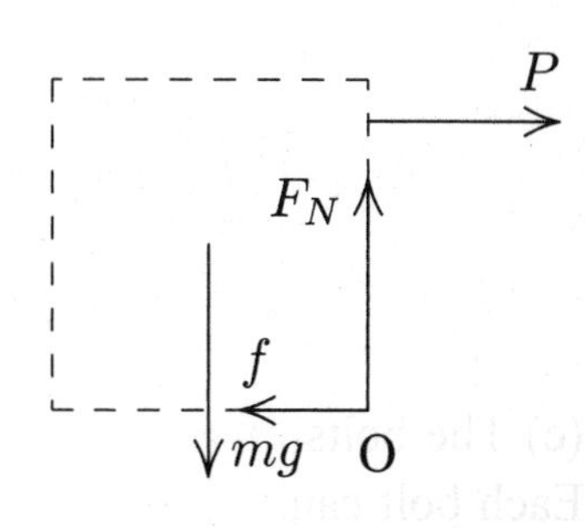

mass is $P - f = 0$ and the vertical component is $T_N - mg = 0$. The net torque about O is $P\ell - mgL/2$, where $\ell$ is the distance between O and the point of application of $\vec{P}$.

(a) The cube slides if $f$ is greater than $\mu_s F_N$, where $\mu_s$ is the coefficient of static friction between the floor and the cube. According to the Newton's second law equations $f = P$ and $F_N = mg$. Thus sliding occurs if $P > \mu_s mg$. Tipping does not occur if $P\ell < mgL/2$. Thus for the cube to slide but not tip as $P$ increases $\mu_s mg$ must be less than $mgL/2\ell$ or $\mu_s$ must be less than $L/2\ell = (8.0\,\text{cm})/2(7.0\,\text{cm}) = 0.57$.

(b) The cube tips before it slides if $P\ell > mgL/2$ and $P < \mu_s mg$. This means $\mu_s mg$ must be greater than $mgL/2\ell$ or $\mu_s$ must be greater than $L/2\ell = 0.57$.

**85**

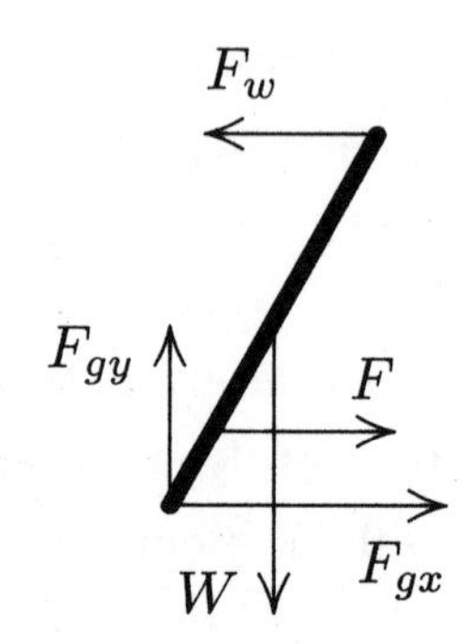


The force diagram for the ladder is shown on the right. $\vec{F}_g$ is the force of the ground on the ladder, $\vec{F}_w$ is the force of the wall, and $W$ is the weight of the ladder. The horizontal component of Newton's second law gives $F + F_{gx} - F_w = 0$ and the vertical component gives $F_{gy} - W = 0$. Set the net torque about the point where the ladder touches the wall equal to zero. The horizontal component of the force of the ground has a lever arm of $h$, the vertical component of the force of the ground has a lever arm of $\ell$, where $\ell$ is the distance from the foot of the ladder to the wall, the applied force has a lever arm of $(1 - d/L)h$, where $L$ is the length of the ladder, and the gravitational force has a lever arm of $\ell/2$. Thus $F_{gx}h - F_{gy}\ell + Fh(1 - d/L) + W\ell/2 = 0$. The $x$ axis was taken to be horizontal with the positive $x$ direction to the right and the $y$ axis was taken to be vertical with the positive direction upward.

The vertical component of the second law gives $Fgy = W = 200\,\text{N}$. The torque equation gives

$$F_{gx} = \frac{\ell}{h}F_gy - \left(1 - \frac{d}{L}\right)F - \frac{d}{2h}W\,.$$

Now $\ell = \sqrt{(10\,\text{m})^2 - (8.0\,\text{m})^2} = 6.0\,\text{m}$, so $(\ell/h)F_{gy} = (6.0\,\text{m})/(8.0\,\text{m})(200\,\text{N}) = 150\,\text{N}$, $1 - d/L = 1 - (2.0\,\text{m})/10\,\text{m}) = 0.80$, and $(d/2h)W = (6.0\,\text{m})(200\,\text{N}/2(8.0\,\text{m}) = 75\,\text{N}$. This means $F_{gx} = 150\,\text{N} - (0.80)F - 75\,\text{N} = 75\,\text{N} - (0.80)F$.

(a) If $F = 50\,\text{N}$, $F_{gx} = 75\,\text{N} - (0.80)(50\,\text{N}) = 35\,\text{N}$ and $\vec{F}_g = (35\,\text{N})\hat{\imath} + (200\,\text{N})\hat{\jmath}$.

(b) If $F = 150\,\text{N}$, $F_{gx} = 75\,\text{N} - (0.80)(150\,\text{N}) = -45\,\text{N}$ and $\vec{F}_g = (-45\,\text{N})\hat{\imath} + (200\,\text{N})\hat{\jmath}$.

(c) When the ladder is on the verge of slipping the frictional force is to the left, in the negative $x$ direction. Its magnitude is $(0.80)F - 75\,\text{N}$. If the ladder does not slip this must be less than $\mu_s F_{gy} = (0.38)(200\,\text{N} = 76\,\text{N}$. The applied force must be less than $(75\,\text{N} + 76\,\text{N})/(0.80) = 1.9 \times 10^2\,\text{N}$. Thus the applied force that will just start the ladder moving is $1.9 \times 10^2\,\text{N}$.

# Chapter 13

**1**

The magnitude of the force of one particle on the other is given by $F = Gm_1m_2/r^2$, where $m_1$ and $m_2$ are the masses, $r$ is their separation, and $G$ is the universal gravitational constant. Solve for $r$:

$$r = \sqrt{\frac{Gm_1m_2}{F}} = \sqrt{\frac{(6.67 \times 10^{-11}\,\mathrm{N \cdot m^2/kg^2})(5.2\,\mathrm{kg})(2.4\,\mathrm{kg})}{2.3 \times 10^{-12}\,\mathrm{N}}} = 19\,\mathrm{m}\,.$$

**7**

At the point where the forces balance $GM_em/r_1^2 = GM_sm/r_2^2$, where $M_e$ is the mass of Earth, $M_s$ is the mass of the Sun, $m$ is the mass of the space probe, $r_1$ is the distance from the center of Earth to the probe, and $r_2$ is the distance from the center of the Sun to the probe. Substitute $r_2 = d - r_1$, where $d$ is the distance from the center of Earth to the center of the Sun, to find

$$\frac{M_e}{r_1^2} = \frac{M_s}{(d - r_1)^2}\,.$$

Take the positive square root of both sides, then solve for $r_1$. A little algebra yields

$$r_1 = \frac{d\sqrt{M_e}}{\sqrt{M_s} + \sqrt{M_e}} = \frac{(150 \times 10^9\,\mathrm{m})\sqrt{5.98 \times 10^{24}\,\mathrm{kg}}}{\sqrt{1.99 \times 10^{30}\,\mathrm{kg}} + \sqrt{5.98 \times 10^{24}\,\mathrm{kg}}} = 2.60 \times 10^8\,\mathrm{m}\,.$$

Values for $M_e$, $M_s$, and $d$ can be found in Appendix C.

**17**

The gravitational acceleration is given by $a_g = GM/r^2$, where $M$ is the mass of Earth and $r$ is the distance from Earth's center. Substitute $r = R + h$, where $R$ is the radius of Earth and $h$ is the altitude, to obtain $a_g = GM/(R + h)^2$. Solve for $h$. You should get $h = \sqrt{GM/a_g} - R$. According to Appendix C of the text, $R = 6.37 \times 10^6\,\mathrm{m}$ and $M = 5.98 \times 10^{24}\,\mathrm{kg}$, so

$$h = \sqrt{\frac{(6.67 \times 10^{-11}\,\mathrm{m^3/s^2 \cdot kg})(5.98 \times 10^{24}\,\mathrm{kg})}{4.9\,\mathrm{m/s^2}}} - 6.37 \times 10^6\,\mathrm{m} = 2.6 \times 10^6\,\mathrm{m}\,.$$

**29**

(a) The density of a uniform sphere is given by $\rho = 3M/4\pi R^3$, where $M$ is its mass and $R$ is its radius. The ratio of the density of Mars to the density of Earth is

$$\frac{\rho_M}{\rho_E} = \frac{M_M}{M_E}\frac{R_E^3}{R_M^3} = 0.11\left(\frac{0.65 \times 10^4\,\mathrm{km}}{3.45 \times 10^3\,\mathrm{km}}\right)^3 = 0.74\,.$$

(b) The value of $a_g$ at the surface of a planet is given by $a_g = GM/R^2$, so the value for Mars is

$$a_{gM} = \frac{M_M}{M_E}\frac{R_E^2}{R_M^2}a_{gE} = 0.11\left(\frac{0.65 \times 10^4\,\text{km}}{3.45 \times 10^3\,\text{km}}\right)^2 (9.8\,\text{m/s}^2) = 3.8\,\text{m/s}^2\,.$$

(c) If $v$ is the escape speed, then, for a particle of mass $m$

$$\tfrac{1}{2}mv^2 = G\frac{mM}{R}$$

and

$$v = \sqrt{\frac{2GM}{R}}\,.$$

For Mars

$$v = \sqrt{\frac{2(6.67 \times 10^{-11}\,\text{m}^3/\text{s}^2 \cdot \text{kg})(0.11)(5.98 \times 10^{24}\,\text{kg})}{3.45 \times 10^6\,\text{m}}} = 5.0 \times 10^3\,\text{m/s}\,.$$

**37**

(a) Use the principle of conservation of energy. Initially the particle is at the surface of the asteroid and has potential energy $U_i = -GMm/R$, where $M$ is the mass of the asteroid, $R$ is its radius, and $m$ is the mass of the particle being fired upward. The initial kinetic energy is $\frac{1}{2}mv^2$. The particle just escapes if its kinetic energy is zero when it is infinitely far from the asteroid. The final potential and kinetic energies are both zero. Conservation of energy yields $-GMm/R + \frac{1}{2}mv^2 = 0$. Replace $GM/R$ with $a_gR$, where $a_g$ is the gravitational acceleration at the surface. Then the energy equation becomes $-a_gR + \frac{1}{2}v^2 = 0$. Solve for $v$:

$$v = \sqrt{2a_gR} = \sqrt{2(3.0\,\text{m/s}^2)(500 \times 10^3\,\text{m})} = 1.7 \times 10^3\,\text{m/s}\,.$$

(b) Initially the particle is at the surface; the potential energy is $U_i = -GMm/R$ and the kinetic energy is $K_i = \frac{1}{2}mv^2$. Suppose the particle is a distance $h$ above the surface when it momentarily comes to rest. The final potential energy is $U_f = -GMm/(R+h)$ and the final kinetic energy is $K_f = 0$. Conservation of energy yields

$$-\frac{GMm}{R} + \frac{1}{2}mv^2 = -\frac{GMm}{R+h}\,.$$

Replace $GM$ with $a_gR^2$ and cancel $m$ in the energy equation to obtain

$$-a_gR + \frac{1}{2}v^2 = -\frac{a_gR^2}{(R+h)}\,.$$

The solution for $h$ is

$$\begin{aligned} h &= \frac{2a_gR^2}{2a_gR - v^2} - R \\ &= \frac{2(3.0\,\text{m/s}^2)(500 \times 10^3\,\text{m})^2}{2(3.0\,\text{m/s}^2)(500 \times 10^3\,\text{m}) - (1000\,\text{m/s})^2} - (500 \times 10^3\,\text{m}) \\ &= 2.5 \times 10^5\,\text{m}\,. \end{aligned}$$

(c) Initially the particle is a distance $h$ above the surface and is at rest. The potential energy is $U_i = -GMm/(R+h)$ and the initial kinetic energy is $K_i = 0$. Just before it hits the asteroid the potential energy is $U_f = -GMm/R$. Write $\frac{1}{2}mv_f^2$ for the final kinetic energy. Conservation of energy yields

$$-\frac{GMm}{R+h} = -\frac{GMm}{R} + \frac{1}{2}mv^2 .$$

Replace $GM$ with $a_g R^2$ and cancel $m$ to obtain

$$-\frac{a_g R^2}{R+h} = -a_g R + \frac{1}{2}v^2 .$$

The solution for $v$ is

$$\begin{aligned} v &= \sqrt{2a_g R - \frac{2a_g R^2}{R+h}} \\ &= \sqrt{2(3.0\,\text{m/s}^2)(500 \times 10^3\,\text{m}) - \frac{2(3.0\,\text{m/s}^2)(500 \times 10^3\,\text{m})^2}{500 \times 10^3\,\text{m} + 1000 \times 10^3\,\text{m}}} \\ &= 1.4 \times 10^3\,\text{m/s} . \end{aligned}$$

**39**

(a) The momentum of the two-star system is conserved, and since the stars have the same mass, their speeds and kinetic energies are the same. Use the principle of conservation of energy. The initial potential energy is $U_i = -GM^2/r_i$, where $M$ is the mass of either star and $r_i$ is their initial center-to-center separation. The initial kinetic energy is zero since the stars are at rest. The final potential energy is $U_f = -2GM^2/r_i$ since the final separation is $r_i/2$. Write $Mv^2$ for the final kinetic energy of the system. This is the sum of two terms, each of which is $\frac{1}{2}Mv^2$. Conservation of energy yields

$$-\frac{GM^2}{r_i} = -\frac{2GM^2}{r_i} + Mv^2 .$$

The solution for $v$ is

$$v = \sqrt{\frac{GM}{r_i}} = \sqrt{\frac{(6.67 \times 10^{-11}\,\text{m}^3/\text{s}^2 \cdot \text{kg})(10^{30}\,\text{kg})}{10^{10}\,\text{m}}} = 8.2 \times 10^4\,\text{m/s} .$$

(b) Now the final separation of the centers is $r_f = 2R = 2 \times 10^5$ m, where $R$ is the radius of either of the stars. The final potential energy is given by $U_f = -GM^2/r_f$ and the energy equation becomes $-GM^2/r_i = -GM^2/r_f + Mv^2$. The solution for $v$ is

$$\begin{aligned} v &= \sqrt{GM\left(\frac{1}{r_f} - \frac{1}{r_i}\right)} \\ &= \sqrt{(6.67 \times 10^{-11}\,\text{m}^3/\text{s}^2 \cdot \text{kg})(10^{30}\,\text{kg})\left(\frac{1}{2 \times 10^5\,\text{m}} - \frac{1}{10^{10}\,\text{m}}\right)} \\ &= 1.8 \times 10^7\,\text{m/s} . \end{aligned}$$

**45**

Let $N$ be the number of stars in the galaxy, $M$ be the mass of the Sun, and $r$ be the radius of the galaxy. The total mass in the galaxy is $NM$ and the magnitude of the gravitational force acting on the Sun is $F = GNM^2/r^2$. The force points toward the galactic center. The magnitude of the Sun's acceleration is $a = v^2/R$, where $v$ is its speed. If $T$ is the period of the Sun's motion around the galactic center then $v = 2\pi R/T$ and $a = 4\pi^2 R/T^2$. Newton's second law yields $GNM^2/R^2 = 4\pi^2 MR/T^2$. The solution for $N$ is

$$N = \frac{4\pi^2 R^3}{GT^2 M} .$$

The period is $2.5 \times 10^8$ y, which is $7.88 \times 10^{15}$ s, so

$$N = \frac{4\pi^2 (2.2 \times 10^{20}\,\text{m})^3}{(6.67 \times 10^{-11}\,\text{m}^3/\text{s}^2 \cdot \text{kg})(7.88 \times 10^{15}\,\text{s})^2 (2.0 \times 10^{30}\,\text{kg})} = 5.1 \times 10^{10} .$$

**47**

(a) The greatest distance between the satellite and Earth's center (the apogee distance) is $R_a = 6.37 \times 10^6\,\text{m} + 360 \times 10^3\,\text{m} = 6.73 \times 10^6\,\text{m}$. The least distance (perigee distance) is $R_p = 6.37 \times 10^6\,\text{m} + 180 \times 10^3\,\text{m} = 6.55 \times 10^6\,\text{m}$. Here $6.37 \times 10^6$ m is the radius of Earth. Look at Fig. 13–13 to see that the semimajor axis is $a = (R_a + R_p)/2 = (6.73 \times 10^6\,\text{m} + 6.55 \times 10^6\,\text{m})/2 = 6.64 \times 10^6\,\text{m}$.

(b) The apogee and perigee distances are related to the eccentricity $e$ by $R_a = a(1 + e)$ and $R_p = a(1 - e)$. Add to obtain $R_a + R_p = 2a$ and $a = (R_a + R_p)/2$. Subtract to obtain $R_a - R_p = 2ae$. Thus

$$e = \frac{R_a - R_p}{2a} = \frac{R_a - R_p}{R_a + R_p} = \frac{6.73 \times 10^6\,\text{m} - 6.55 \times 10^6\,\text{m}}{6.73 \times 10^6\,\text{m} + 6.55 \times 10^6\,\text{m}} = 0.0136 .$$

**61**

(a) Use the law of periods: $T^2 = (4\pi^2/GM)r^3$, where $M$ is the mass of the Sun ($1.99 \times 10^{30}$ kg) and $r$ is the radius of the orbit. The radius of the orbit is twice the radius of Earth's orbit: $r = 2r_e = 2(150 \times 10^9\,\text{m}) = 300 \times 10^9\,\text{m}$. Thus

$$T = \sqrt{\frac{4\pi^2 r^3}{GM}} = \sqrt{\frac{4\pi^2 (300 \times 10^9\,\text{m})^3}{(6.67 \times 10^{-11}\,\text{m}^3/\text{s}^2 \cdot \text{kg})(1.99 \times 10^{30}\,\text{kg})}} = 8.96 \times 10^7\,\text{s} .$$

Divide by (365 d/y)(24 h/d)(60 min/h)(60 s/min) to obtain $T = 2.8$ y.

(b) The kinetic energy of any asteroid or planet in a circular orbit of radius $r$ is given by $K = GMm/2r$, where $m$ is the mass of the asteroid or planet. Notice that it is proportional to $m$ and inversely proportional to $r$. The ratio of the kinetic energy of the asteroid to the kinetic energy of Earth is $K/K_e = (m/m_e)(r_e/r)$. Substitute $m = 2.0 \times 10^{-4} m_e$ and $r = 2r_e$ to obtain $K/K_e = 1.0 \times 10^{-4}$.

**75**

(a) Kepler's law of periods gives

$$T = \sqrt{\frac{4\pi^2 r^3}{GM}} = \sqrt{\frac{4\pi^2(4.20 \times 10^7\,\text{m})^3}{(6.67 \times 10^{-11}\,\text{N} \cdot \text{m}^2/\text{kg}^2)(9.50 \times 10^{25}\,\text{kg})}} = 2.15 \times 10^4\,\text{s}.$$

(b) The craft goes a distance $2\pi r$ in a period, so its speed is $v_0 = 2\pi(4.20 \times 10^7\,\text{m})/(2.15 \times 10^4\,\text{s}) = 1.23 \times 10^4\,\text{m/s}$.

(c) The new speed is $v = 0.98v_0 = (0.98)(1.23 \times 10^4\,\text{m/s}) = 1.21 \times 10^4\,\text{m/s}$.

(d) The kinetic energy of the craft is $K = \frac{1}{2}mv^2 = \frac{1}{2}(3000\,\text{kg})(1.21 \times 10^4\,\text{m/s})^2 = 2.20 \times 10^{11}\,\text{J}$.

(e) The gravitational potential energy of the planet-craft system is

$$U = -G\frac{mM}{r} = -(6.67 \times 10^{-11}\,\text{N} \cdot \text{m}^2/\text{kg}^2)\frac{(3000\,\text{kg})(9.50 \times 10^{25}\,\text{kg})}{4.20 \times 10^7\,\text{m}} = -4.53 \times 10^{11}\,\text{J},$$

where the potential energy was taken to be zero when the craft is far from the planet.

(f) The mechanical energy of the planet-craft system is $E = K + U = 2.20 \times 10^{11}\,\text{J} - 4.53 \times 10^{11}\,\text{J} = -2.33 \times 10^{11}\,\text{J}$.

(g) The mechanical energy of a satellite is given by $E = -GmM/2a$, where $a$ is the semimajor axis. Thus

$$a = -\frac{GmM}{2E} = -\frac{(6.67 \times 10^{-11}\,\text{N} \cdot \text{m}^2/\text{kg}^2)(3000\,\text{kg})(9.50 \times 10^{25}\,\text{kg})}{2(-2.33 \times 10^{11}\,\text{J})} = 4.08 \times 10^7\,\text{m}.$$

(h) and (i) The new period is

$$T' = \sqrt{\frac{4\pi^2 a^3}{GM}} = \sqrt{\frac{4\pi^2(4.08 \times 10^7\,\text{m})^3}{(6.67 \times 10^{-11}\,\text{N} \cdot \text{m}^2/\text{kg}^2)(9.50 \times 10^{25}\,\text{kg})}} = 2.06 \times 10^4\,\text{s}.$$

The change in the period is $2.06 \times 10^4\,\text{s} - 2.15 \times 10^4\,\text{s} = -9 \times 10^3\,\text{s}$. The period for the second orbit is smaller by $9 \times 10^2(4.20 \times 10^7\,\text{m})^3\,\text{s}$.

**79**

Use $F = Gm_s m_m/r^2$, where $m_s$ is the mass of the satellite, $m_m$ is the mass of the meteor, and $r$ is the distance between their centers. The distance between centers is $r = R + d = 15\,\text{m} + 3\,\text{m} = 18\,\text{m}$. Here $R$ is the radius of the satellite and $d$ is the distance from its surface to the center of the meteor. Thus

$$F = \frac{(6.67 \times 10^{-11}\,\text{N} \cdot \text{m}^2/\text{kg}^2)(20\,\text{kg})(7.0\,\text{kg})}{(18\,\text{m})^2} = 2.9 \times 10^{-11}\,\text{N}.$$

**83**

(a) The centripetal acceleration of either star is given by $a = \omega^2 r$, where $\omega$ is the angular speed and $r$ is the radius of the orbit. Since the distance between the stars is $2r$ the gravitational force of one on the other is $Gm^2/(2r)^2$, where $m$ is the mass of either star. Newton's second law gives $Gm^2/(2r)^2 = m\omega^2 r$. Thus

$$\omega = \frac{1}{2}\sqrt{\frac{Gm}{r^3}} = \frac{1}{2}\sqrt{\frac{(6.67 \times 10^{-11}\,\mathrm{N \cdot m/kg^2})(3.0 \times 10^{30}\,\mathrm{kg})^2}{(1.0 \times 10^{11}\,\mathrm{m})^3}} = 2.2 \times 10^{-7}\,\mathrm{rad/s}\,.$$

(b) As the meteoroid goes from the center of the two-star system to far away the kinetic energy changes by $\Delta K = -\frac{1}{2}mv^2$ and the potential energy changes by $\Delta U = 2GmM/r$, where $M$ is the mass of the meteoroid and $v$ is its speed when it is at the center of the two-star system. Since energy is conserved $\Delta K + \Delta U = 0$ and

$$v = \sqrt{\frac{4Gm}{r}} = \sqrt{\frac{4(6.67 \times 10^{-11}\,\mathrm{N \cdot m^2/kg^2})(3.0 \times 10^{30}\,\mathrm{kg})}{1.0 \times 10^{11}\,\mathrm{m}}} = 8.9 \times 10^4\,\mathrm{m/s}\,.$$

**87**

(a) Since energy is conserved it is the same throughout the motion and there is no variation.

(b) The potential energy at the closest distance (perihelion) is

$$\begin{aligned} U_p &= -G\frac{M_E M_S}{r_p} = -(6.67 \times 10^{-11}\,\mathrm{N \cdot m^2/kg^2})\frac{(5.98 \times 10^{24}\,\mathrm{kg})(1.99 \times 10^{30}\,\mathrm{kg})}{1.47 \times 10^{11}\,\mathrm{m}} \\ &= -5.40 \times 10^{33}\,\mathrm{J} \end{aligned}$$

and at the furthest distance (aphelion) is

$$\begin{aligned} U_a &= -G\frac{M_E M_S}{r_a} = -(6.67 \times 10^{-11}\,\mathrm{N \cdot m^2/kg^2})\frac{(5.98 \times 10^{24}\,\mathrm{kg})(1.99 \times 10^{30}\,\mathrm{kg})}{1.52 \times 10^{11}\,\mathrm{m}} \\ &= -5.22 \times 10^{33}\,\mathrm{J}\,. \end{aligned}$$

The difference is $1.8 \times 10^{32}$ J.

(c) Since energy is conserved the variation in the kinetic energy must be the same as the variation in the potential energy, $1.8 \times 10^{32}$ J.

(d) The semimajor axis is $a = (r_p + r_a)/2 = (1.47 \times 10^8\,\mathrm{km} + 1.495 \times 10^8\,\mathrm{km})/2 = 1.50 \times 10^8\,\mathrm{km}$. The kinetic energy at perihelion is

$$K_p = GM_E M_S\left[\frac{1}{r_p} - \frac{1}{2a}\right]\,.$$

Now

$$\left[\frac{1}{r_p} - \frac{1}{2a}\right] = \frac{1}{1.47 \times 10^{11}\,\mathrm{m}} - \frac{1}{2(1.495 \times 10^{11}\,\mathrm{m}} = 3.46 \times 10^{-12}\,\mathrm{m^{-1}}\,,$$

so

$$\begin{aligned} K_p &= (6.67 \times 10^{-11}\,\mathrm{N \cdot m^2/kg^2})(5.98 \times 10^{24}\,\mathrm{kg})(1.99 \times 10^{30}\,\mathrm{kg})(3.46 \times 10^{-12}\,\mathrm{m^{-1}}) \\ &= 2.74 \times 10^{33}\,\mathrm{J} \end{aligned}$$

and the speed is $v_p = \sqrt{2K/M_E} = \sqrt{2(2.74 \times 10^{33}\,\mathrm{J})/(5.98 \times 10^{24}\,\mathrm{kg})} = 3.02 \times 10^4\,\mathrm{m/s}$.

Since angular momentum is conserved $v_p r_p = v_a r_a$ and the speed at aphelion is $v_a = v_p r_p / r_a = (3.02 \times 10^4\,\text{m/s})(1.47 \times 10^{11}\,\text{m})/(1.52 \times 10^{11}\,\text{m}) = 2.93 \times 10^4\,\text{m/s}$. The variation is $3.02 \times 10^4\,\text{m/s} - 2.93 \times 10^4\,\text{m/s} = 9.0 \times 10^2\,\text{m/s}$.

**93**

Each star is a distance $r$ from the central star and a distance $2r$ from the other orbiting star, so it is attracted toward the center of its orbit with a force of magnitude

$$F = \frac{GMm}{r^2} + \frac{Gm^2}{(2r)^2} = \frac{Gm}{4r^2}(4M + m)\,.$$

According to Newton's second law this must equal the product of the mass and centripetal acceleration $v^2/r$. Each star travels a distance $2\pi r$ in a time equal to the period $T$, so $v = 2\pi r/T$, and the centripetal acceleration is $4\pi^2 r/T^2$. Thus

$$\frac{Gm}{4r^2}(4M + m) = m\frac{4\pi^2 r}{T^2}\,.$$

The solution for $T$ is

$$T = \frac{4\pi r^{3/2}}{\sqrt{G(4M + m)}}\,.$$

# Chapter 14

**1**

The air inside pushes outward with a force given by $p_i A$, where $p_i$ is the pressure inside the room and $A$ is the area of the window. Similarly, the air on the outside pushes inward with a force given by $p_o A$, where $p_o$ is the pressure outside. The magnitude of the net force is $F = (p_i - p_o)A$. Since 1 atm = $1.013 \times 10^5$ Pa,

$$F = (1.0\,\text{atm} - 0.96\,\text{atm})(1.013 \times 10^5\,\text{Pa/atm})(3.4\,\text{m})(2.1\,\text{m}) = 2.9 \times 10^4\,\text{N}.$$

**3**

The change in the pressure is the force applied by the nurse divided by the cross-sectional area of the syringe:

$$\Delta p = \frac{F}{A} = \frac{F}{\pi R^2} = \frac{42\,\text{N}}{\pi(1.1 \times 10^{-2}\,\text{m})^2} = 1.1 \times 10^5\,\text{Pa}.$$

**11**

The pressure $p$ at the depth $d$ of the hatch cover is $p_0 + \rho g d$, where $\rho$ is the density of ocean water and $p_0$ is atmospheric pressure. The downward force of the water on the hatch cover is $(p_0 + \rho g d)A$, where $A$ is the area of the cover. If the air in the submarine is at atmospheric pressure then it exerts an upward force of $p_0 A$. The minimum force that must be applied by the crew to open the cover has magnitude $F = (p_0 + \rho g d)A - p_0 A = \rho g d A = (1024\,\text{kg/m}^3)(9.8\,\text{m/s}^2)(100\,\text{m})(1.2\,\text{m})(0.60\,\text{m}) = 7.2 \times 10^5\,\text{N}$.

**19**

When the levels are the same the height of the liquid is $h = (h_1 + h_2)/2$, where $h_1$ and $h_2$ are the original heights. Suppose $h_1$ is greater than $h_2$. The final situation can then be achieved by taking liquid with volume $A(h_1 - h)$ and mass $\rho A(h_1 - h)$, in the first vessel, and lowering it a distance $h - h_2$. The work done by the force of gravity is $W = \rho A(h_1 - h)g(h - h_2)$. Substitute $h = (h_1 + h_2)/2$ to obtain

$$\begin{aligned} W &= \tfrac{1}{4}\rho g A(h_1 - h_2)^2 \\ &= \tfrac{1}{4}(1.30 \times 10^3\,\text{kg/m}^3)(9.8\,\text{m/s}^2)(4.00 \times 10^{-4}\,\text{m}^2)(1.56\,\text{m} - 0.854\,\text{m})^2 \\ &= 0.635\,\text{J}. \end{aligned}$$

**27**

(a) Use the expression for the variation of pressure with height in an incompressible fluid: $p_2 = p_1 - \rho g(y_2 - y_1)$. Take $y_1$ to be at the surface of Earth, where the pressure is $p_1 = 1.01 \times 10^5$ Pa, and $y_2$ to be at the top of the atmosphere, where the pressure is $p_2 = 0$. Take the density to be $1.3\,\mathrm{kg/m^3}$. Then,

$$y_2 - y_1 = \frac{p_1}{\rho g} = \frac{1.01 \times 10^5\,\mathrm{Pa}}{(1.3\,\mathrm{kg/m^3})(9.8\,\mathrm{m/s^2})} = 7.9 \times 10^3\,\mathrm{m} = 7.9\,\mathrm{km}\,.$$

(b) Let $h$ be the height of the atmosphere. Since the density varies with altitude, you must use the integral

$$p_2 = p_1 - \int_0^h \rho g\, dy\,.$$

Take $\rho = \rho_0(1 - y/h)$, where $\rho_0$ is the density at Earth's surface. This expression predicts that $\rho = \rho_0$ at $y = 0$ and $\rho = 0$ at $y = h$. Assume $g$ is uniform from $y = 0$ to $y = h$. Now the integral can be evaluated:

$$p_2 = p_1 - \int_0^h \rho_0 g \left(1 - \frac{y}{h}\right) dy = p_1 - \tfrac{1}{2}\rho_0 g h\,.$$

Since $p_2 = 0$, this means

$$h = \frac{2p_1}{\rho_0 g} = \frac{2(1.01 \times 10^5\,\mathrm{Pa})}{(1.3\,\mathrm{kg/m^3})(9.8\,\mathrm{m/s^2})} = 16 \times 10^3\,\mathrm{m} = 16\,\mathrm{km}\,.$$

<u>31</u>

(a) The anchor is completely submerged. It appears to be lighter than its actual weight because the water is pushing up on it with a buoyant force of $\rho_w g V$, where $\rho_w$ is the density of water and $V$ is the volume of the anchor. Its effective weight (in water) is $W_{\mathrm{eff}} = W - \rho_w g V$, where $W$ is its actual weight (the force of gravity). Thus

$$V = \frac{W - W_{\mathrm{eff}}}{\rho_w g} = \frac{200\,\mathrm{N}}{(998\,\mathrm{kg/m^3})(9.8\,\mathrm{m/s^2})} = 2.045 \times 10^{-2}\,\mathrm{m^3}\,.$$

The density of water was obtained from Table 14–1 of the text.

(b) The mass of the anchor is $m = \rho V$, where $\rho$ is the density of iron. Its weight in air is $W = mg = \rho g V = (7870\,\mathrm{kg/m^3})(9.8\,\mathrm{m/s^2})(2.045 \times 10^{-2}\,\mathrm{m^3}) = 1.58 \times 10^3\,\mathrm{N}$.

<u>35</u>

(a) Let $V$ be the volume of the block. Then, the submerged volume is $V_s = 2V/3$. According to Archimedes' principle the weight of the displaced water is equal to the weight of the block, so $\rho_w V_s = \rho_b V$, where $\rho_w$ is the density of water, and $\rho_b$ is the density of the block. Substitute $V_s = 2V/3$ to obtain $\rho_b = 2\rho_w/3 = 2(998\,\mathrm{kg/m^3})/3 = 6.7 \times 10^2\,\mathrm{kg/m^3}$. The density of water was obtained from Table 14–1 of the text.

(b) If $\rho_o$ is the density of the oil, then Archimedes' principle yields $\rho_o V_s = \rho_b V$. Substitute $V_s = 0.90V$ to obtain $\rho_o = \rho_b/0.90 = 7.4 \times 10^2\,\mathrm{kg/m^3}$.

<u>37</u>

(a) The force of gravity $mg$ is balanced by the buoyant force of the liquid $\rho g V_s$: $mg = \rho g V_s$. Here $m$ is the mass of the sphere, $\rho$ is the density of the liquid, and $V_s$ is the submerged volume.

Thus $m = \rho V_s$. The submerged volume is half the volume enclosed by the outer surface of the sphere, or $V_s = \frac{1}{2}(4\pi/3)r_o^3$, where $r_o$ is the outer radius. This means

$$m = \frac{4\pi}{6}\rho r_o^3 = \left(\frac{4\pi}{6}\right)(800\,\text{kg/m}^3)(0.090\,\text{m})^3 = 1.2\,\text{kg}\,.$$

Air in the hollow sphere, if any, has been neglected.

(b) The density $\rho_m$ of the material, assumed to be uniform, is given by $\rho_m = m/V$, where $m$ is the mass of the sphere and $V$ is its volume. If $r_i$ is the inner radius, the volume is

$$V = \frac{4\pi}{3}\left(r_o^3 - r_i^3\right) = \frac{4\pi}{3}\left[(0.090\,\text{m})^3 - (0.080\,\text{m})^3\right] = 9.09 \times 10^{-4}\,\text{m}^3\,.$$

The density is

$$\rho = \frac{m}{V} = \frac{1.22\,\text{kg}}{9.09 \times 10^{-4}\,\text{m}^3} = 1.3 \times 10^3\,\text{kg/m}^3\,.$$

**49**

Use the equation of continuity. Let $v_1$ be the speed of the water in the hose and $v_2$ be its speed as it leaves one of the holes. Let $A_1$ be the cross-sectional area of the hose. If there are $N$ holes you may think of the water in the hose as $N$ tubes of flow, each of which goes through a single hole. The cross-sectional area of each tube of flow is $A_1/N$. If $A_2$ is the area of a hole the equation of continuity becomes $v_1A_1/N = v_2A_2$. Thus $v_2 = (A_1/NA_2)v_1 = (R^2/Nr^2)v_1$, where $R$ is the radius of the hose and $r$ is the radius of a hole. Thus

$$v_2 = \frac{R^2}{Nr^2}v_1 = \frac{(0.95\,\text{cm})^2}{24(0.065\,\text{cm})^2}(0.91\,\text{m/s}) = 8.1\,\text{m/s}\,.$$

**53**

Suppose that a mass $\Delta m$ of water is pumped in time $\Delta t$. The pump increases the potential energy of the water by $\Delta mgh$, where $h$ is the vertical distance through which it is lifted, and increases its kinetic energy by $\frac{1}{2}\Delta mv^2$, where $v$ is its final speed. The work it does is $\Delta W = \Delta mgh + \frac{1}{2}\Delta mv^2$ and its power is

$$P = \frac{\Delta W}{\Delta t} = \frac{\Delta m}{\Delta t}\left(gh + \frac{1}{2}v^2\right)\,.$$

Now the rate of mass flow is $\Delta m/\Delta t = \rho Av$, where $\rho$ is the density of water and $A$ is the area of the hose. The area of the hose is $A = \pi r^2 = \pi(0.010\,\text{m})^2 = 3.14 \times 10^{-4}\,\text{m}^2$ and $\rho Av = (998\,\text{kg/m}^3)(3.14 \times 10^{-4}\,\text{m}^2)(5.0\,\text{m/s}) = 1.57\,\text{kg/s}$, where the density of water was obtained from Table 14–1 of the text. Thus

$$\begin{aligned}P &= \rho Av\left(gh + \frac{1}{2}v^2\right)\\ &= (1.57\,\text{kg/s})\left[(9.8\,\text{m/s}^2)(3.0\,\text{m}) + \frac{(5.0\,\text{m/s})^2}{2}\right]\\ &= 66\,\text{W}\,.\end{aligned}$$

**55**

(a) Use the equation of continuity: $A_1 v_1 = A_2 v_2$. Here $A_1$ is the cross-sectional area of the pipe at the top and $v_1$ is the speed of the water there; $A_2$ is the cross-sectional area of the pipe at the bottom and $v_2$ is the speed of the water there. Thus $v_2 = (A_1/A_2)v_1 = \left[(4.0\,\text{cm}^2)/(8.0\,\text{cm}^2)\right](5.0\,\text{m/s}) = 2.5\,\text{m/s}$.

(b) Use the Bernoulli equation: $p_1 + \frac{1}{2}\rho v_1^2 + \rho g h_1 = p_2 + \frac{1}{2}\rho v_2^2 + \rho g h_2$, where $\rho$ is the density of water, $h_1$ is its initial altitude, and $h_2$ is its final altitude. Thus

$$\begin{aligned} p_2 &= p_1 + \frac{1}{2}\rho(v_1^2 - v_2^2) + \rho g(h_1 - h_2) \\ &= 1.5 \times 10^5\,\text{Pa} + \frac{1}{2}(998\,\text{kg/m}^3)\left[(5.0\,\text{m/s})^2 - (2.5\,\text{m/s})^2\right] \\ &\qquad + (998\,\text{kg/m}^3)(9.8\,\text{m/s}^2)(10\,\text{m}) \\ &= 2.6 \times 10^5\,\text{Pa}. \end{aligned}$$

The density of water was obtained from Table 14–1 of the text.

**59**

(a) Use the Bernoulli equation: $p_1 + \frac{1}{2}\rho v_1^2 + \rho g h_1 = p_2 + \frac{1}{2}\rho v_2^2 + \rho g h_2$, where $h_1$ is the height of the water in the tank, $p_1$ is the pressure there, and $v_1$ is the speed of the water there; $h_2$ is the altitude of the hole, $p_2$ is the pressure there, and $v_2$ is the speed of the water there. $\rho$ is the density of water. The pressure at the top of the tank and at the hole is atmospheric, so $p_1 = p_2$. Since the tank is large we may neglect the water speed at the top; it is much smaller than the speed at the hole. The Bernoulli equation then becomes $\rho g h_1 = \frac{1}{2}\rho v_2^2 + \rho g h_2$ and

$$v_2 = \sqrt{2g(h_1 - h_2)} = \sqrt{2(9.8\,\text{m/s}^2)(0.30\,\text{m})} = 2.42\,\text{m/s}.$$

The flow rate is $A_2 v_2 = (6.5 \times 10^{-4}\,\text{m}^2)(2.42\,\text{m/s}) = 1.6 \times 10^{-3}\,\text{m}^3/\text{s}$.

(b) Use the equation of continuity: $A_2 v_2 = A_3 v_3$, where $A_3 = A_2/2$ and $v_3$ is the water speed where the cross-sectional area of the stream is half its cross-sectional area at the hole. Thus $v_3 = (A_2/A_3)v_2 = 2v_2 = 4.84\,\text{m/s}$. The water is in free fall and we wish to know how far it has fallen when its speed is doubled to $4.84\,\text{m/s}$. Since the pressure is the same throughout the fall, $\frac{1}{2}\rho v_2^2 + \rho g h_2 = \frac{1}{2}\rho v_3^2 + \rho g h_3$. Thus

$$h_2 - h_3 = \frac{v_3^2 - v_2^2}{2g} = \frac{(4.84\,\text{m/s})^2 - (2.42\,\text{m/s})^2}{2(9.8\,\text{m/s}^2)} = 0.90\,\text{m}.$$

**67**

(a) The continuity equation yields $Av = aV$ and Bernoulli's equation yields $\frac{1}{2}\rho v^2 = \Delta p + \frac{1}{2}\rho V^2$, where $\Delta p = p_2 - p_1$. The first equation gives $V = (A/a)v$. Use this to substitute for $V$ in the second equation. You should obtain $\frac{1}{2}\rho v^2 = \Delta p + \frac{1}{2}\rho(A/a)^2 v^2$. Solve for $v$. The result is

$$v = \sqrt{\frac{2\,\Delta p}{\rho\left(1 - \frac{A^2}{a^2}\right)}} = \sqrt{\frac{2a^2\,\Delta p}{\rho(a^2 - A^2)}}.$$

(b) Substitute values to obtain

$$v = \sqrt{\frac{2(32 \times 10^{-4}\,\text{m}^2)^2(41 \times 10^3\,\text{Pa} - 55 \times 10^3\,\text{Pa})}{(998\,\text{kg/m}^3)\left[(32 \times 10^{-4}\,\text{m}^2)^2 - (64 \times 10^{-4}\,\text{m}^2)^2\right]}} = 3.06\,\text{m/s}\,.$$

The density of water was obtained from Table 14–1 of the text. The flow rate is $Av = (64 \times 10^{-4}\,\text{m}^2)(3.06\,\text{m/s}) = 2.0 \times 10^{-2}\,\text{m}^3/\text{s}$.

75

Let $\rho$ (= $998\,\text{kg/m}^3$) be the density of water and $\rho_\ell$ (= $800\,\text{kg/m}^3$) be the density of the other liquid. Let $d_{wL}$ be the length of the water column on the left side, $d_{wR}$ (= 10.9 cm) be the length of the water column on the right side, and $d_\ell$ (= 8.0 cm) be the length of the column of the other liquid. The pressure at the bottom of the tube is given by $p_0 + \rho_\ell d_\ell + \rho_w d_{wL}$, where $p_0$ is atmospheric pressure, and by $p_0 + \rho_w d_{wR}$. These expressions must be equal, so $p_0 + \rho_\ell d_\ell + \rho_w d_{wL} = p_0 + \rho_w d_{wR}$. The solution for $d_{wL}$ is

$$d_{wL} = \frac{\rho_w d_{wR} - \rho_\ell d_\ell}{\rho_w} = \frac{(998\,\text{kg/m}^3)(10.0\,\text{cm}) - (800\,\text{kg/m}^3)(8.0\,\text{cm})}{998\,\text{kg/m}^3} = 3.59\,\text{cm}\,.$$

Before the other liquid is poured into the tube the length of the water column on the left side is the same as the water column on the right side, namely 10.0 cm. After the liquid is poured it is 3.59 cm. The length decreases by 10 cm – 3.59 cm = 6.41 cm. The volume of water that flows out of the right arm is $\pi(1.50\,\text{cm})^2(6.41\,\text{cm}) = 45.3\,\text{cm}^3$.

# Chapter 15

**3**

(a) The amplitude is half the range of the displacement, or $x_m = 1.0\,\text{mm}$.

(b) The maximum speed $v_m$ is related to the amplitude $x_m$ by $v_m = \omega x_m$, where $\omega$ is the angular frequency. Since $\omega = 2\pi f$, where $f$ is the frequency, $v_m = 2\pi f x_m = 2\pi(120\,\text{Hz})(1.0\times 10^{-3}\,\text{m}) = 0.75\,\text{m/s}$.

(c) The maximum acceleration is $a_m = \omega^2 x_m = (2\pi f)^2 x_m = (2\pi \times 120\,\text{Hz})^2(1.0 \times 10^{-3}\,\text{m}) = 5.7 \times 10^2\,\text{m/s}^2$.

**7**

(a) The motion repeats every 0.500 s so the period must be $T = 0.500\,\text{s}$.

(b) The frequency is the reciprocal of the period: $f = 1/T = 1/(0.500\,\text{s}) = 2.00\,\text{Hz}$.

(c) The angular frequency is $\omega = 2\pi f = 2\pi(2.00\,\text{Hz}) = 12.6\,\text{rad/s}$.

(d) The angular frequency is related to the spring constant $k$ and the mass $m$ by $\omega = \sqrt{k/m}$, so $k = m\omega^2 = (0.500\,\text{kg})(12.57\,\text{rad/s})^2 = 79.0\,\text{N/m}$.

(e) If $x_m$ is the amplitude, the maximum speed is $v_m = \omega x_m = (12.57\,\text{rad/s})(0.350\,\text{m}) = 4.40\,\text{m/s}$.

(f) The maximum force is exerted when the displacement is a maximum and its magnitude is given by $F_m = kx_m = (79.0\,\text{N/m})(0.350\,\text{m}) = 27.6\,\text{N}$.

**9**

The magnitude of the maximum acceleration is given by $a_m = \omega^2 x_m$, where $\omega$ is the angular frequency and $x_m$ is the amplitude. The angular frequency for which the maximum acceleration is $g$ is given by $\omega = \sqrt{g/x_m}$ and the corresponding frequency is given by

$$f = \frac{\omega}{2\pi} = \frac{1}{2\pi}\sqrt{\frac{g}{x_m}} = \frac{1}{2\pi}\sqrt{\frac{9.8\,\text{m/s}^2}{1.00 \times 10^{-6}\,\text{m}}} = 498\,\text{Hz}\,.$$

For frequencies greater than $498z$ Hz the acceleration exceeds $g$ for some part of the motion.

**17**

The maximum force that can be exerted by the surface must be less than $\mu_s F_N$ or else the block will not follow the surface in its motion. Here, $\mu_s$ is the coefficient of static friction and $F_N$ is the normal force exerted by the surface on the block. Since the block does not accelerate vertically, you know that $F_N = mg$, where $m$ is the mass of the block. If the block follows the table and moves in simple harmonic motion, the magnitude of the maximum force exerted on it is given by $F = ma_m = m\omega^2 x_m = m(2\pi f)^2 x_m$, where $a_m$ is the magnitude of the maximum acceleration, $\omega$ is the angular frequency, and $f$ is the frequency. The relationship $\omega = 2\pi f$ was used to obtain the last form.

Substitute $F = m(2\pi f)^2 x_m$ and $F_N = mg$ into $F < \mu_s F_N$ to obtain $m(2\pi f)^2 x_m < \mu_s mg$. The largest amplitude for which the block does not slip is

$$x_m = \frac{\mu_s g}{(2\pi f)^2} = \frac{(0.50)(9.8\,\text{m/s}^2)}{(2\pi \times 2.0\,\text{Hz})^2} = 0.031\,\text{m}\,.$$

A larger amplitude requires a larger force at the end points of the motion. The surface cannot supply the larger force and the block slips.

**19**

(a) Let

$$x_1 = \frac{A}{2}\cos\left(\frac{2\pi t}{T}\right)$$

be the coordinate as a function of time for particle 1 and

$$x_2 = \frac{A}{2}\cos\left(\frac{2\pi t}{T} + \frac{\pi}{6}\right)$$

be the coordinate as a function of time for particle 2. Here $T$ is the period. Note that since the range of the motion is $A$, the amplitudes are both $A/2$. The arguments of the cosine functions are in radians.

Particle 1 is at one end of its path ($x_1 = A/2$) when $t = 0$. Particle 2 is at $A/2$ when $2\pi t/T + \pi/6 = 0$ or $t = -T/12$. That is, particle 1 lags particle 2 by one-twelfth a period. We want the coordinates of the particles at $t = 0.50\,\text{s}$. They are

$$x_1 = \frac{A}{2}\cos\left(\frac{2\pi \times 0.50\,\text{s}}{1.5\,\text{s}}\right) = -0.250A$$

and

$$x_2 = \frac{A}{2}\cos\left(\frac{2\pi \times 0.50\,\text{s}}{1.5\,\text{s}} + \frac{\pi}{6}\right) = -0.433A\,.$$

Their separation at that time is $x_1 - x_2 = -0.250A + 0.433A = 0.183A$.

(b) The velocities of the particles are given by

$$v_1 = \frac{dx_1}{dt} = \frac{\pi A}{T}\sin\left(\frac{2\pi t}{T}\right)$$

and

$$v_2 = \frac{dx_2}{dt} = \frac{\pi A}{T}\sin\left(\frac{2\pi t}{T} + \frac{\pi}{6}\right)\,.$$

Evaluate these expressions for $t = 0.50\,\text{s}$. You will find they are both negative, indicating that the particles are moving in the same direction.

**27**

When the block is at the end of its path and is momentarily stopped, its displacement is equal to the amplitude and all the energy is potential in nature. If the spring potential energy is taken to be zero when the block is at its equilibrium position, then

$$E = \frac{1}{2}kx_m^2 = \frac{1}{2}(1.3 \times 10^2\,\text{N/m})(0.024\,\text{m})^2 = 3.7 \times 10^{-2}\,\text{J}\,.$$

**29**

(a) and (b) The total energy is given by $E = \frac{1}{2}kx_m^2$, where $k$ is the spring constant and $x_m$ is the amplitude. When $x = \frac{1}{2}x_m$ the potential energy is $U = \frac{1}{2}kx^2 = \frac{1}{8}kx_m^2$. The ratio is

$$\frac{U}{E} = \frac{\frac{1}{8}kx_m^2}{\frac{1}{2}kx_m^2} = \frac{1}{4}.$$

The fraction of the energy that is kinetic is

$$\frac{K}{E} = \frac{E-U}{E} = 1 - \frac{U}{E} = 1 - \frac{1}{4} = \frac{3}{4}.$$

(c) Since $E = \frac{1}{2}kx_m^2$ and $U = \frac{1}{2}kx^2$, $U/E = x^2/x_m^2$. Solve $x^2/x_m^2 = 1/2$ for $x$. You should get $x = x_m/\sqrt{2}$.

**39**

(a) Take the angular displacement of the wheel to be $\theta = \theta_m \cos(2\pi t/T)$, where $\theta_m$ is the amplitude and $T$ is the period. Differentiate with respect to time to find the angular velocity: $\Omega = -(2\pi/T)\theta_m \sin(2\pi t/T)$. The symbol $\Omega$ is used for the angular velocity of the wheel so it is not confused with the angular frequency. The maximum angular velocity is

$$\Omega_m = \frac{2\pi\theta_m}{T} = \frac{(2\pi)(\pi\,\text{rad})}{0.500\,\text{s}} = 39.5\,\text{rad/s}.$$

(b) When $\theta = \pi/2$, then $\theta/\theta_m = 1/2$, $\cos(2\pi t/T) = 1/2$, and

$$\sin(2\pi t/T) = \sqrt{1 - \cos^2(2\pi t/T)} = \sqrt{1 - (1/2)^2} = \sqrt{3}/2,$$

where the trigonometric identity $\cos^2 A + \sin^2 A = 1$ was used. Thus

$$\Omega = -\frac{2\pi}{T}\theta_m \sin\left(\frac{2\pi t}{T}\right) = -\left(\frac{2\pi}{0.500\,\text{s}}\right)(\pi\,\text{rad})\left(\frac{\sqrt{3}}{2}\right) = -34.2\,\text{rad/s}.$$

The negative sign is not significant. During another portion of the cycle its angular speed is +34.2 rad/s when its angular displacement is $\pi/2$ rad.

(c) The angular acceleration is

$$\alpha = \frac{d^2\theta}{dt^2} = -\left(\frac{2\pi}{T}\right)^2 \theta_m \cos(2\pi t/T) = -\left(\frac{2\pi}{T}\right)^2 \theta.$$

When $\theta = \pi/4$,

$$\alpha = -\left(\frac{2\pi}{0.500\,\text{s}}\right)^2 \left(\frac{\pi}{4}\right) = -124\,\text{rad/s}^2.$$

Again the negative sign is not significant.

## 43

(a) A uniform disk pivoted at its center has a rotational inertia of $\frac{1}{2}MR^2$, where $M$ is its mass and $R$ is its radius. See Table 10–2. The disk of this problem rotates about a point that is displaced from its center by $R + L$, where $L$ is the length of the rod, so, according to the parallel-axis theorem, its rotational inertia is $\frac{1}{2}MR^2 + M(L+R)^2$. The rod is pivoted at one end and has a rotational inertia of $mL^2/3$, where $m$ is its mass. The total rotational inertia of the disk and rod is $I = \frac{1}{2}MR^2 + M(L+R)^2 + \frac{1}{3}mL^2 = \frac{1}{2}(0.500\,\text{kg})(0.100\,\text{m})^2 + (0.500\,\text{kg})(0.500\,\text{m} + 0.100\,\text{m})^2 + \frac{1}{3}(0.270\,\text{kg})(0.500\,\text{m})^2 = 0.205\,\text{kg}\cdot\text{m}^2$.

(b) Put the origin at the pivot. The center of mass of the disk is $\ell_d = L+R = 0.500\,\text{m}+0.100\,\text{m} = 0.600\,\text{m}$ away and the center of mass of the rod is $\ell_r = L/2 = (0.500\,\text{m})/2 = 0.250\,\text{m}$ away, on the same line. The distance from the pivot point to the center of mass of the disk-rod system is

$$d = \frac{M\ell_d + m\ell_r}{M+m} = \frac{(0.500\,\text{kg})(0.600\,\text{m}) + (0.270\,\text{kg})(0.250\,\text{m})}{0.500\,\text{kg} + 0.270\,\text{kg}} = 0.477\,\text{m}\,.$$

(c) The period of oscillation is

$$T = 2\pi\sqrt{\frac{I}{(M+m)gd}} = 2\pi\sqrt{\frac{0.205\,\text{kg}\cdot\text{m}^2}{(0.500\,\text{kg} + 0.270\,\text{kg})(9.8\,\text{m/s}^2)(0.447\,\text{m})}} = 1.50\,\text{s}\,.$$

## 51

If the torque exerted by the spring on the rod is proportional to the angle of rotation of the rod and if the torque tends to pull the rod toward its equilibrium orientation, then the rod will oscillate in simple harmonic motion. If $\tau = -C\theta$, where $\tau$ is the torque, $\theta$ is the angle of rotation, and $C$ is a constant of proportionality, then the angular frequency of oscillation is $\omega = \sqrt{C/I}$ and the period is $T = 2\pi/\omega = 2\pi\sqrt{I/C}$, where $I$ is the rotational inertia of the rod. The plan is to find the torque as a function of $\theta$ and identify the constant $C$ in terms of given quantities. This immediately gives the period in terms of given quantities.

Let $\ell_0$ be the distance from the pivot point to the wall. This is also the equilibrium length of the spring. Suppose the rod turns through the angle $\theta$, with the left end moving away from the wall. If $L$ is the length of the rod, this end is now $(L/2)\sin\theta$ further from the wall and has moved $(L/2)(1-\cos\theta)$ to the right. The spring length is now $\sqrt{(L/2)^2(1-\cos\theta)^2 + [\ell_0 + (L/2)\sin\theta]^2}$. If the angle $\theta$ is small we may approximate $\cos\theta$ with 1 and $\sin\theta$ with $\theta$ in radians. Then the length of the spring is given by $\ell_0 + L\theta/2$ and its elongation is $\Delta x = L\theta/2$. The force it exerts on the rod has magnitude $F = k\,\Delta x = kL\theta/2$, where $k$ is the spring constant. Since $\theta$ is small we may approximate the torque exerted by the spring on the rod by $\tau = -FL/2$, where the pivot point was taken as the origin. Thus $\tau = -(kL^2/4)\theta$. The constant of proportionality $C$ that relates the torque and angle of rotation is $C = kL^2/4$.

The rotational inertia for a rod pivoted at its center is $I = mL^2/12$, where $m$ is its mass. See Table 10–2. Thus the period of oscillation is

$$T = 2\pi\sqrt{\frac{I}{C}} = 2\pi\sqrt{\frac{mL^2/12}{kL^2/4}} = 2\pi\sqrt{\frac{m}{3k}} = 2\pi\sqrt{\frac{0.600\,\text{kg}}{3(1850\,\text{N/m})}} = 0.0653\,\text{s}\,.$$

**57**

(a) You want to solve $e^{-bt/2m} = 1/3$ for $t$. Take the natural logarithm of both sides to obtain $-bt/2m = \ln(1/3)$. Now solve for $t$: $t = -(2m/b)\ln(1/3) = (2m/b)\ln 3$, where the sign was reversed when the argument of the logarithm was replaced by its reciprocal. Thus

$$t = \frac{2(1.50\,\text{kg})}{0.230\,\text{kg/s}} \ln 3 = 14.3\,\text{s}\,.$$

(b) The angular frequency is

$$\omega' = \sqrt{\frac{k}{m} - \frac{b^2}{4m^2}} = \sqrt{\frac{8.00\,\text{N/m}}{1.50\,\text{kg}} - \frac{(0.230\,\text{kg/s})^2}{4(1.50\,\text{kg})^2}} = 2.31\,\text{rad/s}\,.$$

The period is $T = 2\pi/\omega' = (2\pi)/(2.31\,\text{rad/s}) = 2.72\,\text{s}$ and the number of oscillations is $t/T = (14.3\,\text{s})/(2.72\,\text{s}) = 5.27$.

**75**

(a) The frequency for small amplitude oscillations is $f = (1/2\pi)\sqrt{g/L}$, where $L$ is the length of the pendulum. This gives $f = (1/2\pi)\sqrt{(9.80\,\text{m/s}^2)/(2.0\,\text{m})} = 0.35\,\text{Hz}$.

(b) The forces acing on the pendulum are the tension force $\vec{T}$ of the rod and the force of gravity $m\vec{g}$. Newton's second law yields $\vec{T} + m\vec{g} = m\vec{a}$, where $m$ is the mass and $\vec{a}$ is the acceleration of the pendulum. Let $\vec{a} = \vec{a}_e + \vec{a}'$, where $\vec{a}_e$ is the acceleration of the elevator and $\vec{a}'$ is the acceleration of the pendulum relative to the elevator. Newton's second law can then be written $m(\vec{g} - \vec{a}_e) + \vec{T} = m\vec{a}'$. Relative to the elevator the motion is exactly the same as it would be in an inertial frame where the acceleration due to gravity is $\vec{g} - \vec{a}_e$. Since $\vec{g}$ and $\vec{a}_e$ are along the same line and in opposite directions we can find the frequency for small amplitude oscillations by replacing $g$ with $g + a_e$ in the expression $f = (1/2\pi)\sqrt{g/L}$. Thus

$$f = \frac{1}{2\pi}\sqrt{\frac{g + a_e}{L}} = \frac{1}{2\pi}\sqrt{\frac{9.8\,\text{m/s}^2 + 2.0\,\text{m/s}^2}{2.0\,\text{m}}} = 0.39\,\text{Hz}\,.$$

(c) Now the acceleration due to gravity and the acceleration of the elevator are in the same direction and have the same magnitude. That is, $\vec{g} - \vec{a}_e = 0$. To find the frequency for small amplitude oscillations, replace $g$ with zero in $f = (1/2\pi)\sqrt{g/L}$. The result is zero. The pendulum does not oscillate.

**83**

Use $v_m = \omega x_m = 2\pi f x_m$. The frequency is $180/(60\,\text{s}) = 3.0\,\text{Hz}$ and the amplitude is half the stroke, or $0.38\,\text{m}$. Thus $v_m = 2\pi(3.0\,\text{Hz})(0.38\,\text{m}) = 7.2\,\text{m/s}$.

**89**

(a) The spring stretches until the magnitude of its upward force on the block equals the magnitude of the downward force of gravity: $ky = mg$, where $y$ is the elongation of the spring at equilibrium, $k$ is the spring constant, and $m$ is the mass of the block. Thus $k = mg/y = (1.3\,\text{kg})(9.8\,\text{m/s}^2)/(0.096\,\text{m}) = 133\,\text{N/m}$.

(b) The period is given by $T = 1/f = 2\pi/\omega = 2\pi\sqrt{m/k} = 2\pi\sqrt{(1.3\,\text{kg})/(133\,\text{N/m})} = 0.62\,\text{s}$.

(c) The frequency is $f = 1/T = 1/0.62\,\text{s} = 1.6\,\text{Hz}$.

(d) The block oscillates in simple harmonic motion about the equilibrium point determined by the forces of the spring and gravity. It is started from rest 5.0 cm below the equilibrium point so the amplitude is 5.0 cm.

(e) The block has maximum speed as it passes the equilibrium point. At the initial position, the block is not moving but it has potential energy

$$U_i = -mgy_i + \frac{1}{2}ky_i^2 = -(1.3\,\text{kg})(9.8\,\text{m/s}^2)(0.146\,\text{m}) + \frac{1}{2}(133\,\text{N/m})(0.146\,\text{m})^2 = -0.44\,\text{J}\,.$$

When the block is at the equilibrium point, the elongation of the spring is $y = 9.6\,\text{cm}$ and the potential energy is

$$U_f = -mgy + \frac{1}{2}ky^2 = -(1.3\,\text{kg})(9.8\,\text{m/s}^2)(0.096\,\text{m}) + \frac{1}{2}(133\,\text{N/m})(0.096\,\text{m})^2 = -0.61\,\text{J}\,.$$

Write the equation for conservation of energy as $U_i = U_f + \frac{1}{2}mv^2$ and solve for $v$:

$$v = \sqrt{\frac{2(U_i - U_f)}{m}} = \sqrt{\frac{2(-0.44\,\text{J} + 0.61\,\text{J})}{1.3\,\text{kg}}} = 0.51\,\text{m/s}\,.$$

**91**

(a) The frequency of oscillation is

$$f = \frac{1}{2\pi}\sqrt{\frac{k}{m}} = \frac{1}{2\pi}\sqrt{\frac{480\,\text{N/m}}{1.2\,\text{kg}}} = 3.2\,\text{Hz}\,.$$

(b) Because mechanical energy is conserved the maximum kinetic energy of the block has the same value as the maximum potential energy stored in the spring, so $\frac{1}{2}mv_m^2 = \frac{1}{2}kx_m^2$. Thus

$$x_m = \sqrt{\frac{m}{k}}v_m = \sqrt{\frac{1.2\,\text{kg}}{480\,\text{N/m}}(5.2\,\text{m/s})} = 0.26\,[\text{m}\,.$$

(c) The position of the block is given by $x = x_m\cos(\omega t + \phi)$, where $x_m = 0.26\,\text{m}$ and $\omega = 2\pi f = 2\pi(3.2\,\text{Hz}) = 20\,\text{rad/s}$. Since $x = 0$ at time $t = 0$, the phase constant $\phi$ must be either $+\pi/2$ or $-\pi/2$. The velocity at $t = 0$ is given by $-\omega x_m \sin\phi$ and this is positive, so $\phi$ must be $-\pi/2$, The function is $x = (0.26\,\text{m})\cos[(20\,\text{rad/s})t - \pi/2]$.

# Chapter 16

**15**

The wave speed $v$ is given by $v = \sqrt{\tau/\mu}$, where $\tau$ is the tension in the rope and $\mu$ is the linear mass density of the rope. The linear mass density is the mass per unit length of rope:

$$\mu = \frac{m}{L} = \frac{0.0600\,\text{kg}}{2.00\,\text{m}} = 0.0300\,\text{kg/m}\,.$$

Thus

$$v = \sqrt{\frac{500\,\text{N}}{0.0300\,\text{kg/m}}} = 129\,\text{m/s}\,.$$

**17**

(a) In the expression given for $y$, the quantity $y_m$ is the amplitude and so is 0.12 mm.

(b) The wave speed is given by $v = \sqrt{\tau/\mu}$, where $\tau$ is the tension in the string and $\mu$ is the linear mass density of the string, so the wavelength is

$$\lambda = \frac{v}{f} = \frac{\sqrt{\tau/\mu}}{f}$$

and the angular wave number is

$$k = \frac{2\pi}{\lambda} = 2\pi f\sqrt{\frac{\mu}{\tau}} = 2\pi(100\,\text{Hz})\sqrt{\frac{0.50\,\text{kg/m}}{10\,\text{N}}} = 141\,\text{m}^{-1}\,.$$

(c) The frequency is $f = 100\,\text{Hz}$, so the angular frequency is $\omega = 2\pi f = 2\pi(100\,\text{Hz}) = 628\,\text{rad/s}$.

(d) The positive sign is used since the wave is traveling in the negative $x$ direction.

**21**

(a) Read the amplitude from the graph. It is the displacement at the peak and is about 5.0 cm.

(b) Read the wavelength from the graph. The curve crosses $y = 0$ at about $x = 15$ cm and again with the same slope at about $x = 55$ cm, so $\lambda = 55\,\text{cm} - 15\,\text{cm} = 40\,\text{cm} = 0.40\,\text{m}$.

(c) The wave speed is

$$v = \sqrt{\frac{\tau}{\mu}}\,,$$

where $\tau$ is the tension in the string and $\mu$ is the linear mass density of the string. Thus

$$v = \sqrt{\frac{3.6\,\text{N}}{25\times10^{-3}\,\text{kg/m}}} = 12\,\text{m/s}\,.$$

(d) The frequency is

$$f = \frac{v}{\lambda} = \frac{12\,\text{m/s}}{0.40\,\text{m}} = 30\,\text{Hz}$$

and the period is

$$T = \frac{1}{f} = \frac{1}{30\,\text{Hz}} = 0.033\,\text{s}\,.$$

(e) The maximum string speed is

$$u_m = \omega y_m = 2\pi f y_m = 2\pi(30\,\text{Hz})(5.0\,\text{cm}) = 940\,\text{cm/s} = 9.4\,\text{m/s}\,.$$

(f) The angular wave number is

$$k = \frac{2\pi}{\lambda} = \frac{2\pi}{0.40\,\text{m}} = 16\,\text{m}^{-1}\,.$$

(g) The angular frequency is $\omega = 2\pi f = 2\pi(30\,\text{Hz}) = 1.9 \times 10^2\,\text{rad/s}$.

(h) According to the graph, the displacement at $x = 0$ and $t = 0$ is $4.0 \times 10^{-2}\,\text{m}$. The formula for the displacement gives $y(0,0) = y_m \sin\phi$. We wish to select $\phi$ so that $5.0 \times 10^{-2} \sin\phi = 4.0 \times 10^{-2}$. The solution is either 0.93 rad or 2.21 rad. In the first case the function has a positive slope at $x = 0$ and matches the graph. In the second case it has negative slope and does not match the graph. We select $\phi = 0.93\,\text{rad}$.

(i) A positive sign appears in front of $\omega$ because the wave is moving in the negative $x$ direction.

**31**

The displacement of the string is given by

$$y = y_m \sin(kx - \omega t) + y_m \sin(kx - \omega t + \phi) = 2y_m \cos(\tfrac{1}{2}\phi)\sin(kx - \omega t + \tfrac{1}{2}\phi)\,,$$

where $\phi = \pi/2$. The amplitude is

$$A = 2y_m \cos(\tfrac{1}{2}\phi) = 2y_m \cos(\pi/4) = 1.41 y_m\,.$$

**35**

The phasor diagram is shown to the right: $y_{1m}$ and $y_{2m}$ represent the original waves and $y_m$ represents the resultant wave. The phasors corresponding to the two constituent waves make an angle of 90° with each other, so the triangle is a right triangle. The Pythagorean theorem gives

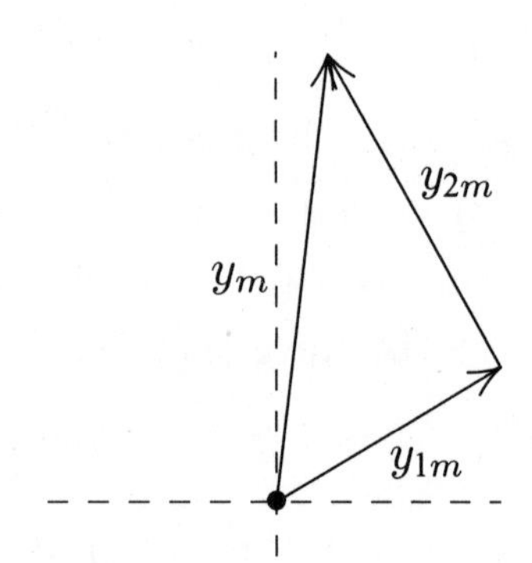


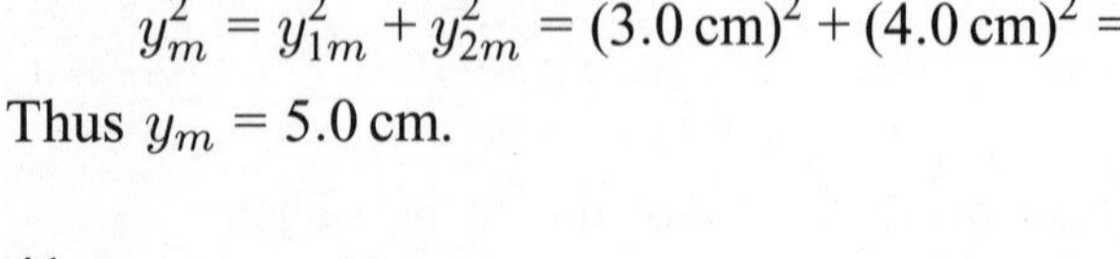

$$y_m^2 = y_{1m}^2 + y_{2m}^2 = (3.0\,\text{cm})^2 + (4.0\,\text{cm})^2 = 25\,\text{cm}^2\,.$$

Thus $y_m = 5.0\,\text{cm}$.

**41**

Possible wavelengths are given by $\lambda = 2L/n$, where $L$ is the length of the wire and $n$ is an integer. The corresponding frequencies are given by $f = v/\lambda = nv/2L$, where $v$ is the wave speed. The wave speed is given by $v = \sqrt{\tau/\mu} = \sqrt{\tau L/M}$, where $\tau$ is the tension in the wire, $\mu$ is the linear mass density of the wire, and $M$ is the mass of the wire. $\mu = M/L$ was used to obtain the last form. Thus

$$f = \frac{n}{2L}\sqrt{\frac{\tau L}{M}} = \frac{n}{2}\sqrt{\frac{\tau}{LM}} = \frac{n}{2}\sqrt{\frac{250\,\text{N}}{(10.0\,\text{m})(0.100\,\text{kg})}} = n(7.91\,\text{Hz})\,.$$

(a) For $n = 1$, $f = 7.91\,\text{Hz}$.

(b) For $n = 2$, $f = 15.8\,\text{Hz}$.

(c) For $n = 3$, $f = 23.7\,\text{Hz}$.

**43**

(a) The wave speed is given by $v = \sqrt{\tau/\mu}$, where $\tau$ is the tension in the string and $\mu$ is the linear mass density of the string. Since the mass density is the mass per unit length, $\mu = M/L$, where $M$ is the mass of the string and $L$ is its length. Thus

$$v = \sqrt{\frac{\tau L}{M}} = \sqrt{\frac{(96.0\,\text{N})(8.40\,\text{m})}{0.120\,\text{kg}}} = 82.0\,\text{m/s}\,.$$

(b) The longest possible wavelength $\lambda$ for a standing wave is related to the length of the string by $L = \lambda/2$, so $\lambda = 2L = 2(8.40\,\text{m}) = 16.8\,\text{m}$.

(c) The frequency is $f = v/\lambda = (82.0\,\text{m/s})/(16.8\,\text{m}) = 4.88\,\text{Hz}$.

**47**

(a) The resonant wavelengths are given by $\lambda = 2L/n$, where $L$ is the length of the string and $n$ is an integer, and the resonant frequencies are given by $f = v/\lambda = nv/2L$, where $v$ is the wave speed. Suppose the lower frequency is associated with the integer $n$. Then since there are no resonant frequencies between, the higher frequency is associated with $n+1$. That is, $f_1 = nv/2L$ is the lower frequency and $f_2 = (n+1)v/2L$ is the higher. The ratio of the frequencies is

$$\frac{f_2}{f_1} = \frac{n+1}{n}\,.$$

The solution for $n$ is

$$n = \frac{f_1}{f_2 - f_1} = \frac{315\,\text{Hz}}{420\,\text{Hz} - 315\,\text{Hz}} = 3\,.$$

The lowest possible resonant frequency is $f = v/2L = f_1/n = (315\,\text{Hz})/3 = 105\,\text{Hz}$.

(b) The longest possible wavelength is $\lambda = 2L$. If $f$ is the lowest possible frequency then $v = \lambda f = 2Lf = 2(0.75\,\text{m})(105\,\text{Hz}) = 158\,\text{m/s}$.

**53**

The waves have the same amplitude, the same angular frequency, and the same angular wave number, but they travel in opposite directions.

(a) The amplitude of each of the constituent waves is half the amplitude of the standing wave or 0.50 cm.

(b) Since the standing wave has three loops the string is three half-wavelengths long. If $L$ is the length of the string and $\lambda$ is the wavelength, then $L = 3\lambda/2$, or $\lambda = 2L/3 = 2(3.0\,\text{m})/3 = 2.0\,\text{m}$. The angular wave number is $k = 2\pi/\lambda = 2\pi/(2.0\,\text{m}) = 3.1\,\text{m}^{-1}$.

(c) If $v$ is the wave speed, then the frequency is

$$f = \frac{v}{\lambda} = \frac{3v}{2L} = \frac{3(100\,\text{m/s})}{2(3.0\,\text{m})} = 50\,\text{Hz}\,.$$

The angular frequency is $\omega = 2\pi f = 2\pi(50\,\text{Hz}) = 3.1 \times 10^2\,\text{rad/s}$.

(d) Since the first wave travels in the negative $x$ direction, the second wave must travel in the positive $x$ direction and the sign in front of $\omega$ must be a negative sign.

**61**

(a) The phasor diagram is shown to the right: $y_1$, $y_2$, and $y_3$ represent the original waves and $y_m$ represents the resultant wave. The horizontal component of the resultant is $y_{mh} = y_1 - y_3 = y_1 - y_1/3 = 2y_1/3$. The vertical component is $y_{mv} = y_2 = y_1/2$. The amplitude of the resultant is

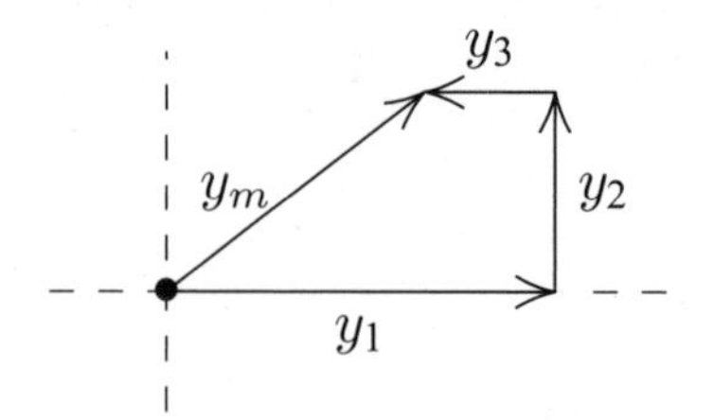


$$y_m = \sqrt{y_{mh}^2 + y_{mv}^2} = \sqrt{\left(\frac{2y_1}{3}\right)^2 + \left(\frac{y_1}{2}\right)^2} = \frac{5}{6}y_1 = 0.83y_1\,.$$

(b) The phase constant for the resultant is

$$\phi = \tan^{-1}\frac{y_{mv}}{y_{mh}} = \tan^{-1}\left(\frac{y_1/2}{2y_1/3}\right) = \tan^{-1}\frac{3}{4} = 0.644\,\text{rad} = 37^\circ\,.$$

(c) The resultant wave is

$$y = \frac{5}{6}y_1 \sin(kx - \omega t + 0.644\,\text{rad})\,.$$

The graph below shows the wave at time $t = 0$. As time goes on it moves to the right with speed $v = \omega/k$.

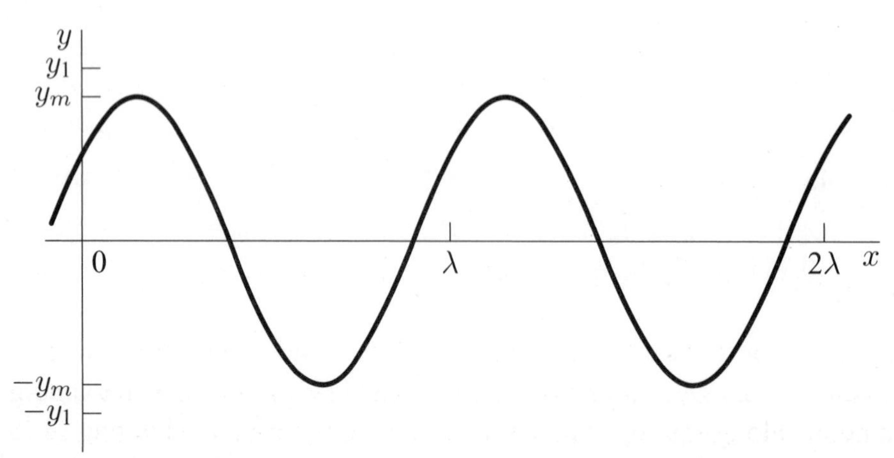

## 69

(a) Take the form of the displacement to be $y(x,t) = y_m \sin(kx - \omega t)$. The speed of a point on the cord is $u(x,t) = \partial y/\partial t = -\omega y_m \cos(kx - \omega t)$ and its maximum value is $u_m = \omega y_m$. The wave speed, on the other hand, is given by $v = \lambda/T = \omega/k$. The ratio is

$$\frac{u_m}{v} = \frac{\omega y_m}{\omega/k} = k y_m = \frac{2\pi y_m}{\lambda}.$$

(b) The ratio of the speeds depends only on the ratio of the amplitude to the wavelength. Different waves on different cords have the same ratio of speeds if they have the same amplitude and wavelength, regardless of the wave speeds , linear densities of the cords, and the tensions in the cords.

## 77

(a) If $\tau$ is the tension in the wire and $\mu$ is its linear mass density, then the wave speed is $v = \sqrt{\tau/\mu}$. The linear mass density is $\mu = m/L$, where $m$ is the mass of the wire and $L$ is its length. Thus

$$v = \sqrt{\frac{\tau L}{m}} = \sqrt{\frac{(120\,\text{N})(1.50\,\text{m})}{8.70 \times 10^{-3}\,\text{kg}}} = 144\,\text{m/s}.$$

(b) A one-loop standing wave has two nodes, one a each end, and these are half a wavelength apart, so the wavelength is $\lambda = 2L = 2(1.50\,\text{m}) = 3.00\,\text{m}$.

A two-loop standing wave has three nodes, one at each end and one at the midpoint. Since the nodes are half a wavelength apart, the wavelength is $\lambda = L = 1.50\,\text{m}$.

(c) and (d) The frequency is $f = v/\lambda$. For the one-loop wave $f = (144\,\text{m/s})/(3.00\,\text{m}) = 48.0\,\text{Hz}$ and for the two-loop wave $f = (144\,\text{m/s})/(1.50\,\text{m}) = 96.6\,\text{Hz}$.

## 87

(a) The transverse rope velocity is given by $u = \omega y_m$, where $\omega$ is the angular frequency and $y_m$ is the amplitude. The angular frequency is $\omega = 2\pi f$, where $f$ is the frequency. Thus

$$y_m = \frac{u}{\omega} = \frac{u}{2\pi f} = \frac{5.0\,\text{m/s}}{2\pi(5.0\,\text{Hz})} = 0.16\,\text{m}.$$

(b) The wave speed is $v = \sqrt{\tau/\mu}$, where $\tau$ is the tension in the rope and $\mu$ is the linear mass density of the rope. The linear mass density is $\mu = m/L$, where $m$ is the mass of the rope and $L$ is its length. The wave speed is $\lambda f$, where $\lambda$ is the wavelength. Since the rope is vibrating in its fundamental mode $\lambda = 2L$. Thus

$$\tau = \mu v^2 = \frac{m}{L}(\lambda f)^2 = \frac{m}{L}(2Lf)^2 = 4mLf^2 = 4(1.2\,\text{kg})(2.0\,\text{m})(5.0\,\text{Hz})^2 = 2.4 \times 10^2\,\text{N}.$$

(c) The general form for the displacement at coordinate $x$ and time $t$ for a standing wave that has nodes at $x = 0$ and $x = L$ is $y = y_m \sin(2\pi x/\lambda)\sin(2\pi f t$. Here $y_m$ is the maximum displacement of any of the points along the rope. Since, in this case, the rope is vibrating in its fundamental mode $y_m$ is the maximum displacement of the point at its center and thus has the value calculated in part (a). The displacement at any coordinate $x$ is

$$y(x,t) = (0.16\,\text{m})\sin\left[\frac{2\pi}{4.0\,\text{m}}x\right]\sin\left[2\pi(5.0\,\text{Hz})t\right] = (0.16\,\text{m})\sin[(1.6\,\text{m}^{-1})x]\sin[(31\,\text{s}^{-1})t]\,.$$

**89**

(a) The wave speed is given by $v = \sqrt{\tau/\mu}$, where $\tau$ is the tension in the rubber band and $\mu$ is the linear mass density of the rubber band. According to Hooke's law the tension is $\tau = k\,\Delta\ell$. The length of the stretched rubber band is $\ell + \Delta\ell$, so the linear mass density is $\mu = m/(\ell + \Delta\ell)$. The wave speed is

$$v = \sqrt{\frac{k\,\Delta\ell(\ell + \Delta\ell)}{m}}\,.$$

(b) The time for a pulse to travel the length of the rubber band is

$$t = \frac{\ell + \Delta\ell}{v} = (\ell + \Delta\ell)\sqrt{\frac{m}{k\,\Delta\ell(\ell + \Delta\ell)}} = \sqrt{\frac{m(\ell + \Delta\ell)}{k\,\Delta\ell}}\,.$$

If $\Delta\ell$ is much less than $\ell$ we may neglect the $\Delta\ell$ in the numerator. Then

$$v = \sqrt{\frac{m\ell}{k\,\Delta\ell}}\,,$$

which is proportional to $1/\sqrt{\Delta\ell}$.

(c) If $\Delta\ell$ is much greater than $\ell$ we may neglect $\ell$ in the numerator. Then

$$v = \sqrt{\frac{m\,\Delta\ell}{k\,\Delta\ell}} = \sqrt{\frac{m}{k}}\,,$$

which is independent of $\Delta\ell$.

# Chapter 17

<u>5</u>

Let $t_f$ be the time for the stone to fall to the water and $t_s$ be the time for the sound of the splash to travel from the water to the top of the well. Then the total time elapsed from dropping the stone to hearing the splash is $t = t_f + t_s$. If $d$ is the depth of the well, then the kinematics of free fall gives $d = \frac{1}{2}gt_f^2$, or $t_f = \sqrt{2d/g}$. The sound travels at a constant speed $v_s$, so $d = v_s t_s$, or $t_s = d/v_s$. Thus the total time is

$$t = \sqrt{\frac{2d}{g}} + \frac{d}{v_s}\,.$$

This equation is to be solved for $d$. Rewrite it as

$$\sqrt{\frac{2d}{g}} = t - \frac{d}{v_s}$$

and square both sides to obtain

$$\frac{2d}{g} = t^2 - 2\frac{t}{v_s}d + \frac{1}{v_s^2}d^2\,.$$

Now multiply by $gv_s^2$ and rearrange to get

$$gd^2 - 2v_s(gt + v_s)d + gv_s^2t^2 = 0\,.$$

This is a quadratic equation for $d$. Its solutions are

$$d = \frac{2v_s(gt + v_s) \pm \sqrt{4v_s^2(gt + v_s)^2 - 4g^2v_s^2t^2}}{2g}\,.$$

The physical solution must yield $d = 0$ for $t = 0$, so we take the solution with the negative sign in front of the square root. Once values are substituted the result $d = 40.7\,\text{m}$ is obtained.

<u>7</u>

If $d$ is the distance from the location of the earthquake to the seismograph and $v_s$ is the speed of the S waves, then the time for these waves to reach the seismograph is $t_s = d/v_s$. Similarly, the time for P waves to reach the seismograph is $t_p = d/v_p$. The time delay is

$$\Delta t = \frac{d}{v_s} - \frac{d}{v_p} = \frac{d(v_p - v_s)}{v_s v_p}\,,$$

so

$$d = \frac{v_s v_p \, \Delta t}{(v_p - v_s)} = \frac{(4.5\,\text{km/s})(8.0\,\text{km/s})(3.0\,\text{min})(60\,\text{s/min})}{8.0\,\text{km/s} - 4.5\,\text{km/s}} = 1.9 \times 10^3\,\text{km}\,.$$

Notice that values for the speeds were substituted as given, in km/s, but that the value for the time delay was converted from minutes to seconds.

**9**

(a) Use $\lambda = v/f$, where $v$ is the speed of sound in air and $f$ is the frequency. Thus

$$\lambda = \frac{343\,\text{m/s}}{4.5 \times 10^6\,\text{Hz}} = 7.62 \times 10^{-5}\,\text{m}\,.$$

(b) Now $\lambda = v/f$, where $v$ is the speed of sound in tissue. The frequency is the same for air and tissue. Thus

$$\lambda = \frac{1500\,\text{m/s}}{4.5 \times 10^6\,\text{Hz}} = 3.33 \times 10^{-4}\,\text{m}\,.$$

**19**

Let $L_1$ be the distance from the closer speaker to the listener. The distance from the other speaker to the listener is $L_2 = \sqrt{L_1^2 + d^2}$, where $d$ is the distance between the speakers. The phase difference at the listener is

$$\phi = \frac{2\pi(L_2 - L_1)}{\lambda}\,,$$

where $\lambda$ is the wavelength.

(a) For a minimum in intensity at the listener, $\phi = (2n + 1)\pi$, where $n$ is an integer. Thus $\lambda = 2(L_2 - L_1)/(2n + 1)$. The frequency is

$$f = \frac{v}{\lambda} = \frac{(2n+1)v}{2\left[\sqrt{L_1^2 + d^2} - L_1\right]} = \frac{(2n+1)(343\,\text{m/s})}{2\left[\sqrt{(3.75\,\text{m})^2 + (2.00\,\text{m})^2} - 3.75\,\text{m}\right]} = (2n+1)(343\,\text{Hz})\,.$$

To obtain the lowest frequency for which a minimum occurs set $n$ equal to 0. The frequency is $f_{\text{min},\,1} = 343$ Hz.

(b) To obtain the second lowest frequency set $n$ equal to 1. This means multiply $F_{\text{min},\,1}$ by 3.

(c) To obtain the third lowest frequency set $n$ equal to 2. This means multiply $F_{\text{min},\,1}$ by 5.

For a maximum in intensity at the listener, $\phi = 2n\pi$, where $n$ is any positive integer. Thus

$$\lambda = \frac{1}{n}\left[\sqrt{L_1^2 + d^2} - L_1\right]$$

and

$$f = \frac{v}{\lambda} = \frac{nv}{\sqrt{L_1^2 + d^2} - L_1} = \frac{n(343\,\text{m/s})}{\sqrt{(3.75\,\text{m})^2 + (2.00\,\text{m})^2} - 3.75\,\text{m}} = n(686\,\text{Hz})\,.$$

(d) To obtain the lowest frequency for which a maximum occurs set $n$ equal to 1. The frequency is $f_{\text{max},\,1} = 686$ Hz.

(e) To obtain the second lowest frequency set $n$ equal to 2. This means multiply $F_{\text{min},\,1}$ by 2.

(f) To obtain the third lowest frequency set $n$ equal to 3. This means multiply $F_{\text{min},\,1}$ by 3.

**25**

The intensity is the rate of energy flow per unit area perpendicular to the flow. The rate at which energy flows across every sphere centered at the source is the same, regardless of the sphere radius, and is the same as the power output of the source. If $P$ is the power output and $I$ is the intensity a distance $r$ from the source, then $P = IA = 4\pi r^2 I$, where $A$ ($= 4\pi r^2$) is the surface area of a sphere of radius $r$. Thus $P = 4\pi(2.50\,\text{m})^2(1.91 \times 10^{-4}\,\text{W/m}^2) = 1.50 \times 10^{-2}\,\text{W}$.

**29**

(a) Let $I_1$ be the original intensity and $I_2$ be the final intensity. The original sound level is $\beta_1 = (10\,\text{dB})\log(I_1/I_0)$ and the final sound level is $\beta_2 = (10\,\text{dB})\log(I_2/I_0)$, where $I_0$ is the reference intensity. Since $\beta_2 = \beta_1 + 30\,\text{dB}$,

$$(10\,\text{dB})\log(I_2/I_0) = (10\,\text{dB})\log(I_1/I_0) + 30\,\text{dB}\,,$$

or

$$(10\,\text{dB})\log(I_2/I_0) - (10\,\text{dB})\log(I_1/I_0) = 30\,\text{dB}\,.$$

Divide by 10 dB and use $\log(I_2/I_0) - \log(I_1/I_0) = \log(I_2/I_1)$ to obtain $\log(I_2/I_1) = 3$. Now use each side as an exponent of 10 and recognize that

$$10^{\log(I_2/I_1)} = I_2/I_1\,.$$

The result is $I_2/I_1 = 10^3$. The intensity is multiplied by a factor of $1.0 \times 10^3$.

(b) The pressure amplitude is proportional to the square root of the intensity so it is multiplied by a factor of $\sqrt{1000} = 32$.

**43**

(a) The string is fixed at both ends and, when vibrating at its lowest resonant frequency, exactly half a wavelength fits between the ends. If $L$ is the length of the string and $\lambda$ is the wavelength, then $\lambda = 2L$. The frequency is $f = v/\lambda = v/2L$, where $v$ is the speed of waves on the string. Thus $v = 2Lf = 2(0.220\,\text{m})(920\,\text{Hz}) = 405\,\text{m/s}$.

(b) The wave speed is given by $v = \sqrt{\tau/\mu}$, where $\tau$ is the tension in the string and $\mu$ is the linear mass density of the string. If $M$ is the mass of the string, then $\mu = M/L$ since the string is uniform. Thus

$$\tau = \mu v^2 = \frac{M}{L}v^2 = \frac{800 \times 10^{-6}\,\text{kg}}{0.220\,\text{m}}(405\,\text{m/s})^2 = 596\,\text{N}\,.$$

(c) The wavelength is $\lambda = 2L = 2(0.220\,\text{m}) = 0.440\,\text{m}$.

(d) The frequency of the sound wave in air is the same as the frequency of oscillation of the string. The wavelength is different because the wave speed is different. If $v_a$ is the speed of sound in air the wavelength in air is

$$\lambda_a = \frac{v_a}{f} = \frac{343\,\text{m/s})}{920\,\text{Hz}} = 0.373\,\text{m}\,.$$

<u>45</u>

(a) Since the pipe is open at both ends there are displacement antinodes at both ends and an integer number of half-wavelengths fit into the length of the pipe. If $L$ is the pipe length and $\lambda$ is the wavelength then $\lambda = 2L/n$, where $n$ is an integer. If $v$ is the speed of sound then the resonant frequencies are given by $f = v/\lambda = nv/2L$. Now $L = 0.457\,\text{m}$, so $f = n(344\,\text{m/s})/2(0.457\,\text{m}) = 376.4n\,\text{Hz}$. To find the resonant frequencies that lie between 1000 Hz and 2000 Hz, first set $f = 1000\,\text{Hz}$ and solve for $n$, then set $f = 2000\,\text{Hz}$ and again solve for $n$. You should get 2.66 and 5.32. This means $n = 3$, 4, and 5 are the appropriate values of $n$. There are three resonance frequencies in the given range.

(b) For $n = 3$, $f = 3(376.4\,\text{Hz}) = 1129\,\text{Hz}$.

(c) For $n = 4$, $f = 4(376.4\,\text{Hz}) = 1506\,\text{Hz}$.

<u>47</u>

The string is fixed at both ends so the resonant wavelengths are given by $\lambda = 2L/n$, where $L$ is the length of the string and $n$ is an integer. The resonant frequencies are given by $f = v/\lambda = nv/2L$, where $v$ is the wave speed on the string. Now $v = \sqrt{\tau/\mu}$, where $\tau$ is the tension in the string and $\mu$ is the linear mass density of the string. Thus $f = (n/2L)\sqrt{\tau/\mu}$. Suppose the lower frequency is associated with $n = n_1$ and the higher frequency is associated with $n = n_1 + 1$. There are no resonant frequencies between so you know that the integers associated with the given frequencies differ by 1. Thus $f_1 = (n_1/2L)\sqrt{\tau/\mu}$ and

$$f_2 = \frac{n_1+1}{2L}\sqrt{\frac{\tau}{\mu}} = \frac{n_1}{2L}\sqrt{\frac{\tau}{\mu}} + \frac{1}{2L}\sqrt{\frac{\tau}{\mu}} = f_1 + \frac{1}{2L}\sqrt{\frac{\tau}{\mu}}\,.$$

This means $f_2 - f_1 = (1/2L)\sqrt{\tau/\mu}$ and

$$\begin{aligned}\tau &= 4L^2\mu(f_2 - f_1)^2 \\ &= 4(0.300\,\text{m})^2(0.650 \times 10^{-3}\,\text{kg/m})(1320\,\text{Hz} - 880\,\text{Hz})^2 \\ &= 45.3\,\text{N}\,.\end{aligned}$$

<u>53</u>

Each wire is vibrating in its fundamental mode so the wavelength is twice the length of the wire ($\lambda = 2L$) and the frequency is

$$f = \frac{v}{\lambda} = \frac{1}{2L}\sqrt{\frac{\tau}{\mu}}\,,$$

where $v$ ($= \sqrt{\tau/\mu}$) is the wave speed for the wire, $\tau$ is the tension in the wire, and $\mu$ is the linear mass density of the wire.

Suppose the tension in one wire is $\tau$ and the oscillation frequency of that wire is $f_1$. The tension in the other wire is $\tau + \Delta\tau$ and its frequency is $f_2$. You want to calculate $\Delta\tau/\tau$ for $f_1 = 600\,\text{Hz}$ and $f_2 = 606\,\text{Hz}$.

Now

$$f_1 = \frac{1}{2L}\sqrt{\frac{\tau}{\mu}}$$

and

$$f_2 = \frac{1}{2L}\sqrt{\frac{\tau + \Delta\tau}{\mu}},$$

so

$$\frac{f_2}{f_1} = \sqrt{\frac{\tau + \Delta\tau}{\tau}} = \sqrt{1 + \frac{\Delta\tau}{\tau}}.$$

This means

$$\frac{\Delta\tau}{\tau} = \left(\frac{f_2}{f_1}\right)^2 - 1 = \left(\frac{606\,\text{Hz}}{600\,\text{Hz}}\right)^2 - 1 = 0.020\,.$$

**65**

(a) The expression for the Doppler shifted frequency is

$$f' = f\,\frac{v \pm v_D}{v \mp v_S},$$

where $f$ is the unshifted frequency, $v$ is the speed of sound, $v_D$ is the speed of the detector (the uncle), and $v_S$ is the speed of the source (the locomotive). All speeds are relative to the air. The uncle is at rest with respect to the air, so $v_D = 0$. The speed of the source is $v_S = 10\,\text{m/s}$. Since the locomotive is moving away from the uncle the frequency decreases and we use the positive sign in the denominator. Thus

$$f' = f\,\frac{v}{v + v_S} = (500.0\,\text{Hz})\left(\frac{343\,\text{m/s}}{343\,\text{m/s} + 10.00\,\text{m/s}}\right) = 485.8\,\text{Hz}\,.$$

(b) The girl is now the detector. Relative to the air she is moving with speed $v_D = 10.00\,\text{m/s}$ toward the source. This tends to increase the frequency and we use the positive sign in the numerator. The source is moving at $v_S = 10.00\,\text{m/s}$ away from the girl. This tends to decrease the frequency and we use the positive sign in the denominator. Thus $(v + v_D) = (v + v_S)$ and $f' = f = 500.0\,\text{Hz}$.

(c) Relative to the air the locomotive is moving at $v_S = 20.00\,\text{m/s}$ away from the uncle. Use the positive sign in the denominator. Relative to the air the uncle is moving at $v_D = 10.00\,\text{m/s}$ toward the locomotive. Use the positive sign in the numerator. Thus

$$f' = f\,\frac{v + v_D}{v + v_S} = (500.0\,\text{Hz})\left(\frac{343\,\text{m/s} + 10.00\,\text{m/s}}{343\,\text{m/s} + 20.00\,\text{m/s}}\right) = 486.2\,\text{Hz}\,.$$

(d) Relative to the air the locomotive is moving at $v_S = 20.00\,\text{m/s}$ away from the girl and the girl is moving at $v_D = 20.00\,\text{m/s}$ toward the locomotive. Use the positive signs in both the numerator and the denominator. Thus $(v + v_D) = (v + v_S)$ and $f' = f = 500.0\,\text{Hz}$.

**69**

(a) The half angle $\theta$ of the Mach cone is given by $\sin\theta = v/v_S$, where $v$ is the speed of sound and $v_S$ is the speed of the plane. Since $v_S = 1.5v$, $\sin\theta = v/1.5v = 1/1.5$. This means $\theta = 42°$.
(b) Let $h$ be the altitude of the plane and suppose the Mach cone intersects Earth's surface a distance $d$ behind the plane. The situation is shown on the diagram to the right, with P indicating the plane and O indicating the observer. The cone angle is related to $h$ and $d$ by $\tan\theta = h/d$, so $d = h/\tan\theta$. The shock wave reaches O in the time the plane takes to fly the distance $d$:

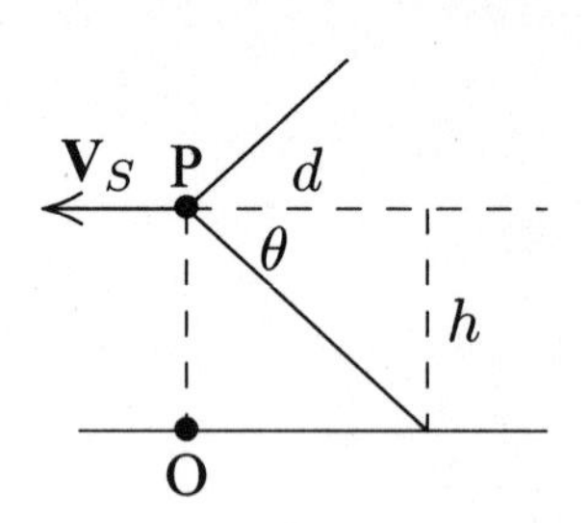


$$t = d/v = h/v\tan\theta = (5000\,\text{m})/1.5(331\,\text{m/s})\tan 42° = 11\,\text{s}\,.$$

**77**

(a) Use the Doppler shift equation, which gives the detected frequency $f'$:

$$f' = f\frac{v \pm v_D}{v \pm v_S}\,,$$

where $f$ is the emitted frequency, $v$ is the speed of sound, $V_D$ is the speed of the detector, and $v_S$ is the speed of the source. You are the detector and are stationary, so $v_D = 0$. The siren is the source and is moving away from you, which lowers the frequency, so the positive sign is used in the denominator. Thus

$$f' = (1000\,\text{Hz})\frac{330\,\text{m/s}}{330\,\text{m/s} + 10\,\text{m/s}} = 9.7 \times 10^2\,\text{Hz}\,.$$

(b) The detector is the cliff, which is stationary. The source is the siren, which is moving toward the cliff. This increases the frequency and the minus sign is used in the denominator. Thus

$$f' = (1000\,\text{Hz})\frac{330\,\text{m/s}}{330\,\text{m/s} - 10\,\text{m/s}} = 1.0 \times 10^3\,\text{Hz}\,.$$

(c) The beat frequency is $f_{\text{beat}} = 1.0 \times 10^3\,\text{Hz} - 9.7 \times 10^2\,\text{Hz} = 60\,\text{Hz}$.

**81**

(a) The rate with which sound energy is passing through the surface of a sphere with radius $r$ is $P = 4\pi r^2 I$, where $I$ is the intensity. Since energy is conserved this must be the power of the source. Thus $P = 4\pi(0.0080\,\text{W/m}^2)(10\,\text{m})^2 = 10\,\text{W}$.
(b) The sound intensity is $I = P/4\pi r^2 = (10\,\text{W}/4\pi(5.0\,\text{m})^2 = 0.032\,\text{W/m}^2$.
(c) The sound level is $\beta = (10\,\text{dB}\log(I/I_0)$, where $I_0$ is the standard reference intensity, which is given as $10^{-12}\,\text{W/m}^2$ in the text. At a point 10 m from the source it is

$$\beta = (10\,\text{dB})log\frac{0.0080\,\text{W/m}^2}{10^{-12}} = 99\,\text{dB}\,.$$

**85**

(a) The intensity is given by $I = \frac{1}{2}\rho v \omega^2 s_m^2$, where $\rho$ is the density of the medium, $v$ is the speed of sound, $\omega$ is the angular frequency, and $s_m$ is the displacement amplitude. The displacement and pressure amplitudes are related by $\Delta p_m = \rho v \omega s_m$, so $s_m = \Delta p_m / \rho v \omega$ and $I = (\Delta p_m)^2 / 2\rho v$. For waves of the same frequency the ratio of the intensity for propagation in water to the intensity for propagation in air is

$$\frac{I_w}{I_a} = \left(\frac{\Delta p_{mw}}{\Delta p_{ma}}\right)^2 \frac{\rho_a v_a}{\rho_w v_w},$$

where the subscript $a$ denotes air and the subscript $w$ denotes water.

Since $I_a = I_w$,

$$\frac{\Delta p_{mw}}{\Delta p_{ma}} = \sqrt{\frac{\rho_w v_w}{\rho_a v_a}} = \sqrt{\frac{(0.998 \times 10^3\,\text{kg/m}^3)(1482\,\text{m/s})}{(1.21\,\text{kg/m}^3)(343\,\text{m/s})}} = 59.7\,.$$

The speeds of sound are given in Table 17–1 and the densities are given in Table 14–1.

(b) Now $\Delta p_{mw} = \Delta p_{ma}$, so

$$\frac{I_w}{I_a} = \frac{\rho_a v_a}{\rho_w v_w} = \frac{(1.21\,\text{kg/m}^3)(343\,\text{m/s})}{(0.998 \times 10^3\,\text{kg/m}^3)(1482\,\text{m/s})} = 2.81 \times 10^{-4}\,.$$

**87**

(a) When the right side of the instrument is pulled out a distance $d$ the path length for sound waves increases by $2d$. Since the interference pattern changes from a minimum to the next maximum, this distance must be half a wavelength of the sound. So $2d = \lambda/2$, where $\lambda$ is the wavelength. Thus $\lambda = 4d$ and, if $v$ is the speed of sound, the frequency is $f = v/\lambda = v/4d = (343\,\text{m/s})/4(0.0165\,\text{m}) = 5.2 \times 10^3$ Hz.

(b) The displacement amplitude is proportional to the square root of the intensity (see Eq. 17–27). Write $\sqrt{I} = Cs_m$, where $I$ is the intensity, $s_m$ is the displacement amplitude, and $C$ is a constant of proportionality. At the minimum, interference is destructive and the displacement amplitude is the difference in the amplitudes of the individual waves: $s_m = s_{SAD} - s_{SBD}$, where the subscripts indicate the paths of the waves. At the maximum, the waves interfere constructively and the displacement amplitude is the sum of the amplitudes of the individual waves: $s_m = s_{SAD} + s_{SBD}$. Solve $\sqrt{100} = C(s_{SAD} - s_{SBD})$ and $\sqrt{900} = C(s_{SAD} + s_{SBD})$ for $s_{SAD}$ and $s_{SBD}$. Add the equations to obtain $s_{SAD} = (\sqrt{100} + \sqrt{900})/2C = 20/C$, then subtract them to obtain $s_{SBD} = (\sqrt{900} - \sqrt{100})/2C = 10/C$. The ratio of the amplitudes is $s_{SAD}/s_{SBD} = 2$.

(c) Any energy losses, such as might be caused by frictional forces of the walls on the air in the tubes, result in a decrease in the displacement amplitude. Those losses are greater on path B since it is longer than path A.

**101**

(a) The frequency is increased by reflection from the flowing blood, so the blood must be flowing to the right, with a positive velocity component in the direction of the original source of the ultrasound.

(b) Use the Doppler shift equation twice. It is

$$f' = f\frac{v \pm v_D}{v \pm v_S},$$

where $f$ is the emitted frequency, $v$ is the speed of sound, $V_D$ is the speed of the detector, and $v_S$ is the speed of the source. First, take the source to be the ultrasound generator, which is stationary, and the detector to be the blood. Thus $v_S = 0$ and $v_D = v_b \cos\theta$, where $v_b$ is the speed of the blood. Since the detected frequency is greater than the generator frequency, use the plus sign in the numerator. Thus

$$f' = f\frac{v + v_b \cos\theta}{v}.$$

In the next step the blood is the source and $f'$ is the emitted frequency. The generator is the detector and is stationary. Take $v_D = 0$ and $v_S = v_b \cos\theta$. The detected frequency is higher, so we use the minus sign in the denominator of the Doppler shift equation. The detected frequency is

$$f'' = f'\frac{v}{v - v_b\cos\theta} = f\left(\frac{v + v_b\cos\theta}{v}\right)\left(\frac{v}{v - v_b\cos\theta}\right) = f\frac{v + v_b\cos\theta}{v - v_b\cos\theta},$$

where $f'$ was replaced by the expression developed previously. The solution for $v_b$ is

$$v_b = \left(\frac{f'' - f}{f'' + f}\right)\left(\frac{v}{\cos\theta}\right) = \left(\frac{5495\,\text{Hz}}{5.000\,000 \times 10^6\,\text{Hz} + 5495\,\text{Hz}}\right)\left(\frac{1540\,\text{m/s}}{\cos 20^\circ}\right) = 0.90\,\text{m/s}.$$

(c) If $\theta$ increases, $\cos\theta$ decreases. This means the numerator of the expression for $f''$ decreases and the denominator increases. Both changes result in a decrease in $f''$.

# Chapter 18

<u>9</u>

Since a volume is the product of three lengths, the change in volume due to a temperature change $\Delta T$ is given by $\Delta V = 3\alpha V\,\Delta T$, where $V$ is the original volume and $\alpha$ is the coefficient of linear expansion. See Eq. 18–11. Since $V = (4\pi/3)R^3$, where $R$ is the original radius of the sphere,

$$\Delta V = 3\alpha\left(\frac{4\pi}{3}R^3\right)\Delta T = (23\times10^{-6}\,/\text{C}^\circ)(4\pi)(10\,\text{cm})^3(100\,\text{C}^\circ) = 29\,\text{cm}^3\,.$$

The value for the coefficient of linear expansion was obtained from Table 18–2.

<u>15</u>

If $V_c$ is the original volume of the cup, $\alpha_a$ is the coefficient of linear expansion of aluminum, and $\Delta T$ is the temperature increase, then the change in the volume of the cup is $\Delta V_c = 3\alpha_a V_c\,\Delta T$. See Eq. 18–11. If $\beta$ is the coefficient of volume expansion for glycerin then the change in the volume of glycerin is $\Delta V_g = \beta V_c\,\Delta T$. Note that the original volume of glycerin is the same as the original volume of the cup. The volume of glycerin that spills is

$$\begin{aligned}\Delta V_g - \Delta V_c &= (\beta - 3\alpha_a)V_c\,\Delta T\\ &= \left[(5.1\times10^{-4}\,/\text{C}^\circ) - 3(23\times10^{-6}\,/\text{C}^\circ)\right](100\,\text{cm}^3)(6\,\text{C}^\circ) = 0.26\,\text{cm}^3\,.\end{aligned}$$

<u>21</u>

Consider half the bar. Its original length is $\ell_0 = L_0/2$ and its length after the temperature increase is $\ell = \ell_0 + \alpha\ell_0\,\Delta T$. The old position of the half-bar, its new position, and the distance $x$ that one end is displaced form a right triangle, with a hypotenuse of length $\ell$, one side of length $\ell_0$, and the other side of length $x$. The Pythagorean theorem yields $x^2 = \ell^2 - \ell_0^2 = \ell_0^2(1+\alpha\,\Delta T)^2 - \ell_0^2$. Since the change in length is small we may approximate $(1+\alpha\,\Delta T)^2$ by $1 + 2\alpha\,\Delta T$, where the small term $(\alpha\,\Delta T)^2$ was neglected. Then

$$x^2 = \ell_0^2 + 2\ell_0^2\alpha\,\Delta T - \ell_0^2 = 2\ell_0^2\alpha\,\Delta T$$

and

$$x = \ell_0\sqrt{2\alpha\,\Delta T} = \frac{3.77\,\text{m}}{2}\sqrt{2(25\times10^{-6}\,/\text{C}^\circ)(32\,\text{C}^\circ)} = 7.5\times10^{-2}\,\text{m}\,.$$

<u>25</u>

The melting point of silver is 1235 K, so the temperature of the silver must first be raised from 15.0°C (= 288 K) to 1235 K. If $m$ is the mass of the silver and $c$ is its specific heat, this requires energy

$$Q = cm(T_f - T_i) = (236\,\text{J/kg}\cdot\text{K})(0.130\,\text{kg})(1235^\circ\text{C} - 288^\circ\text{C}) = 2.91\times10^4\,\text{J}\,.$$

Now the silver at its melting point must be melted. If $L_F$ is the heat of fusion for silver this requires

$$Q = mL_F = (0.130\,\text{kg})(105 \times 10^3\,\text{J/kg}) = 1.36 \times 10^4\,\text{J}.$$

The total energy required as heat is $2.91 \times 10^4\,\text{J} + 1.36 \times 10^4\,\text{J} = 4.27 \times 10^4\,\text{J}$. The specific heat of silver can be found in Table 18–3 and its heat of fusion can be found in Table 18–4.

**27**

Mass $m$ (= 0.100 kg) of water, with specific heat $c$ (= 4190 J/kg · K), is raised from an initial temperature $T_i$ (= 23°C) to its boiling point $T_f$ (= 100°C). The heat input is given by $Q = cm(T_f - T_i)$. This must be the power output of the heater $P$ multiplied by the time $t$; $Q = Pt$. Thus

$$t = \frac{Q}{P} = \frac{cm(T_f - T_i)}{P} = \frac{(4190\,\text{J/kg}\cdot\text{K})(0.100\,\text{kg})(100^\circ\text{C} - 23^\circ\text{C})}{200\,\text{W}} = 160\,\text{s}.$$

**41**

(a) There are three possibilities:

1. None of the ice melts and the water-ice system reaches thermal equilibrium at a temperature that is at or below the melting point of ice.

2. The system reaches thermal equilibrium at the melting point of ice, with some of the ice melted.

3. All of the ice melts and the system reaches thermal equilibrium at a temperature at or above the melting point of ice.

First suppose that no ice melts. The temperature of the water decreases from $T_{Wi}$ (= 25°C) to some final temperature $T_f$ and the temperature of the ice increases from $T_{Ii}$ (= −15°C) to $T_f$. If $m_W$ is the mass of the water and $c_W$ is its specific heat then the water loses energy

$$Q = c_W m_W (T_{Wi} - T_f).$$

If $m_I$ is the mass of the ice and $c_I$ is its specific heat then the ice absorbs energy

$$Q = c_I m_I (T_f - T_{Ii}).$$

Since no energy is lost these two energies must be the same and

$$c_W m_W (T_{Wi} - T_f) = c_I m_I (T_f - T_{Ii}).$$

The solution for the final temperature is

$$\begin{aligned} T_f &= \frac{c_W m_W T_{Wi} + c_I m_I T_{Ii}}{c_W m_W + c_I m_I} \\ &= \frac{(4190\,\text{J/kg}\cdot\text{K})(0.200\,\text{kg})(25^\circ\text{C}) + (2220\,\text{J/kg}\cdot\text{K})(0.100\,\text{kg})(-15^\circ\text{C})}{(4190\,\text{J/kg}\cdot\text{K})(0.200\,\text{kg}) + (2220\,\text{J/kg}\cdot\text{K})(0.100\,\text{kg})} \\ &= 16.6^\circ\text{C}. \end{aligned}$$

This is above the melting point of ice, so at least some of the ice must have melted. The calculation just completed does not take into account the melting of the ice and is in error.

Now assume the water and ice reach thermal equilibrium at $T_f = 0°\text{C}$, with mass $m$ ($< m_I$) of the ice melted. The magnitude of the energy lost by the water is

$$Q = c_W m_W T_{Wi}\,,$$

and the energy absorbed by the ice is

$$Q = c_I m_I (0 - T_{Ii}) + m L_F\,,$$

where $L_F$ is the heat of fusion for water. The first term is the energy required to warm all the ice from its initial temperature to 0°C and the second term is the energy required to melt mass $m$ of the ice. The energy lost by the water equals the energy gained by the ice, so

$$c_W m_W T_{Wi} = -c_I m_I T_{Ii} + m L_F\,.$$

This equation can be solved for the mass $m$ of ice melted:

$$\begin{aligned} m &= \frac{c_W m_W T_{Wi} + c_I m_I T_{Ii}}{L_F} \\ &= \frac{(4190\,\text{J/kg}\cdot\text{K})(0.200\,\text{kg})(25°\text{C}) + (2220\,\text{J/kg}\cdot\text{K})(0.100\,\text{kg})(-15°\text{C})}{333\times 10^3\,\text{J/kg}} \\ &= 5.3\times 10^{-2}\,\text{kg} = 53\,\text{g}\,. \end{aligned}$$

Since the total mass of ice present initially was 100 g, there is enough ice to bring the water temperature down to 0°C. This is the solution: the ice and water reach thermal equilibrium at a temperature of 0°C with 53 g of ice melted.

(b) Now there is less than 53 g of ice present initially. All the ice melts and the final temperature is above the melting point of ice. The energy lost by the water is

$$Q = c_W m_W (T_{Wi} - T_f)$$

and the energy absorbed by the ice and the water it becomes when it melts is

$$Q = c_I m_I (0 - T_{Ii}) + c_W m_I (T_f - 0) + m_I L_F\,.$$

The first term is the energy required to raise the temperature of the ice to 0°C, the second term is the energy required to raise the temperature of the melted ice from 0°C to $T_f$, and the third term is the energy required to melt all the ice. Since the two energies are equal,

$$c_W m_W (T_{Wi} - T_f) = c_I m_I (-T_{Ii}) + c_W m_I T_f + m_I L_F\,.$$

The solution for $T_f$ is

$$T_f = \frac{c_W m_W T_{Wi} + c_I m_I T_{Ii} - m_I L_F}{c_W (m_W + m_I)}\,.$$

Substitute given values to obtain $T_f = 2.5°\text{C}$.

**43**

The internal energy is the same at the beginning and end of a cycle, so the energy $Q$ absorbed as heat equals the work done: $Q = W$. Over the portion of the cycle from A to B the pressure $p$ is a linear function of the volume $V$ and we may write $p = a + bV$, where $a = (10/3)\,\text{Pa}$ and $b = (20/3\,\text{Pa/m}^3$. The coefficients $a$ and $b$ were chosen so that $p = 10\,\text{Pa}$ when $V = 1.0\,\text{m}^3$ and $p = 30\,\text{Pa}$ when $V = 4.0\,\text{m}^3$. The work done by the gas during this portion of the cycle is

$$W_{AB} = \int_{V_A}^{V_B} p\,\mathrm{d}V = \int_{V_A}^{V_B} (a + bV)\,\mathrm{d}V = a(V_B - V_A) + \tfrac{1}{2}b(V_B^2 - V_A^2)$$
$$= (10/3)[4.0\,\text{m}^3 - 1.0\,\text{m}^3] + (20/6)[(4.0\,\text{m}^3)^2 - (1.0\,\text{m}^3)^2] = 60\,\text{J}\,.$$

The BC portion of the cycle is at constant pressure and the work done by the gas is $W_{BC} = p\,\Delta V = (30\,\text{Pa})(1.0\,\text{m}^3 - 4.0\,\text{m}^3) = -90\,\text{J}$. The CA portion of the cycle is at constant volume, so no work is done. The total work done by the gas is $W = W_{AB} + W_{BC} + W_{CA} = 60\,\text{J} - 90\,\text{J} + 0 = -30\,\text{J}$ and the total energy absorbed as heat is $Q = W = -30\,\text{J}$. This means the gas loses 30 J of energy in the form of heat.

**49**

(a) The change in internal energy $\Delta E_{\text{int}}$ is the same for path $iaf$ and path $ibf$. According to the first law of thermodynamics, $\Delta E_{\text{int}} = Q - W$, where $Q$ is the energy absorbed as heat and $W$ is the work done by the system. Along $iaf$ $\Delta E_{\text{int}} = Q - W = 50\,\text{cal} - 20\,\text{cal} = 30\,\text{cal}$. Along $ibf$ $W = Q - \Delta E_{\text{int}} = 36\,\text{cal} - 30\,\text{cal} = 6\,\text{cal}$.

(b) Since the curved path is traversed from $f$ to $i$ the change in internal energy is $-30\,\text{cal}$ and $Q = \Delta E_{\text{int}} + W = -30\,\text{cal} - 13\,\text{cal} = -43\,\text{cal}$.

(c) Let $\Delta E_{\text{int}} = E_{\text{int},\,f} - E_{\text{int},\,i}$. Then $E_{\text{int},\,f} = \Delta E_{\text{int}} + E_{\text{int},\,i} = 30\,\text{cal} + 10\,\text{cal} = 40\,\text{cal}$.

(d) and (e) The work $W_{bf}$ for the path $bf$ is zero, so $Q_{bf} = E_{\text{int},\,f} - E_{\text{int},\,b} = 40\,\text{cal} - 22\,\text{cal} = 18\,\text{cal}$. For the path $ibf$ $Q = 36\,\text{cal}$ so $Q_{ib} = Q - Q_{bf} = 36\,\text{cal} - 18\,\text{cal} = 18\,\text{cal}$.

**51**

The rate of heat flow is given by

$$P_{\text{cond}} = kA\,\frac{T_H - T_C}{L}\,,$$

where $k$ is the thermal conductivity of copper $(401\,\text{W/m}\cdot\text{K})$, $A$ is the cross-sectional area (in a plane perpendicular to the flow), $L$ is the distance along the direction of flow between the points where the temperature is $T_H$ and $T_C$. Thus

$$P_{\text{cond}} = \frac{(401\,\text{W/m}\cdot\text{K})(90.0\times 10^{-4}\,\text{m}^2)(125^\circ\text{C} - 10.0^\circ\text{C})}{0.250\,\text{m}} = 1.66\times 10^3\,\text{J/s}\,.$$

The thermal conductivity can be found in Table 18–6 of the text.

**65**

Let $h$ be the thickness of the slab and $A$ be its area. Then the rate of heat flow through the slab is

$$P_{\text{cond}} = \frac{kA(T_H - T_C)}{h},$$

where $k$ is the thermal conductivity of ice, $T_H$ is the temperature of the water (0°C), and $T_C$ is the temperature of the air above the ice (−10°C). The energy leaving the water freezes it, the energy required to freeze mass $m$ of water being $Q = L_F m$, where $L_F$ is the heat of fusion for water. Differentiate with respect to time and recognize that $\mathrm{d}Q/\mathrm{d}t = P_{\text{cond}}$ to obtain

$$P_{\text{cond}} = L_F \frac{\mathrm{d}m}{\mathrm{d}t}.$$

Now the mass of the ice is given by $m = \rho A h$, where $\rho$ is the density of ice and $h$ is the thickness of the ice slab, so $\mathrm{d}m/\mathrm{d}t = \rho A(\mathrm{d}h/\mathrm{d}t)$ and

$$P_{\text{cond}} = L_F \rho A \frac{\mathrm{d}h}{\mathrm{d}t}.$$

Equate the two expressions for $P_{\text{cond}}$ and solve for $\mathrm{d}h/\mathrm{d}t$:

$$\frac{\mathrm{d}h}{\mathrm{d}t} = \frac{k(T_H - T_C)}{L_F \rho h}.$$

Since 1 cal = 4.186 J and 1 cm = $1 \times 10^{-2}$ m, the thermal conductivity of ice has the SI value $k = (0.0040\,\text{cal/s}\cdot\text{cm}\cdot\text{K})(4.186\,\text{J/cal})/(1 \times 10^{-2}\,\text{m/cm}) = 1.674\,\text{W/m}\cdot\text{K}$. The SI value for the density of ice is $\rho = 0.92\,\text{g/cm}^3 = 0.92 \times 10^3\,\text{kg/m}^3$. Thus

$$\frac{\mathrm{d}h}{\mathrm{d}t} = \frac{(1.674\,\text{W/m}\cdot\text{K})(0°\text{C} + 10°\text{C})}{(333 \times 10^3\,\text{J/kg})(0.92 \times 10^3\,\text{kg/m}^3)(0.050\,\text{m})} = 1.1 \times 10^{-6}\,\text{m/s} = 0.40\,\text{cm/h}.$$

**73**

(a) The work done in process 1 is $W_1 = 4.0p_iV_i$, so according to the first law of thermodynamics the change in the internal energy is $\Delta E_{\text{int}} = Q - W_1 = 10p_iV_i - 4.0p_iV_i = 6.0p_iV_i$. The work done in process 2 can be computed as the sum of the areas of a triangle (with altitude $3p_i/2 - p_i = p_i/2$ and base $4.0V_i$) and rectangle (with sides $p_i$ and $4.0V_i$): $W_2 = \frac{1}{2}(p_i/2)(4.0V_i) + 4.0p_1V_i = 5.0p_iV_i$. The change in the internal energy is the same so the energy transferred to the gas a heat is $Q = \Delta E_{\text{int}} + W_2 = 6.0p_1V_i + 5.0p_iV_i = 11p_iV_i$.

(b) The change in internal energy is the same for all processes that start at state a and end at state b, namely $6.0p_iV_i$.

**75**

At the colder temperature the volume of the disk is $V = \pi R^2 \ell$, where $R$ is its radius and $\ell$ is its thickness. After it is heated its volume is $\pi(R + \Delta R)r(\ell + \Delta\ell) = \pi R^2\ell + \pi R^2\,\Delta\ell + 2\pi R\ell\,\Delta R + \ldots$, where terms that are proportional to the products of small quantities are neglected. Thus the change in the volume is $\Delta V = \pi R^2\,\Delta\ell + 2\pi R\ell\,\Delta R$. Now $\Delta R = R\alpha\,\Delta T$ and $\Delta\ell = \ell\alpha\,\Delta T$, where $\Delta T$ is the change in temperature and $\alpha$ (= $3.2 \times 10^{-6} \times 10^{-6}/\text{C}^\circ$ from Table 18–2) is the coefficient of linear expansion for Pyrex. Thus

$$\begin{aligned}\Delta V &= 3\pi R^2\ell\alpha\,\Delta T = 3\pi(0.0800\,\text{m})^2(0.00500\,\text{m})(3.2\times10^{-6}\,\text{C}^\circ)(60.0^\circ\text{C} - 10.0^\circ\text{C})\\ &= 4.83\times10^{-8}\,\text{m}^3\,.\end{aligned}$$

**77**

Let $p_i$ be the initial pressure, $V_i$ be the initial volume, and $V_f$ be the final volume. Then the work done by the gas is

$$\int_{V_i}^{V_f} p\,dV = \int_{V_i}^{V_f} aV^2\,dV = \frac{a}{3}(V_f^3 - V_i^3) = \frac{10\,\text{m}^8}{(2.0\,\text{m}^3)^3 - (1.0\,\text{m}^3)^3} = 23\,\text{J}\,.$$

**81**

The magnitude of the energy transferred as heat from the aluminum is given by $Q = m_A c_A(T_A - T)$ and the magnitude of the energy transferred as heat into the water is given by $Q = m_w c_w(T - T_w)$, where $c_A$ is the specific heat of aluminum (900 J/kg·K from table 18–3), $c_w$ is the specific heat of water (4190 J/kg·K from the same table), $T_A$ is the initial temperature of the aluminum, $T_w$ is the initial temperature of the water, and $T$ is the common final temperature. The solution for $T$ is

$$\begin{aligned}T &= \frac{m_A c_A T_A + m_w c_w T_w}{m_A c_A + m_w c_w}\\ &= \frac{(2.50\,\text{kg})(900\,\text{J/kg}\cdot\text{K})(92.0^\circ\text{C}) + (8.00\,\text{kg})(4190\,\text{J/kg}\cdot\text{K})(5.00^\circ\text{C})}{(2.50\,\text{kg})(900\,\text{J/kg}\cdot\text{K}) + (8.00\,\text{kg})(4190\,\text{J/kg}\cdot\text{K})}\\ &= 10.5^\circ\text{C}\,.\end{aligned}$$

Note that it is not necessary to convert the temperatures to the Kelvin scale.

**83**

Let $m$ be the mass of the ice cube and $c_{\text{ice}}$ be its specific heat. Then the energy required to bring the ice cube temperature to 0°C is

$$Q = mc_{\text{ice}}\,\Delta T = (0.700\,\text{kg})(2220\,\text{J/kg}\cdot\text{K})(150\,\text{K}) = 2.331\times10^5\,\text{J}\,.$$

This is less than the $6.993 \times 10^5$ J that are supplied, so the ice cube is brought to 0°C and all or part of it melts. The specific heat for ice can be found in Table 18–3.
If $L_F$ is the neat of fusion for water, then the energy required to melt all the ice is

$$mL_F = (0.700\,\text{kg})(333\times10^3\,\text{J/kg}) = 2.331\times10^5\,\text{J}\,.$$

This less than the $6.993 \times 10^5\,\text{J} - 2.331 \times 10^5\,\text{J} = 4.662 \times 10^5$ J that are available, so all the ice melts. The heat of fusion for water can be found in Table 18–4. Now energy $E = 4.662 \times 10^5\,\text{J} - 2.331 \times 10^5\,\text{J} = 2.331 \times 10^5$ J is used to raise the temperature of the water from 0°C. The final temperature is

$$T = \frac{E}{mc_{\text{water}}} = \frac{2.331\times10^5\,\text{J}}{(0.700\,\text{kg})(4190\,\text{J/kg}\cdot\text{K})} = 79.5^\circ\text{C}\,.$$

# Chapter 19

<u>7</u>

(a) Solve the ideal gas law $pV = nRT$ for $n$. First, convert the temperature to the Kelvin scale: $T = 40.0 + 273.15 = 313.15\,\text{K}$. Also convert the volume to $\text{m}^3$: $1000\,\text{cm}^3 = 1000 \times 10^{-6}\,\text{m}^3$. Then

$$n = \frac{pV}{RT} = \frac{(1.01 \times 10^5\,\text{Pa})(1000 \times 10^{-6}\,\text{m}^3)}{(8.31\,\text{J/mol}\cdot\text{K})(313.15\,\text{K})} = 3.88 \times 10^{-2}\,\text{mol}\,.$$

(b) Solve the ideal gas law $pV = nRT$ for $T$:

$$T = \frac{pV}{nR} = \frac{(1.06 \times 10^5\,\text{Pa})(1500 \times 10^{-6}\,\text{m}^3)}{(3.88 \times 10^{-2}\,\text{mol})(8.31\,\text{J/mol}\cdot\text{K})} = 493\,\text{K} = 220^\circ\,\text{C}\,.$$

<u>13</u>

Suppose the gas expands from volume $V_i$ to volume $V_f$ during the isothermal portion of the process. The work it does is

$$W = \int_{V_i}^{V_f} p\,\text{d}V = nRT \int_{V_i}^{V_f} \frac{\text{d}V}{V} = nRT \ln \frac{V_f}{V_i}\,,$$

where the ideal gas law $pV = nRT$ was used to replace $p$ with $nRT/V$. Now $V_i = nRT/p_i$ and $V_f = nRT/p_f$, so $V_f/V_i = p_i/p_f$. Also replace $nRT$ with $p_iV_i$ to obtain

$$W = p_iV_i \ln \frac{p_i}{p_f}\,.$$

Since the initial gauge pressure is $1.03 \times 10^5$ Pa, $p_i = 1.03 \times 10^5\,\text{Pa} + 1.013 \times 10^5\,\text{Pa} = 2.04 \times 10^5\,\text{Pa}$. The final pressure is atmospheric pressure: $p_f = 1.013 \times 10^5$ Pa. Thus

$$W = (2.04 \times 10^5\,\text{Pa})(0.140\,\text{m}^3) \ln \frac{2.04 \times 10^5\,\text{Pa}}{1.013 \times 10^5\,\text{Pa}} = 2.00 \times 10^4\,\text{J}\,.$$

During the constant pressure portion of the process the work done by the gas is $W = p_f(V_i - V_f)$. Notice that the gas starts in a state with pressure $p_f$, so this is the pressure throughout this portion of the process. Also note that the volume decreases from $V_f$ to $V_i$. Now $V_f = p_iV_i/p_f$, so

$$\begin{aligned} W &= p_f\left(V_i - \frac{p_iV_i}{p_f}\right) = (p_f - p_i)V_i \\ &= (1.013 \times 10^5\,\text{Pa} - 2.04 \times 10^5\,\text{Pa})(0.140\,\text{m}^3) = -1.44 \times 10^4\,\text{J}\,. \end{aligned}$$

The total work done by the gas over the entire process is $W = 2.00 \times 10^4\,\text{J} - 1.44 \times 10^4\,\text{J} = 5.60 \times 10^3\,\text{J}$.

**19**

According to kinetic theory, the rms speed is

$$v_{\text{rms}} = \sqrt{\frac{3RT}{M}},$$

where $T$ is the temperature on the Kelvin scale and $M$ is the molar mass. See Eq. 19–34. According to Table 19–1, the molar mass of molecular hydrogen is $2.02\,\text{g/mol} = 2.02 \times 10^{-3}\,\text{kg/mol}$, so

$$v_{\text{rms}} = \sqrt{\frac{3(8.31\,\text{J/mol}\cdot\text{K})(2.7\,\text{K})}{2.02 \times 10^{-3}\,\text{kg/mol}}} = 1.8 \times 10^{2}\,\text{m/s}.$$

**29**

(a) According to Eq. 19–25, the mean free path for molecules in a gas is given by

$$\lambda = \frac{1}{\sqrt{2}\pi d^2 N/V},$$

where $d$ is the diameter of a molecule and $N$ is the number of molecules in volume $V$. Substitute $d = 2.0 \times 10^{-10}\,\text{m}$ and $N/V = 1 \times 10^{6}\,\text{molecules/m}^3$ to obtain

$$\lambda = \frac{1}{\sqrt{2}\pi(2.0 \times 10^{-10}\,\text{m})^2(1 \times 10^{6}\,\text{m}^{-3})} = 6 \times 10^{12}\,\text{m}.$$

(b) At this altitude most of the gas particles are in orbit around Earth and do not suffer randomizing collisions. The mean free path has little physical significance.

**35**

(a) The average speed is

$$\overline{v} = \frac{\sum v}{N},$$

where the sum is over the speeds of the particles and $N$ is the number of particles. Thus

$$\overline{v} = \frac{(2.0 + 3.0 + 4.0 + 5.0 + 6.0 + 7.0 + 8.0 + 9.0 + 10.0 + 11.0)\,\text{km/s}}{10} = 6.5\,\text{km/s}.$$

(b) The rms speed is given by

$$v_{\text{rms}} = \sqrt{\frac{\sum v^2}{N}}.$$

Now

$$\sum v^2 = \Big[(2.0)^2 + (3.0)^2 + (4.0)^2 + (5.0)^2 + (6.0)^2 + (7.0)^2 + (8.0)^2 + (9.0)^2 + (10.0)^2 + (11.0)^2\Big]\,\text{km}^2/\text{s}^2 = 505\,\text{km}^2/\text{s}^2,$$

so

$$v_{\text{rms}} = \sqrt{\frac{505\,\text{km}^2/\text{s}^2}{10}} = 7.1\,\text{km/s}\,.$$

**41**

(a) The distribution function gives the fraction of particles with speeds between $v$ and $v + dv$, so its integral over all speeds is unity: $\int P(v)\,dv = 1$. Evaluate the integral by calculating the area under the curve in Fig. 19–24. The area of the triangular portion is half the product of the base and altitude, or $\frac{1}{2}av_0$. The area of the rectangular portion is the product of the sides, or $av_0$. Thus $\int P(v)\,dv = \frac{1}{2}av_0 + av_0 = \frac{3}{2}av_0$, so $\frac{3}{2}av_0 = 1$ and $av_0 = 2/3$.

(b) The average speed is given by

$$v_{\text{avg}} = \int vP(v)\,dv\,.$$

For the triangular portion of the distribution $P(v) = av/v_0$ and the contribution of this portion is

$$\frac{a}{v_0}\int_0^{v_0} v^2\,dv = \frac{a}{3v_0}v_0^3 = \frac{av_0^2}{3} = \frac{2}{9}v_0\,,$$

where $2/3v_0$ was substituted for $a$. $P(v) = a$ in the rectangular portion and the contribution of this portion is

$$a\int_{v_0}^{2v_0} v\,dv = \frac{a}{2}(4v_0^2 - v_0^2) = \frac{3a}{2}v_0^2 = v_0\,.$$

Thus $v_{\text{avg}} = \frac{2}{9}v_0 + v_0 = 1.2v_0$ and $v_{\text{avg}}/v_0 = 1.2$.

(c) The mean-square speed is given by

$$v_{\text{rms}}^2 = \int v^2P(v)\,dv\,.$$

The contribution of the triangular section is

$$\frac{a}{v_0}\int_0^{v_0} v^3\,dv = \frac{a}{4v_0}v_0^4 = \frac{1}{6}v_0^2\,.$$

The contribution of the rectangular portion is

$$a\int_{v_0}^{2v_0} v^2\,dv = \frac{a}{3}(8v_0^3 - v_0^3) = \frac{7a}{3}v_0^3 = \frac{14}{9}v_0^2\,.$$

Thus $v_{\text{rms}} = \sqrt{\frac{1}{6}v_0^2 + \frac{14}{9}v_0^2} = 1.31v_0$ and $v_{\text{rms}}/v_0 = 1.3$.

(d) The number of particles with speeds between $1.5v_0$ and $2v_0$ is given by $N\int_{1.5v_0}^{2v_0} P(v)\,dv$. The integral is easy to evaluate since $P(v) = a$ throughout the range of integration. Thus the number of particles with speeds in the given range is $Na(2.0v_0 - 1.5v_0) = 0.5Nav_0 = N/3$, where $2/3v_0$ was substituted for $a$. The fraction of particles in the given range is 0.33.

**45**

When the temperature changes by $\Delta T$ the internal energy of the first gas changes by $n_1C_1\,\Delta T$, the internal energy of the second gas changes by $n_2C_2\,\Delta T$, and the internal energy of the third gas changes by $n_3C_3\,\Delta T$. The change in the internal energy of the composite gas is $\Delta E_{\text{int}} = (n_1C_1 + n_2C_2 + n_3C_3)\,\Delta T$. This must be $(n_1 + n_2 + n_3)C\,\Delta T$, where $C$ is the molar specific heat of the mixture. Thus

$$\begin{aligned} C &= \frac{n_1C_1 + n_2C_2 + n_3C_3}{n_1 + n_2 + n_3} \\ &= \frac{(2.40\,\text{mol})(12.0\,\text{J/mol}\cdot\text{K}) + (1.50\,\text{mol})(12.8\,\text{J/mol}\cdot\text{K}) + (3.20\,\text{mol})(20.0\,\text{J/mol}\cdot\text{K})}{2.40\,\text{mol} + 1.50\,\text{mol} + 3.20\,\text{mol}} \\ &= 15.8\,\text{J/mol}\cdot\text{K}\,. \end{aligned}$$

**53**

(a) Since the process is at constant pressure energy transferred as heat to the gas is given by $Q = nC_p\,\Delta T$, where $n$ is the number of moles in the gas, $C_p$ is the molar specific heat at constant pressure, and $\Delta T$ is the increase in temperature. For a diatomic ideal gas with rotating molecules $C_p = \frac{7}{2}R$. Thus

$$Q = \frac{7}{2}nR\,\Delta T = \frac{7}{2}(4.00\,\text{mol})(8.314\,\text{J/mol}\cdot\text{K})(60.0\,\text{K}) = 6.98 \times 10^3\,\text{J}\,.$$

See Table 19–3 for the expression for $C_p$.

(b) The change in the internal energy is given by $\Delta E_{\text{int}} = nC_V\,\Delta T$, where $C_V$ is the specific heat at constant volume. For a diatomic ideal gas with rotating molecules $C_V = \frac{5}{2}R$, so

$$\Delta E_{\text{int}} = \frac{5}{2}nR\,\Delta T = \frac{5}{2}(4.00\,\text{mol})(8.314\,\text{J/mol}\cdot\text{K})(60.0\,\text{K}) = 4.99 \times 10^3\,\text{J}\,.$$

See Table 19–3 for the expression for $C_V$.

(c) According to the first law of thermodynamics, $\Delta E_{\text{int}} = Q - W$, so

$$W = Q - \Delta E_{\text{int}} = 6.98 \times 10^3\,\text{J} - 4.99 \times 10^3\,\text{J} = 1.99 \times 10^3\,\text{J}\,.$$

(d) The change in the total translational kinetic energy is

$$\Delta K = \frac{3}{2}nR\,\Delta T = \frac{3}{2}(4.00\,\text{mol})(8.314\,\text{J/mol}\cdot\text{K})(60.0\,\text{K}) = 2.99 \times 10^3\,\text{J}\,.$$

**67**

Let $\rho_a$ be the density of air surrounding the balloon and $\rho_H$ be the density of hot air in the balloon. The buoyant force on the balloon has magnitude $\rho_a gV$, where $V$ is the volume of the envelope, and it is upward. The magnitude of the force of gravity is $W + \rho_h gV$, where $W$ is the weight of the basket and envelope combined, and it is downward. The second term is the weight of the hot air. If $F_{\text{net}}$ (= 2.67 kN) is the net force on the balloon then $F_{\text{net}} = \rho_a gV - W - \rho_h gV$, so

$$\rho_h g = \frac{\rho_a gV - W - F_{\text{net}}}{V} = \frac{(11.9\,\text{N/m}^3)(2.18 \times 10^3\,\text{m}^3) - 2.45 \times 10^3\,\text{N} - 2.67 \times 10^3\,\text{N}}{2.18 \times 10^3\,\text{m}^3}$$
$$= 9.55\,\text{N/m}^3 .$$

The ideal gas law tell us that the number of moles of hot air per unit volume is $n/V = p/RT$, where $p$ is the pressure and $T$ is the temperature. Multiply by the molar mass $M$ of air to obtain $\rho_h = pM/RT$. The pressure is atmospheric pressure ($1.01 \times 10^5$ Pa) so the temperature should be

$$T = \frac{pM}{R\rho_h} = \frac{(1.01 \times 10^5\,\text{Pa})(0.028\,\text{kg/mol})}{(8.31\,\text{J/mol}\cdot\text{K})(9.55\,\text{N/m}^3)} = 356\,\text{K} .$$

**69**

The molar specific heat of a monatomic ideal gas is $C_V = (3/2)R$, where $R$ is the universal gas constant. Let $n$ be the number of moles of gas and $\Delta T$ be the change in temperature. Then the change in the internal energy is

$$\Delta E_{\text{int}} = nC_V\,\Delta T = (3/2)nR\,\Delta T = (3/2)(2.00\,\text{mol})(8.31\,\text{J/mol}\cdot\text{K})(15.0\,\text{K}) = 374\,\text{J} .$$

Since the process is adiabatic the energy transferred as heat is $Q = 0$.

According to the first law of thermodynamics the work done by the gas is $W = Q - \Delta E_{\text{int}} = -374\,\text{J}$.

Since the gas is monatomic the internal energy is translational kinetic energy. The number of atoms is the product of the number of moles and the Avogadro constant: $N = nN_A = (2.00\,\text{mol})(6.02 \times 10^{23}\,\text{atoms/mol}^{-1}) = 1.20 \times 10^{24}$ atoms and the change in the kinetic energy per atom is $(374\,\text{J})/(1.20 \times 10^{24}\,\text{atoms}) = 3.11 \times 10^{22}\,\text{J/atom}$.

**71**

The mean free path is given by $\lambda = 1/\sqrt{2}\pi d^2(N/V)$, where $d$ is the diameter of a molecule and $N$ is the number of molecules in volume $V$. According to the ideal gas law $N/V = p/kT$, where $p$ is the pressure, $T$ is the temperature on the Kelvin scale, and $k$ is the Boltzmann constant. Thus

$$\lambda = \frac{kT}{\sqrt{2}\pi d^2 p} = \frac{(1.38 \times 10^{-23}\,\text{J/K})(400\,\text{K})}{\sqrt{2}\pi(290 \times 10^{-12}\,\text{m})^2(2.00\,\text{atm})(1.01 \times 10^5\,\text{Pa/atm})} = 7.31 \times 10^{-8}\,\text{m} .$$

The average time between collisions is $\tau = \lambda/v_{\text{avg}}$, where $v_{\text{avg}}$ is the average speed of the molecules. This is given by $v_{\text{avg}} = \sqrt{8RT/\pi M}$, where $R$ is the universal gas constant and $M$ is the molar mass (see Eq. 19-30). The molar mass of oxygen is $32.0 \times 10^{-3}$ kg/mol (see table 19—1), so

$$v_{\text{avg}} = \sqrt{\frac{8(8.31\,\text{J/ml}\cdot\text{K})(400\,\text{K})}{\pi(32.0 \times 10^{-3}\,\text{kg/mol})}} = 514\,\text{m/s} .$$

The average time between collisions is $(7.31 \times 10^{-8}\,\text{m})/(514\,\text{m/s}) = 1.42 \times 10^{-9}\,\text{s}$ and the frequency of collision is the reciprocal of this or $7.04 \times 10^{9}$ collisions/s.

77

(a) Let $p_i$ be the initial pressure, $V_i$ be the initial volume, $p_f$ be the final pressure, and $V_f$ be the final volume. According to the ideal gas law $p_iV_i = p_fV_f$ since the initial and final temperatures are the same. Thus

$$p_f = \frac{V_i}{V_f}p_i = \frac{1.0\,\text{L}}{4.0\,\text{L}}(32\,\text{atm}) = 8.0\,\text{atm}\,.$$

(b) The final temperature is the same as the initial temperature: 300 K.

(c) Since $p = nRT/V$, the work done by the gas is

$$W = \int_{V_i}^{V_f} p\,dV = \int_{V_i}^{V_f} \frac{nRT}{V}\,dV = nRT\ln\left(\frac{V_f}{V_i}\right) = p_iV_i\ln\left(\frac{V_f}{V_i}\right)$$

$$= (32\,\text{atm})(1.01 \times 10^5\,\text{Pa/atm})(1.0\,\text{L})(1.00 \times 10^{-3}\,\text{m}^3/\text{L})\ln\left(\frac{4.0\,\text{L}}{1.0\,\text{L}}\right) = 4.5 \times 10^3\,\text{J}\,.$$

(d) Since the process is now adiabatic, $p_iV_i^\gamma = p_fV_f^\gamma$, where $\gamma$ is the ratio of the heat capacity at constant pressure to the heat capacity at constant volume. Since the gas is monatomic $\gamma = 1.667$. The final pressure is

$$p_f = \left(\frac{V_i}{V_f}\right)^\gamma p_i = \left(\frac{1.0\,\text{L}}{4.0\,\text{L}}\right)^{1.667}(32\,\text{atm}) = 3.2\,\text{atm}\,.$$

(e) Let $T_i$ be the initial temperature and $T_f$ be the final temperature. Then according to the ideal gas law $nR = p_iV_i/T_i = p_fV_f/T_f$ and

$$T_f = \frac{p_fV_f}{p_iV_i}T_i = \frac{(3.2\,\text{atm})(4.0\,\text{L})}{(32\,\text{atm})(1.0\,\text{L})}(300\,\text{K}) = 120\,\text{K}\,.$$

(f) The first law of thermodynamics tells us that the change in the internal energy $\Delta E_{\text{int}}$ is equal to the negative of the work done by the gas during an adiabatic process. The change in the internal energy is $\Delta E_{\text{int}} = nC_V(T_f - T_i) = (C_V/R)(p_f - p_i)$, where $C_V$ is the molar specific heat for constant volume processes. The ideal gas law was used to write the equation in the last form. For a monatomic ideal gas $C_V = (3/2)R$, so $W = -\Delta E_{\text{int}} = -(3/2)(p_fV_f - p_iV_i)$. Now $p_fV_f = (3.2\,\text{atm})(4.0\,\text{L})(1.01 \times 10^5\,\text{Pa/atm})(1.0 \times 10^{-3}\,\text{m}^3/\text{L}) = 1.29 \times 10^3\,\text{J}$ and $p_iV_i = (32\,\text{atm})(1.0\,\text{L})(1.01 \times 10^5\,\text{Pa/atm})(1.0 \times 10^{-3}\,\text{m}^3/\text{L}) = 3.23 \times 10^3\,\text{J}$, so

$$W = -\frac{3}{2}(1.29 \times 10^3\,\text{J} - 3.23 \times 10^3\,\text{J}) = 2.9 \times 10^3\,\text{J}\,.$$

(g) Now $\gamma = 1.4$, so

$$p_f = \left(\frac{V_i}{V_f}\right)^\gamma p_i = \left(\frac{1.0\,\text{L}}{4.0\,\text{L}}\right)^{1.4}(32\,\text{atm}) = 4.6\,\text{atm}\,.$$

(h) The final temperature is now

$$T_f = \frac{p_f V_f}{p_i V_i} T_i = \frac{(4.6\,\text{atm})(4.0\,\text{L}}{(32\,\text{atm})(1.0\,\text{L})}(300\,\text{K}) = 170\,\text{K}.$$

(i) The molar specific heat is $C_V = 5/2$, so the work done by the gas is $W = -\frac{5}{2}(p_f V_f - p_i V_i)$. Now $p_f V_f = (4.6\,\text{atm})(4.0\,\text{L})(1.01 \times 10^5\,\text{Pa/atm})(1.0 \times 10^{-3}\,\text{m}^3/\text{L}) = 1.86 \times 10^3\,\text{J}$ and $p_i V_i$ is still $3.23 \times 10^3$ J, so

$$W = -\frac{3}{2}(1.86 \times 10^3\,\text{J} - 3.23 \times 10^3\,\text{J}) = 3.4 \times 10^3\,\text{J}.$$

**81**

(a) Since a molecule must have some speed, the integral of $P(v)$ over all value of $v$ must be 1. Thus

$$\int_0^{v_0} Cv^2\,dv = (1/3)Cv_0^3 = 1$$

and $C = 3v_0^{-3}$.

(b) The average speed of the particles is

$$\int_0^{v_0} P(v)v\,dv = 3v_0^{-3}\int_0^{v_0} v^3\,dv = (3/4)v_0\,.$$

(c) The average of the square of the speed is

$$(v^2)_{\text{avg}} = \int_0^{v_0} P(v)v^2\,dv = 3v_0^{-3}\int_0^{v_0} v^4\,dv = (3/5)v_0^2$$

so the rms speed is $v_{\text{rms}} = \sqrt{(v^2)_{\text{avg}}} = \sqrt{(3/5)v_0^2} = 0.775v_0$.

**83**

(a) According to the ideal gas law $p = nRT/V$, so the work done by the gas as the volume goes from $V_i$ to $V_f$ is

$$\begin{aligned} W &= \int_{V_i}^{V_f} p\,dV = \int_{V_i}^{V_f} \frac{nRT}{V}\,dV = nRT\ln\left(\frac{V_f}{V_i}\right) \\ &= (3.50\,\text{mol})(8.31\,\text{J/mol}\cdot\text{K})(283\,\text{K})\ln\left(\frac{3.00\,\text{m}^3}{4.00\,\text{m}^3}\right) \\ &= -2.37 \times 10^3\,\text{J}. \end{aligned}$$

The temperature was converted from degrees Celsius to kelvins.

(b) The internal energy of an ideal gas does not change unless the temperature changes, so according to the first law of thermodynamics the energy transferred as heat is $Q = W = -2.37 \times 10^3$ J.

# Chapter 20

5

(a) Since the gas is ideal, its pressure $p$ is given in terms of the number of moles $n$, the volume $V$, and the temperature $T$ by $p = nRT/V$. The work done by the gas during the isothermal expansion is

$$W = \int_{V_1}^{V_2} p\,dV = nRT \int_{V_1}^{V_2} \frac{dV}{V} = nRT \ln \frac{V_2}{V_1}.$$

Substitute $V_2 = 2.00V_1$ to obtain

$$W = nRT \ln 2 = (4.00\,\text{mol})(8.314\,\text{J/mol}\cdot\text{K})(400\,\text{K}) \ln 2 = 9.22 \times 10^3\,\text{J}.$$

(b) Since the expansion is isothermal, the change in entropy is given by $\Delta S = \int(1/T)\,dQ = Q/T$, where $Q$ is the energy absorbed as heat. According to the first law of thermodynamics, $\Delta E_{\text{int}} = Q - W$. Now the internal energy of an ideal gas depends only on the temperature and not on the pressure and volume. Since the expansion is isothermal, $\Delta E_{\text{int}} = 0$ and $Q = W$. Thus

$$\Delta S = \frac{W}{T} = \frac{9.22 \times 10^3\,\text{J}}{400\,\text{K}} = 23.1\,\text{J/K}.$$

(c) $\Delta S = 0$ for all reversible adiabatic processes.

7

(a) The energy that leaves the aluminum as heat has magnitude $Q = m_a c_a (T_{ai} - T_f)$, where $m_a$ is the mass of the aluminum, $c_a$ is the specific heat of aluminum, $T_{ai}$ is the initial temperature of the aluminum, and $T_f$ is the final temperature of the aluminum-water system. The energy that enters the water as heat has magnitude $Q = m_w c_w (T_f - T_{wi})$, where $m_w$ is the mass of the water, $c_w$ is the specific heat of water, and $T_{wi}$ is the initial temperature of the water. The two energies are the same in magnitude since no energy is lost. Thus $m_a c_a (T_{ai} - T_f) = m_w c_w (T_f - T_{wi})$ and

$$T_f = \frac{m_a c_a T_{ai} + m_w c_w T_{wi}}{m_a c_a + m_w c_w}.$$

The specific heat of aluminum is 900 J/kg · K and the specific heat of water is 4190 J/kg · K. Thus

$$T_f = \frac{(0.200\,\text{kg})(900\,\text{J/kg}\cdot\text{K})(100°\,\text{C}) + (0.0500\,\text{kg})(4190\,\text{J/kg}\cdot\text{K})(20°\,\text{C})}{(0.200\,\text{kg})(900\,\text{J/kg}\cdot\text{K}) + (0.0500\,\text{kg})(4190\,\text{J/kg}\cdot\text{K})}$$
$$= 57.0°\,\text{C}.$$

This is equivalent to 330 K.

(b) Now temperatures must be given in kelvins: $T_{ai} = 393\,\text{K}$, $T_{wi} = 293\,\text{K}$, and $T_f = 330\,\text{K}$. For the aluminum, $dQ = m_a c_a\, dT$ and the change in entropy is

$$\Delta S_a = \int \frac{dQ}{T} = m_a c_a \int_{T_{ai}}^{T_f} \frac{dT}{T} = m_a c_a \ln \frac{T_f}{T_{ai}}$$
$$= (0.200\,\text{kg})(900\,\text{J/kg}\cdot\text{K}) \ln \frac{330\,\text{K}}{373\,\text{K}} = -22.1\,\text{J/K}\,.$$

(c) The entropy change for the water is

$$\Delta S_w = \int \frac{dQ}{T} = m_w c_w \int_{T_{wi}}^{T_f} \frac{dT}{T} = m_w c_w \ln \frac{T_f}{T_{wi}}$$
$$= (0.0500\,\text{kg})(4190\,\text{J/kg}\cdot\text{K}) \ln \frac{330\,\text{K}}{293\,\text{K}} = +24.9\,\text{J/K}\,.$$

(d) The change in the total entropy of the aluminum-water system is $\Delta S = \Delta S_a + \Delta S_w = -22.1\,\text{J/K} + 24.9\,\text{J/K} = +2.8\,\text{J/K}$.

**25**

(a) The efficiency is

$$\varepsilon = \frac{T_H - T_C}{T_H} = \frac{(235 - 115)\,\text{K}}{(235 + 273)\,\text{K}} = 0.236\,.$$

Note that a temperature difference has the same value on the Kelvin and Celsius scales. Since the temperatures in the equation must be in kelvins, the temperature in the denominator was converted to the Kelvin scale.

(b) Since the efficiency is given by $\varepsilon = |W|/|Q_H|$, the work done is given by $|W| = \varepsilon|Q_H| = 0.236(6.30 \times 10^4\,\text{J}) = 1.49 \times 10^4\,\text{J}$.

**29**

(a) Energy is added as heat during the portion of the process from $a$ to $b$. This portion occurs at constant volume ($V_b$), so $Q_{\text{in}} = nC_V\,\Delta T$, where $C_V$ is the molar specific heat for constant volume processes. The gas is a monatomic ideal gas, so $C_V = \frac{3}{2}R$ and the ideal gas law gives $\Delta T = (1/nR)(p_b V_b - p_a V_a) = (1/nR)(p_b - p_a)V_b$. Thus $Q_{\text{in}} = \frac{3}{2}(p_b - p_a)V_b$. $V_b$ and $p_b$ are given. We need to find $p_a$. Now $p_a$ is the same as $p_c$ and points $c$ and $b$ are connected by an adiabatic process. Thus $p_c V_c^\gamma = p_b V_b^\gamma$ and

$$p_a = p_c = \left(\frac{V_b}{V_c}\right)^\gamma p_b = \left(\frac{1}{8.00}\right)^{5/3} (1.013 \times 10^6\,\text{Pa}) = 3.167 \times 10^4\,\text{Pa}\,.$$

The energy added as heat is

$$Q_{\text{in}} = \frac{3}{2}(1.013 \times 10^6\,\text{Pa} - 3.167 \times 10^4\,\text{Pa})(1.00 \times 10^{-3}\,\text{m}^3) = 1.47 \times 10^3\,\text{J}\,.$$

(b) Energy leaves the gas as heat during the portion of the process from $c$ to $a$. This is a constant pressure process, so

$$Q_{\text{out}} = nC_p\,\Delta T = \frac{5}{2}(p_aV_a - p_cV_c) = \frac{5}{2}p_a(V_a - V_c)$$
$$= \frac{5}{2}(3.167 \times 10^4\,\text{Pa})(-7.00)(1.00 \times 10^{-3}\,\text{m}^3) = -5.54 \times 10^2\,\text{J}\,,$$

where $C_p$ is the molar specific heat for constant pressure processes. The substitutions $V_a - V_c = V_a - 8.00V_a = -7.00V_a$ and $C_p = \frac{5}{2}R$ were made.

(c) For a complete cycle, the change in the internal energy is zero and $W = Q = 1.47 \times 10^3\,\text{J} - 5.54 \times 10^2\,\text{J} = 9.18 \times 10^2\,\text{J}$.

(d) The efficiency is $\varepsilon = W/Q_{\text{in}} = (9.18 \times 10^2\,\text{J})/(1.47 \times 10^3\,\text{J}) = 0.624$.

## 37

An ideal refrigerator working between a hot reservoir at temperature $T_H$ and a cold reservoir at temperature $T_C$ has a coefficient of performance $K$ that is given by $K = T_C/(T_H - T_C)$. For the refrigerator of this problem, $T_H = 96°\,\text{F} = 309\,\text{K}$ and $T_C = 70°\,\text{F} = 294\,\text{K}$, so $K = (294\,\text{K})/(309\,\text{K} - 294\,\text{K}) = 19.6$. The coefficient of performance is the energy $Q_C$ drawn from the cold reservoir as heat divided by the work done: $K = |Q_C|/|W|$. Thus $|Q_C| = K|W| = (19.6)(1.0\,\text{J}) = 20\,\text{J}$.

## 39

The coefficient of performance for a refrigerator is given by $K = |Q_C|/|W|$, where $Q_C$ is the energy absorbed from the cold reservoir as heat and $W$ is the work done by the refrigerator, a negative value. The first law of thermodynamics yields $Q_H + Q_C - W = 0$ for an integer number of cycles. Here $Q_H$ is the energy ejected to the hot reservoir as heat. Thus $Q_C = W - Q_H$. $Q_H$ is negative and greater in magnitude than $W$, so $|Q_C| = |Q_H| - |W|$. Thus

$$K = \frac{|Q_H| - |W|}{|W|}\,.$$

The solution for $|W|$ is $|W| = |Q_H|/(K + 1)$. In one hour,

$$|W| = \frac{7.54\,\text{MJ}}{3.8 + 1} = 1.57\,\text{MJ}\,.$$

The rate at which work is done is $(1.57 \times 10^6\,\text{J})/(3600\,\text{s}) = 440\,\text{W}$.

## 47

(a) Suppose there are $n_L$ molecules in the left third of the box, $n_C$ molecules in the center third, and $n_R$ molecules in the right third. There are $N!$ arrangements of the $N$ molecules, but $n_L!$ are simply rearrangements of the $n_L$ molecules in the right third, $n_C!$ are rearrangements of the $n_C$ molecules in the center third, and $n_R!$ are rearrangements of the $n_R$ molecules in the right third. These rearrangements do not produce a new configuration. Thus the multiplicity is

$$W = \frac{N!}{n_L!\,n_C!\,n_R!}\,.$$

(b) If half the molecules are in the right half of the box and the other half are in the left half of the box, then the multiplicity is

$$W_B = \frac{N!}{(N/2)!\,(N/2)!}\,.$$

If one-third of the molecules are in each third of the box, then the multiplicity is

$$W_A = \frac{N!}{(N/3)!\,(N/3)!\,(N/3)!}\,.$$

The ratio is

$$\frac{W_A}{W_B} = \frac{(N/2)!\,(N/2)!}{(N/3)!\,(N/3)!\,(N/3)!}\,.$$

(c) For $N = 100$,

$$\frac{W_A}{W_B} = \frac{50!\,50!}{33!\,33!\,34!} = 4.16 \times 10^{16}\,.$$

**49**

(a) and (b) The most probable speed is given by $v_P = \sqrt{2RT/M}$ and the rms speed is given by $v_{\text{rms}} = \sqrt{3RT/M}$, where $T$ is the temperature on the Kelvin scale, $M$ is the molar mass, and $R$ is the universal gas constant. See Eqs. 19–34 and 19–35. Thus $\Delta v = \left(\sqrt{3} - \sqrt{2}\right)\sqrt{RT/M}$. According to Table 19–1 the molar mass of nitrogen is 0.028 kg/mol. For $T = 250\,\text{K}$,

$$\Delta v = \left(\sqrt{3} - \sqrt{2}\right)\sqrt{\frac{(8.31\,\text{J/mol}\cdot\text{K})(250\,\text{K})}{0.028\,\text{kg/mol}}} = 87\,\text{m/s}$$

and for $T = 500\,\text{K}$,

$$\Delta v = \left(\sqrt{3} - \sqrt{2}\right)\sqrt{\frac{(8.31\,\text{J/mol}\cdot\text{K})(500\,\text{K})}{0.028\,\text{kg/mol}}} = 1.2 \times 10^2\,\text{m/s}$$

(c) The energy transferred as heat when the temperature changes by the infinitesimal $dT$ at constant volume is $dQ = nC_V\,dT$, where $n$ is the number of moles and $C_V$ is the molar specific heat for constant volume processes. Thus the entropy change is

$$\Delta S = \int \frac{dQ}{T} = \int_{T_i}^{T_f} \frac{nC_V}{T}\,dT = nC_V \ln\left(\frac{T_f}{T_i}\right)\,.$$

Here $T_i$ is the initial temperature and $T_f$ is the final temperature. Since nitrogen is diatomic with rotating molecules $C_V = 5R/2$ (see Table 19-3),

$$\Delta S = (5/2)nR\ln\left(\frac{T_f}{T_i}\right) = (5/2)(1.5\,\text{mol})(8.31\,\text{J/mol}\cdot\text{K})\ln\left(\frac{500\,\text{K}}{250\,\text{K}}\right) = 22\,\text{J/K}\,.$$

**55**

The temperature of the ice is raised to 0°C, then the ice melts and the temperature of the resulting water is raised to 40°C. As the temperature of the ice is raised the infinitesimal $dT$ the energy transferred to it as heat is $dQ = mc_{\text{ice}}\,dT$, where $c_{\text{ice}}$ is the specific heat of ice and $m$ is the mass of the ice. The entropy change is

$$\Delta S_1 = \int \frac{dQ}{T} = \int_{T_i}^{T_f} \frac{mc_{\text{ice}}}{T}\,dT = mc_{\text{ice}} \ln\left(\frac{T_f}{T_i}\right).$$

Table 18–3 gives the specific heat of ice as 2220 J/kg · K. The initial temperature on the Kelvin scale is $T_i = -20° + 273° = 253\,\text{K}$ and the final temperature is $T_f = 273\,\text{K}$, so

$$\Delta S_1 = (0.600\,\text{kg})(2220\,\text{J/kg}\cdot\text{K}) \ln\left(\frac{273\,\text{K}}{253\,\text{K}}\right) = 101\,\text{J/K}.$$

The heat of fusion of water is $L_f = 333 \times 10^3$ J/kg, so the entropy change on melting is

$$\Delta S_2 = \frac{mL_f}{T} = \frac{(0.600\,\text{kg})(333 \times 10^3\,\text{J/kg})}{273\,\text{K}} = 732\,\text{J/K}.$$

The initial temperature of the water is $T_i = 273\,\text{K}$ and its final temperature is $T_f = 40° + 273° = 313\,\text{K}$. The specific heat of water is $c_{\text{water}} = 4190\,\text{J/kg}\cdot\text{K}$, so the change in the entropy of the water as its temperature is raised is

$$\Delta S_3 = mc_{\text{water}} \ln\left(\frac{T_f}{T_i}\right) = (0.600\,\text{kg})(4190\,\text{J/kg}\cdot\text{K}) \ln\left(\frac{313\,\text{K}}{273\,\text{K}}\right) = 344\,\text{J/K}.$$

The change in entropy for the complete process is $\Delta S = \Delta S_1 + \Delta S_2 + \Delta S_3 = 101\,\text{J/K} + 732\,\text{J/K} + 344\,\text{J/K} = 1.18 \times 10^3\,\text{J/K}$.

**63**

(a) The coefficient of performance of a refrigerator is given by

$$K = \frac{|Q_L|}{|Q_H| - |Q_L|},$$

where $|Q_L|$ is the energy extracted as heat from the low temperature reservoir and $|Q_H|$ is the energy transferred as heat to the high temperature reservoir. In this case the low temperature reservoir is the interior of the refrigerator and the high temperature reservoir is the room. The solution for $|Q_H|$ is

$$|Q_H| = |Q_L|\frac{K+1}{K} = (35.0\,\text{kJ})\frac{4.60+1}{4.60} = 42.6\,\text{kJ}.$$

(b) Over a cycle the change in the internal energy of the system is zero, so according to the first law of thermodynamics the work done per cycle is $|W| = |Q_H| - |Q_L|$. Thus $K = |Q_L|/|W|$ and

$$|W| = \frac{|Q_L|}{K} = \frac{35.0\,\text{kJ}}{4.60} = 7.61\,\text{kJ}.$$

**67**

(a) and (b) If there are $N$ particles in all, with $n$ in one box and $N - n$ in the other, the multiplicity is $W = N!/n!(N-n)!$. The least multiplicity occurs for $n = 0$ or $n = N$ and is $W = N!/N!0! = 1$. The greatest multiplicity occurs for $n = (N-1)/2$ or $n = (N+1)/2$ and is

$$W = \frac{N!}{[(1/2)(N-1)]![(1/2)(N+1)]!}.$$

For $N = 3$ this is $W = 3!/1!2! = 3$ and for $N = 5$ this is $W = 5!/2!3! = 10$.

(c) and (d) The entropy is given by $S = k \ln W$, where $k$ is the Boltzmann constant. The greatest entropy occurs if the multiplicity is the greatest. For $N = 3$ it is $S = (1.38 \times 10^{-23}\,\mathrm{J/K}) \ln 3 = 1.5 \times 10^{-23}\,\mathrm{J/K}$ and for $N = 5$ it is $S = (1.38 \times 10^{-23}\,\mathrm{J/K}) \ln 10 = 3.2 \times 10^{-23}\,\mathrm{J/K}$.

# Chapter 21

1

The magnitude of the force that either charge exerts on the other is given by

$$F = \frac{1}{4\pi\epsilon_0}\frac{|q_1||q_2|}{r^2},$$

where $r$ is the distance between them. Thus

$$\begin{aligned} r &= \sqrt{\frac{|q_1||q_2|}{4\pi\epsilon_0 F}} \\ &= \sqrt{\frac{(8.99 \times 10^9\,\text{N}\cdot\text{m}^2/\text{C}^2)(26.0 \times 10^{-6}\,\text{C})(47.0 \times 10^{-6}\,\text{C})}{5.70\,\text{N}}} = 1.38\,\text{m}. \end{aligned}$$

5

The magnitude of the force of either of the charges on the other is given by

$$F = \frac{1}{4\pi\epsilon_0}\frac{q(Q-q)}{r^2},$$

where $r$ is the distance between the charges. You want the value of $q$ that maximizes the function $f(q) = q(Q-q)$. Set the derivative $df/dq$ equal to zero. This yields $Q - 2q = 0$, or $q = Q/2$.

7

Assume the spheres are far apart. Then the charge distribution on each of them is spherically symmetric and Coulomb's law can be used. Let $q_1$ and $q_2$ be the original charges and choose the coordinate system so the force on $q_2$ is positive if it is repelled by $q_1$. Take the distance between the charges to be $r$. Then the force on $q_2$ is

$$F_a = -\frac{1}{4\pi\epsilon_0}\frac{q_1 q_2}{r^2}.$$

The negative sign indicates that the spheres attract each other.

After the wire is connected, the spheres, being identical, have the same charge. Since charge is conserved, the total charge is the same as it was originally. This means the charge on each sphere is $(q_1 + q_2)/2$. The force is now one of repulsion and is given by

$$F_b = \frac{1}{4\pi\epsilon_0}\frac{(q_1 + q_2)^2}{4r^2}.$$

Solve the two force equations simultaneously for $q_1$ and $q_2$. The first gives

$$q_1 q_2 = -4\pi\epsilon_0 r^2 F_a = -\frac{(0.500\,\text{m})^2(0.108\,\text{N})}{8.99 \times 10^9\,\text{N}\cdot\text{m}^2/\text{C}^2} = -3.00 \times 10^{-12}\,\text{C}^2$$

and the second gives

$$q_1 + q_2 = 2r\sqrt{4\pi\epsilon_0 F_b} = 2(0.500\,\text{m})\sqrt{\frac{0.0360\,\text{N}}{8.99 \times 10^9\,\text{N}\cdot\text{m}^2/\text{C}^2}} = 2.00 \times 10^{-6}\,\text{C}\,.$$

Thus

$$q_2 = \frac{-(3.00 \times 10^{-12}\,\text{C}^2)}{q_1}$$

and substitution into the second equation gives

$$q_1 + \frac{-3.00 \times 10^{-12}\,\text{C}^2}{q_1} = 2.00 \times 10^{-6}\,\text{C}\,.$$

Multiply by $q_1$ to obtain the quadratic equation

$$q_1^2 - (2.00 \times 10^{-6}\,\text{C})q_1 - 3.00 \times 10^{-12}\,\text{C}^2 = 0\,.$$

The solutions are

$$q_1 = \frac{2.00 \times 10^{-6}\,\text{C} \pm \sqrt{(-2.00 \times 10^{-6}\,\text{C})^2 + 4(3.00 \times 10^{-12}\,\text{C}^2)}}{2}\,.$$

If the positive sign is used, $q_1 = 3.00 \times 10^{-6}\,\text{C}$ and if the negative sign is used, $q_1 = -1.00 \times 10^{-6}\,\text{C}$. Use $q_2 = (-3.00 \times 10^{-12})/q_1$ to calculate $q_2$. If $q_1 = 3.00 \times 10^{-6}\,\text{C}$, then $q_2 = -1.00 \times 10^{-6}\,\text{C}$ and if $q_1 = -1.00 \times 10^{-6}\,\text{C}$, then $q_2 = 3.00 \times 10^{-6}\,\text{C}$. Since the spheres are identical, the solutions are essentially the same: one sphere originally had charge $-1.00 \times 10^{-6}\,\text{C}$ and the other had charge $+3.00 \times 10^{-6}\,\text{C}$.

**19**

If the system of three particles is to be in equilibrium, the force on each particle must be zero. Let the charge on the third particle be $q_0$. The third particle must lie on the $x$ axis since otherwise the two forces on it would not be along the same line and could not sum to zero. Thus the $y$ coordinate of the particle must be zero. The third particle must lie between the other two since otherwise the forces acting on it would be in the same direction and would not sum to zero. Suppose the third particle is a distance $x$ from the particle with charge $q$, as shown on the diagram to the right. The force acting on it is then given by

$\leftarrow x \rightarrow\leftarrow L - x \rightarrow$

$q$ $q_0$ $4.00q$

$$F_0 = \frac{1}{4\pi\epsilon_0}\left[\frac{qq_0}{x^2} - \frac{4.00qq_0}{(L-x)^2}\right] = 0\,,$$

where the positive direction was taken to be toward the right. Solve this equation for $x$. Canceling common factors yields $1/x^2 = 4.00/(L-x)^2$ and taking the square root yields $1/x = 2.00/(L-x)$. The solution is $x = 0.333L$.

The force on $q$ is

$$F_q = \frac{1}{4\pi\epsilon_0}\left[\frac{qq_0}{x^2} + \frac{4.00q^2}{L^2}\right] = 0\,.$$

Solve for $q_0$: $q_0 = -4.00qx^2/L^2 = -0.444q$, where $x = 0.333L$ was used.
The force on the particle with charge $4.00q$ is

$$F_{4q} = \frac{1}{4\pi\epsilon_0}\left[\frac{4.00q^2}{L^2} + \frac{4.00qq_0}{(L-x)^2}\right] = \frac{1}{4\pi\epsilon_0}\left[\frac{4.00q^2}{L^2} + \frac{4.00(0.444)q^2}{(0.444)L^2}\right]$$
$$= \frac{1}{4\pi\epsilon_0}\left[\frac{4.00q^2}{L^2} - \frac{4.00q^2}{L^2}\right] = 0\,.$$

With $q_0 = -0.444q$ and $x = 0.333L$, all three charges are in equilibrium.

**25**

(a) The magnitude of the force between the ions is given by

$$F = \frac{q^2}{4\pi\epsilon_0 r^2}\,,$$

where $q$ is the charge on either of them and $r$ is the distance between them. Solve for the charge:

$$q = r\sqrt{4\pi\epsilon_0 F} = (5.0 \times 10^{-10}\,\text{m})\sqrt{\frac{3.7 \times 10^{-9}\,\text{N}}{8.99 \times 10^9\,\text{N}\cdot\text{m}^2/\text{C}^2}} = 3.2 \times 10^{-19}\,\text{C}\,.$$

(b) Let $N$ be the number of electrons missing from each ion. Then $Ne = q$ and

$$N = \frac{q}{e} = \frac{3.2 \times 10^{-19}\,\text{C}}{1.60 \times 10^{-19}\,\text{C}} = 2\,.$$

**35**

(a) Every cesium ion at a corner of the cube exerts a force of the same magnitude on the chlorine ion at the cube center. Each force is a force of attraction and is directed toward the cesium ion that exerts it, along the body diagonal of the cube. We can pair every cesium ion with another, diametrically positioned at the opposite corner of the cube. Since the two ions in such a pair exert forces that have the same magnitude but are oppositely directed, the two forces sum to zero and, since every cesium ion can be paired in this way, the total force on the chlorine ion is zero.
(b) Rather than remove a cesium ion, superpose charge $-e$ at the position of one cesium ion. This neutralizes the ion and, as far as the electrical force on the chlorine ion is concerned, it is equivalent to removing the ion. The forces of the eight cesium ions at the cube corners sum to zero, so the only force on the chlorine ion is the force of the added charge.
The length of a body diagonal of a cube is $\sqrt{3}a$, where $a$ is the length of a cube edge. Thus the distance from the center of the cube to a corner is $d = (\sqrt{3}/2)a$. The force has magnitude

$$F = \frac{1}{4\pi\epsilon_0}\frac{e^2}{d^2} = \frac{1}{4\pi\epsilon_0}\frac{e^2}{(3/4)a^2}$$
$$= \frac{(8.99 \times 10^9\,\text{N}\cdot\text{m}^2/\text{C}^2)(1.60 \times 10^{-19}\,\text{C})^2}{(3/4)(0.40 \times 10^{-9}\,\text{m})^2} = 1.9 \times 10^{-9}\,\text{N}\,.$$

Since both the added charge and the chlorine ion are negative, the force is one of repulsion. The chlorine ion is pulled away from the site of the missing cesium ion.

**37**

None of the reactions given include a beta decay, so the number of protons, the number of neutrons, and the number of electrons are each conserved. Atomic numbers (numbers of protons and numbers of electrons) and molar masses (combined numbers of protons and neutrons) can be found in Appendix F of the text.

(a) $^{1}$H has 1 proton, 1 electron, and 0 neutrons and $^{9}$Be has 4 protons, 4 electrons, and $9 - 4 = 5$ neutrons, so X has $1 + 4 = 5$ protons, $1 + 4 = 5$ electrons, and $0 + 5 - 1 = 4$ neutrons. One of the neutrons is freed in the reaction. X must be boron with a molar mass of $5\,\text{g/mol} + 4\,\text{g/mol} = 9\,\text{g/mol}$: $^{9}$B.

(b) $^{12}$C has 6 protons, 6 electrons, and $12 - 6 = 6$ neutrons and $^{1}$H has 1 proton, 1 electron, and 0 neutrons, so X has $6 + 1 = 7$ protons, $6 + 1 = 7$ electrons, and $6 + 0 = 6$ neutrons. It must be nitrogen with a molar mass of $7\,\text{g/mol} + 6\,\text{g/mol} = 13\,\text{g/mol}$: $^{13}$N.

(c) $^{15}$N has 7 protons, 7 electrons, and $15 - 7 = 8$ neutrons; $^{1}$H has 1 proton, 1 electron, and 0 neutrons; and $^{4}$He has 2 protons, 2 electrons, and $4 - 2 = 2$ neutrons; so X has $7 + 1 - 2 = 6$ protons, 6 electrons, and $8 + 0 - 2 = 6$ neutrons. It must be carbon with a molar mass of $6\,\text{g/mol} + 6\,\text{g/mol} = 12\,\text{g/mol}$: $^{12}$C.

**39**

The magnitude of the force of particle 1 on particle 2 is

$$F = \frac{1}{4\pi\epsilon_0}\frac{|q_1||q_2|}{d_1^2 + d_2^2}\,.$$

The signs of the charges are the same, so the particles repel each other along the line that runs through them. This line makes an angle $\theta$ with the $x$ axis such that $\cos\theta = d_2/\sqrt{d_1^2 + d_2^2}$, so the $x$ component of the force is

$$\begin{aligned} F_x &= \frac{1}{4\pi\epsilon_0}\frac{|q_1||q_2|}{d_1^2 + d_2^2}\cos\theta = \frac{1}{4\pi\epsilon_0}\frac{|q_1||q_2|d_2}{(d_1^2 + d_2^2)^{3/2}} \\ &= (8.99 \times 10^{9}\,\text{C}^2/\text{N}\cdot\text{m}^2)\frac{24(1.60 \times 10^{-19}\,\text{C})^2(6.00 \times 10^{-3}\,\text{m})}{[(2.00 \times 10^{-3}\,\text{m})^2 + (6.00 \times 10^{-3}\,\text{m})^2]^{3/2}} = 1.31 \times 10^{-22}\,\text{N}\,. \end{aligned}$$

**50**

The magnitude of the gravitational force on a proton near Earth's surface is $mg$, where $m$ is the mass of the proton ($1.67 \times 10^{-27}$ kg from Appendix B). The electrostatic force between two protons is $F = (1/4\pi\epsilon_0)(e^2/d^2)$, where $d$ is their separation. Equate these forces to each other and solve for $d$. The result is

$$d = \sqrt{\frac{1}{4\pi\epsilon_0}\frac{e^2}{mg}} = \sqrt{(8.99 \times 10^{-9}\,\text{N}\cdot\text{m}^2/\text{C}^2)\frac{(1.60 \times 10^{-19}\,\text{C})^2}{(1.67 \times 10{-}27\,\text{kg})(9.8\,\text{m/s}^2)}} = 0.119\,\text{m}\,.$$

## 60

The magnitude of the force of particle 1 on particle 4 is

$$F_1 = \frac{1}{4\pi\epsilon_0}\frac{|q_1||q_4|}{d_1^2} = (8.99 \times 10^9\,\text{N}\cdot\text{m}^2/\text{C}^2)\frac{(3.20 \times 10^{-19}\,\text{C})(3.20 \times 10^{-19}\,\text{C})}{(0.0300\,\text{m})^2} = 1.02 \times 10^{-24}\,\text{N}\,.$$

The charges have opposite signs, so the particles attract each other and the vector force is

$$\begin{aligned}\vec{F}_1 &= -(1.02^{-24}\,\text{N})(\cos 35.0°)\,\hat{\imath} - (1.02 \times 10^{-24}\,\text{N})(\sin 35.0°)\,\hat{\jmath} \\ &= -(8.36 \times 10^{-25}\,\text{N})\,\hat{\imath} - (5.85 \times 10^{-25}\,\text{N})\,\hat{\jmath}\,.\end{aligned}$$

Particles 2 and 3 repel each other. The force of particle 2 on particle 4 is

$$\begin{aligned}\vec{F}_2 &= -\frac{1}{4\pi\epsilon_0}\frac{|q_2||q_4|}{d_2^2}\,\hat{\jmath} \\ &= -(8.99 \times 10^9\,\text{N}\cdot\text{m}^2/\text{C}^2)\frac{(3.20 \times 10^{-19}\,\text{C})(3.20 \times 10^{-19}\,\text{C})}{(0.0200\,\text{m})^2}\,\hat{\jmath} = -(2.30 \times 10^{-24}\,\text{N})\,\hat{\jmath}\,.\end{aligned}$$

Particles 3 and 4 repel each other and the force of particle 3 on particle 4 is

$$\begin{aligned}\vec{F}_3 &= -\frac{1}{4\pi\epsilon_0}\frac{|2q_3||q_4|}{d_3^2}\,\hat{\imath} \\ &= -(8.99 \times 10^9\,\text{N}\cdot\text{m}^2/\text{C}^2)\frac{(6.40 \times 10^{-19}\,\text{C})(3.20 \times 10^{-19}\,\text{C})}{(0.0200\,\text{m})^2} = -(4.60 \times 10^{-24}\,\text{C})\,\hat{\imath}\,.\end{aligned}$$

The net force is the vector sum of the three forces. The $x$ component is $F_x = -18.36 \times 10^{-25}\,\text{N} - 4.60\,\text{N} = -5.44 \times 10^{-24}\,\text{N}$ and the $y$ component is $F_y = -5.85 \times 10^{-25}\,\text{N} - 2.30 \times 10^{-24}\,\text{N} = -2.89 \times 10^{-24}\,\text{N}$. The magnitude of the force is

$$F = \sqrt{F_x^2 + F_y^2} = \sqrt{(-5.44 \times 10^{-24}\,\text{N})^2 + (-2.89 \times 10^{-24}\,\text{N})^2} = 6.16 \times 10^{24}\,\text{N}\,.$$

The tangent of the angle $\theta$ between the net force and the positive $x$ axis is $tan\theta = F_y/F_x = (-2.89 \times 10^{-24}\,\text{N})/(-5.44 \times 10^{-24}\,\text{N}) = 0.531$ and the angle is either 28° or 208°. The later angle is associated with a vector that has negative $x$ and $y$ components and so is the correct angle.

## 69

The net force on particle 3 is the vector sum of the forces of particles 1 and 2 and for this to be zero the two forces must be along the same line. Since electrostatic forces are along the lines that join the particles, particle 3 must be on the $x$ axis. Its $y$ coordinate is zero.

Particle 3 is repelled by one of the other charges and attracted by the other. As a result, particle 3 cannot be between the other two particles and must be either to the left of particle 1 or to the right of particle 2. Since the magnitude of $q_1$ is greater than the magnitude of $q_2$, particle 3 must be closer to particle 2 than to particle 1 and so must be to the right of particle 2. Let $x$ be the coordinate of particle 3. The the $x$ component of the force on it is

$$F_x = \frac{1}{4\pi\epsilon_0}\left[\frac{q_1 q_3}{x^2} + \frac{q_2 q_3}{(x-L)^2}\right].$$

If $F_x = 0$ the solution for $x$ is

$$x = \frac{\sqrt{-q_1/q_2}}{\sqrt{-q_1/q_2} - 1}L = \frac{\sqrt{-(-5.00q)/(2.00q)}}{\sqrt{-(-5.00q)/(2.00q)} - 1}L = 2.72L\,.$$

# Chapter 22

3

Since the magnitude of the electric field produced by a point particle with charge $q$ is given by $E = |q|/4\pi\epsilon_0 r^2$, where $r$ is the distance from the particle to the point where the field has magnitude $E$, the magnitude of the charge is

$$|q| = 4\pi\epsilon_0 r^2 E = \frac{(0.50\,\text{m})^2(2.0\,\text{N/C})}{8.99 \times 10^9\,\text{N}\cdot\text{m}^2/\text{C}^2} = 5.6 \times 10^{-11}\,\text{C}\,.$$

5

Since the charge is uniformly distributed throughout a sphere, the electric field at the surface is exactly the same as it would be if the charge were all at the center. That is, the magnitude of the field is

$$E = \frac{q}{4\pi\epsilon_0 R^2}\,,$$

where $q$ is the magnitude of the total charge and $R$ is the sphere radius. The magnitude of the total charge is $Ze$, so

$$E = \frac{Ze}{4\pi\epsilon_0 R^2} = \frac{(8.99 \times 10^9\,\text{N}\cdot\text{m}^2/\text{C}^2)(94)(1.60 \times 10^{-19}\,\text{C})}{(6.64 \times 10^{-15}\,\text{m})^2} = 3.07 \times 10^{21}\,\text{N/C}\,.$$

The field is normal to the surface and since the charge is positive it points outward from the surface.

7

At points between the particles, the individual electric fields are in the same direction and do not cancel. Charge $q_2$ has a greater magnitude than charge $q_1$, so a point of zero field must be closer to $q_1$ than to $q_2$. It must be to the right of $q_1$ on the diagram.

d P
$q_2$ $q_1$

Put the origin at the particle with charge $q_2$ and let $x$ be the coordinate of $P$, the point where the field vanishes. Then the total electric field at $P$ is given by

$$E = \frac{1}{4\pi\epsilon_0}\left[\frac{q_2}{x^2} - \frac{q_1}{(x-d)^2}\right],$$

where $q_1$ and $q_2$ are the magnitudes of the charges. If the field is to vanish,

$$\frac{q_2}{x^2} = \frac{q_1}{(x-d)^2}\,.$$

Take the square root of both sides to obtain $\sqrt{q_2}/x = \sqrt{q_1}/(x-d)$. The solution for $x$ is

$$x = \left(\frac{\sqrt{q_2}}{\sqrt{q_2}-\sqrt{q_1}}\right)d = \left(\frac{\sqrt{4.0q_1}}{\sqrt{4.0q_1}-\sqrt{q_1}}\right)d$$
$$= \left(\frac{2.0}{2.0-1.0}\right)d = 2.0d = (2.0)(50\text{ cm}) = 100\text{ cm}.$$

The point is 50 cm to the right of $q_1$.

**9**

Choose the coordinate axes as shown on the diagram to the right. At the center of the square, the electric fields produced by the particles at the lower left and upper right corners are both along the $x$ axis and each points away from the center and toward the particle that produces it. Since each particle is a distance $d = \sqrt{2}a/2 = a/\sqrt{2}$ away from the center, the net field due to these two particles is

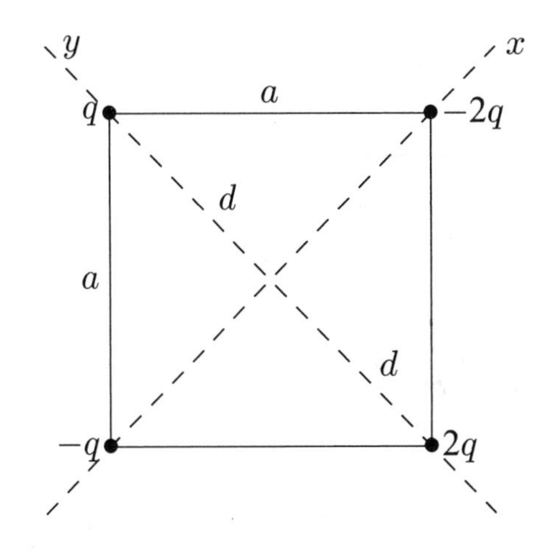


$$E_x = \frac{1}{4\pi\epsilon_0}\left[\frac{2q}{a^2/2} - \frac{q}{a^2/2}\right]$$
$$= \frac{1}{4\pi\epsilon_0}\frac{q}{a^2/2} = \frac{(8.99\times10^9\text{ N}\cdot\text{m}^2/\text{C}^2)(1.0\times10^{-8}\text{ C})}{(0.050\text{ m})^2/2} = 7.19\times10^4\text{ N/C}.$$

At the center of the square, the field produced by the particles at the upper left and lower right corners are both along the $y$ axis and each points away from the particle that produces it. The net field produced at the center by these particles is

$$E_y = \frac{1}{4\pi\epsilon_0}\left[\frac{2q}{a^2/2} - \frac{q}{a^2/2}\right] = \frac{1}{4\pi\epsilon_0}\frac{q}{a^2/2} = 7.19\times10^4\text{ N/C}.$$

The magnitude of the net field is

$$E = \sqrt{E_x^2+E_y^2} = \sqrt{2(7.19\times10^4\text{ N/C})^2} = 1.02\times10^5\text{ N/C}$$

and the angle it makes with the $x$ axis is

$$\theta = \tan^{-1}\frac{E_y}{E_x} = \tan^{-1}(1) = 45°.$$

It is upward in the diagram, from the center of the square toward the center of the upper side.

**21**

Think of the quadrupole as composed of two dipoles, each with dipole moment of magnitude $p = qd$. The moments point in opposite directions and produce fields in opposite directions at points on the quadrupole axis. Consider the point P on the axis, a distance $z$ to the right of the quadrupole center and take a rightward pointing field to be positive. Then the field produced by the right dipole of the pair is $qd/2\pi\epsilon_0(z-d/2)^3$ and the field produced by the left dipole is $-qd/2\pi\epsilon_0(z+d/2)^3$. Use the binomial expansions $(z-d/2)^{-3} \approx z^{-3} - 3z^{-4}(-d/2)$ and $(z+d/2)^{-3} \approx z^{-3} - 3z^{-4}(d/2)$ to obtain

$$E = \frac{qd}{2\pi\epsilon_0}\left[\frac{1}{z^3} + \frac{3d}{2z^4} - \frac{1}{z^3} + \frac{3d}{2z^4}\right] = \frac{6qd^2}{4\pi\epsilon_0 z^4}.$$

Let $Q = 2qd^2$. Then

$$E = \frac{3Q}{4\pi\epsilon_0 z^4}.$$

**27**

(a) The linear charge density $\lambda$ is the charge per unit length of rod. Since the charge is uniformly distributed on the rod, $\lambda = -q/L = -(4.23 \times 10^{-15}\,\mathrm{C})/(0.0815\,\mathrm{m}) = -5.19 \times 10^{-14}\,\mathrm{C/m}$.

(b) and (c) Position the origin at the left end of the rod, as shown in the diagram. Let $dx$ be an infinitesimal length of rod at $x$. The charge in this segment is $dq = \lambda\,dx$. Since the segment may be taken to be a point particle, the electric field it produces at point P has only an $x$ component and this component is given by

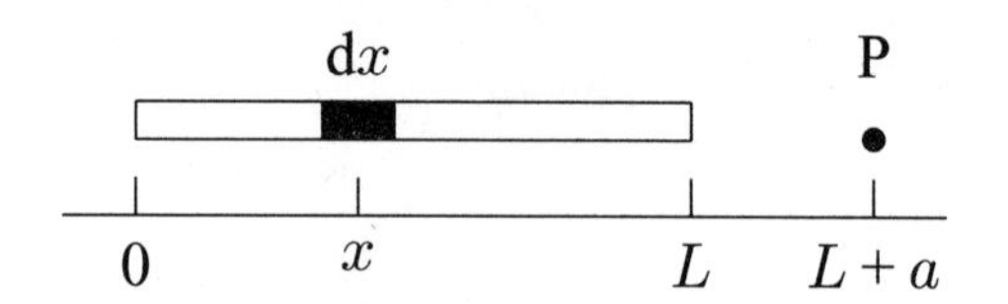


$$dE_x = \frac{1}{4\pi\epsilon_0}\frac{\lambda\,dx}{(L+a-x)^2}.$$

The total electric field produced at P by the whole rod is the integral

$$E_x = \frac{\lambda}{4\pi\epsilon_0}\int_0^L \frac{dx}{(L+a-x)^2} = \frac{\lambda}{4\pi\epsilon_0}\left.\frac{1}{L+a-x}\right|_0^L$$
$$= \frac{\lambda}{4\pi\epsilon_0}\left[\frac{1}{a} - \frac{1}{L+a}\right] = \frac{\lambda}{4\pi\epsilon_0}\frac{L}{a(L+a)}.$$

When $-q/L$ is substituted for $\lambda$ the result is

$$E_x = -\frac{1}{4\pi\epsilon_0}\frac{q}{a(L+a)} = -\frac{(8.99\times10^9\,\mathrm{N\cdot m^2/C^2})(4.23\times10^{-15}\,\mathrm{C})}{(0.120\,\mathrm{m})(0.0815\,\mathrm{m}+0.120\,\mathrm{m})} = -1.57\times10^{-3}\,\mathrm{N/C}.$$

The negative sign indicates that the field is toward the rod and makes an angle of 180° with the positive $x$ direction.

(d) Now

$$E_x = -\frac{1}{4\pi\epsilon_0}\frac{q}{a(L+a)} = -\frac{(8.99\times10^9\,\mathrm{N\cdot m^2/C^2})(4.23\times10^{-15}\,\mathrm{C})}{(50\,\mathrm{m})(0.0815\,\mathrm{m}+50\,\mathrm{m})} = -1.52\times10^{-8}\,\mathrm{N/C}.$$

(e) The field of a point particle at the origin is

$$E_x = -\frac{q}{4\pi\epsilon_0 a^2} = -\frac{(8.99 \times 10^9\,\mathrm{N\cdot m^2/C^2})(4.23 \times 10^{-15}\,\mathrm{C})}{(50\,\mathrm{m})^2} = -1.52 \times 10^{-8}\,\mathrm{N/C}\,.$$

**35**

At a point on the axis of a uniformly charged disk a distance $z$ above the center of the disk, the magnitude of the electric field is

$$E = \frac{\sigma}{2\epsilon_0}\left[1 - \frac{z}{\sqrt{z^2 + R^2}}\right],$$

where $R$ is the radius of the disk and $\sigma$ is the surface charge density on the disk. See Eq. 22–26. The magnitude of the field at the center of the disk ($z = 0$) is $E_c = \sigma/2\epsilon_0$. You want to solve for the value of $z$ such that $E/E_c = 1/2$. This means

$$\frac{E}{E_c} = 1 - \frac{z}{\sqrt{z^2 + R^2}} = \frac{1}{2}$$

or

$$\frac{z}{\sqrt{z^2 + R^2}} = \frac{1}{2}\,.$$

Square both sides, then multiply them by $z^2 + R^2$ to obtain $z^2 = (z^2/4) + (R^2/4)$. Thus $z^2 = R^2/3$ and $z = R/\sqrt{3} = (0.600\,\mathrm{m})/\sqrt{3} = 0.346\,\mathrm{m}$.

**39**

The magnitude of the force acting on the electron is $F = eE$, where $E$ is the magnitude of the electric field at its location. The acceleration of the electron is given by Newton's second law:

$$a = \frac{F}{m} = \frac{eE}{m} = \frac{(1.60 \times 10^{-19}\,\mathrm{C})(2.00 \times 10^4\,\mathrm{N/C})}{9.11 \times 10^{-31}\,\mathrm{kg}} = 3.51 \times 10^{15}\,\mathrm{m/s^2}\,.$$

**43**

(a) The magnitude of the force on the particle is given by $F = qE$, where $q$ is the magnitude of the charge carried by the particle and $E$ is the magnitude of the electric field at the location of the particle. Thus

$$E = \frac{F}{q} = \frac{3.0 \times 10^{-6}\,\mathrm{N}}{2.0 \times 10^{-9}\,\mathrm{C}} = 1.5 \times 10^3\,\mathrm{N/C}\,.$$

The force points downward and the charge is negative, so the field points upward.

(b) The magnitude of the electrostatic force on a proton is

$$F_e = eE = (1.60 \times 10^{-19}\,\mathrm{C})(1.5 \times 10^3\,\mathrm{N/C}) = 2.4 \times 10^{-16}\,\mathrm{N}\,.$$

(c) A proton is positively charged, so the force is in the same direction as the field, upward.

(d) The magnitude of the gravitational force on the proton is

$$F_g = mg = (1.67 \times 10^{-27}\,\text{kg})(9.8\,\text{m/s}^2) = 1.64 \times 10^{-26}\,\text{N}\,.$$

The force is downward.

(e) The ratio of the force magnitudes is

$$\frac{F_e}{F_g} = \frac{2.4 \times 10^{-16}\,\text{N}}{1.64 \times 10^{-26}\,\text{N}} = 1.5 \times 10^{10}\,.$$

**45**

(a) The magnitude of the force acting on the proton is $F = eE$, where $E$ is the magnitude of the electric field. According to Newton's second law, the acceleration of the proton is $a = F/m = eE/m$, where $m$ is the mass of the proton. Thus

$$a = \frac{(1.60 \times 10^{-19}\,\text{C})(2.00 \times 10^4\,\text{N/C})}{1.67 \times 10^{-27}\,\text{kg}} = 1.92 \times 10^{12}\,\text{m/s}^2\,.$$

(b) Assume the proton starts from rest and use the kinematic equation $v^2 = v_0^2 + 2ax$ (or else $x = \frac{1}{2}at^2$ and $v = at$) to show that

$$v = \sqrt{2ax} = \sqrt{2(1.92 \times 10^{12}\,\text{m/s}^2)(0.0100\,\text{m})} = 1.96 \times 10^5\,\text{m/s}\,.$$

**57**

(a) If $q$ is the positive charge in the dipole and $d$ is the separation of the charged particles, the magnitude of the dipole moment is $p = qd = (1.50 \times 10^{-9}\,\text{C})(6.20 \times 10^{-6}\,\text{m}) = 9.30 \times 10^{-15}\,\text{C}\cdot\text{m}$.

(b) If the initial angle between the dipole moment and the electric field is $\theta_0$ and the final angle is $\theta$, then the change in the potential energy as the dipole swings from $\theta = 0$ to $\theta = 180°$ is

$$\begin{aligned}\Delta U &= -pE(\cos\theta - \cos\theta_0) = -(9.30 \times 10^{-15}\,\text{C}\cdot\text{m})(1100\,\text{N/C})(\cos 180° - \cos 0)\\ &= 2.05 \times 10^{-11}\,\text{J}\,.\end{aligned}$$

**79**

(a) and (b) Since the field at the point on the $x$ axis with coordinate $x = 2.0\,\text{cm}$ is in the positive $x$ direction you know that the charged particle is on the $x$ axis. The line through the point with coordinates $x = 3.0\,\text{cm}$ and $y = 3.0\,\text{cm}$ and parallel to the field at that point must pass through the position of the particle. Such a line has slope $(3.0)/(4.0) = 0.75$ and its equation is $y = 0.57 + (0.75)x$. The solution for $y = 0$ is $x = -1.0\,\text{cm}$, so the particle is located at the point with coordinates $x = -1.0\,\text{cm}$ and $y = 0$.

(c) The magnitude of the field at the point on the $x$ axis with coordinate $x = 2.0\,\text{cm}$ is given by $E = (1/4\pi\epsilon_0)q/(2.0\,\text{cm} - x)^2$, so

$$q = 4\pi\epsilon_0 x^2 E = \frac{(0.020\,\text{m} + 0.010\,\text{m})^2(100\,\text{N/C})}{8.99 \times 10^9\,\text{N}\cdot\text{m}^2/\text{C}^2} = 1.0 \times 10^{-11}\,\text{C}.$$

**81**

(a) The potential energy of an electric dipole with dipole moment $\vec{p}$ in an electric field $\vec{E}$ is

$$\begin{aligned} U = -\vec{p}\cdot\vec{E} &= (1.24 \times 10^{-30}\,\text{C}\cdot\text{m})(3.00\,\hat{\text{i}} + 4.00\,\hat{\text{j}}) \cdot (4000\,\text{N/C})\,\hat{\text{i}} \\ &= -(1.24 \times 10^{-30}\,\text{C}\cdot\text{m})(3.00)(4000\,\text{N/C}) = -1.49 \times 10^{-26}\,\text{J}. \end{aligned}$$

Here we used $\vec{a}\cdot\vec{b} = a_x b_x + a_y b_y + a_z b_z$ to evaluate the scalar product.

(b) The torque is

$$\begin{aligned} \vec{\tau} = \vec{p}\times\vec{E} &= (p_x\,\hat{\text{i}} + p_y\,\hat{\text{j}}) \times (E_x\,\hat{\text{i}}) = -p_y E_x\,\hat{\text{k}} \\ &= -(4.00)(1.24 \times 10^{-30}\,\text{C}\cdot\text{m})(4000\,\text{N/C}) = -(1.98 \times 10^{-26}\,\text{N}\cdot\text{m})\,\hat{\text{k}}. \end{aligned}$$

(c) The work done by the agent is equal to the change in the potential energy of the dipole. The initial potential energy is $U_i = -1.49 \times 10^{-26}\,\text{J}$, as computed in part (a). The final potential energy is

$$\begin{aligned} U_f &= (1.24 \times 10^{-30}\,\text{C}\cdot\text{m})(-4.00\,\hat{\text{i}} + 3.00\,\hat{\text{j}}) \cdot (4000\,\text{N/C})\,\hat{\text{i}} \\ &= -(1.24 \times 10^{-30}\,\text{C}\cdot\text{m})(-4.00)(4000\,\text{N/C}) = +1.98 \times 10^{-26}\,\text{J}. \end{aligned}$$

The work done by the agent is $W = (1.98 \times 10^{-26}\,\text{J}) - (-1.49 \times 10^{-26}\,\text{J}) = 3.47 \times 10^{-26}\,\text{J}$.

# Chapter 23

1

The vector area $\vec{A}$ and the electric field $\vec{E}$ are shown on the diagram to the right. The angle $\theta$ between them is $180° - 35° = 145°$, so the electric flux through the area is $\Phi = \vec{E} \cdot \vec{A} = EA\cos\theta = (1800\,\mathrm{N/C})(3.2 \times 10^{-3}\,\mathrm{m})^2 \cos 145° = -1.5 \times 10^{-2}\,\mathrm{N \cdot m^2/C}$.

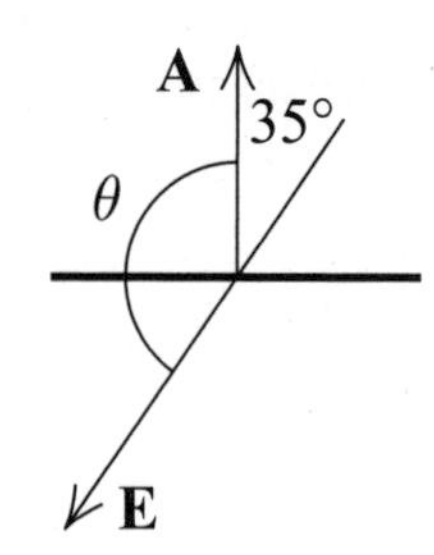


9

Let $A$ be the area of one face of the cube, $E_u$ be the magnitude of the electric field at the upper face, and $E_\ell$ be the magnitude of the field at the lower face. Since the field is downward, the flux through the upper face is negative and the flux through the lower face is positive. The flux through the other faces is zero, so the total flux through the cube surface is $\Phi = A(E_\ell - E_u)$. The net charge inside the cube is given by Gauss' law:

$$q = \epsilon_0\Phi = \epsilon_0 A(E_\ell - E_u) = (8.85 \times 10^{-12}\,\mathrm{C^2/N \cdot m^2})(100\,\mathrm{m})^2(100\,\mathrm{N/C} - 60.0\,\mathrm{N/C})$$
$$= 3.54 \times 10^{-6}\,\mathrm{C} = 3.54\,\mu\mathrm{C}\,.$$

19

(a) The charge on the surface of the sphere is the product of the surface charge density $\sigma$ and the surface area of the sphere ($4\pi r^2$, where $r$ is the radius). Thus

$$q = 4\pi r^2\sigma = 4\pi\left(\frac{1.2\,\mathrm{m}}{2}\right)^2(8.1 \times 10^{-6}\,\mathrm{C/m^2}) = 3.7 \times 10^{-5}\,\mathrm{C}\,.$$

(b) Choose a Gaussian surface in the form a sphere, concentric with the conducting sphere and with a slightly larger radius. The flux through the surface is given by Gauss' law:

$$\Phi = \frac{q}{\epsilon_0} = \frac{3.7 \times 10^{-5}\,\mathrm{C}}{8.85 \times 10^{-12}\,\mathrm{C^2/N \cdot m^2}} = 4.1 \times 10^6\,\mathrm{N \cdot m^2/C}\,.$$

23

The magnitude of the electric field produced by a uniformly charged infinite line is $E = \lambda/2\pi\epsilon_0 r$, where $\lambda$ is the linear charge density and $r$ is the distance from the line to the point where the field is measured. See Eq. 23–12. Thus

$$\lambda = 2\pi\epsilon_0 Er = 2\pi(8.85 \times 10^{-12}\,\mathrm{C^2/N \cdot m^2})(4.5 \times 10^4\,\mathrm{N/C})(2.0\,\mathrm{m}) = 5.0 \times 10^{-6}\,\mathrm{C/m}\,.$$

**27**

Assume the charge density of both the conducting rod and the shell are uniform. Neglect fringing. Symmetry can be used to show that the electric field is radial, both between the rod and the shell and outside the shell. It is zero, of course, inside the rod and inside the shell since they are conductors.

(a) and (b) Take the Gaussian surface to be a cylinder of length $L$ and radius $r$, concentric with the conducting rod and shell and with its curved surface outside the shell. The area of the curved surface is $2\pi rL$. The field is normal to the curved portion of the surface and has uniform magnitude over it, so the flux through this portion of the surface is $\Phi = 2\pi rLE$, where $E$ is the magnitude of the field at the Gaussian surface. The flux through the ends is zero. The charge enclosed by the Gaussian surface is $Q_1 - 2.00Q_1 = -Q_1$. Gauss' law yields $2\pi r\epsilon_0 LE = -Q_1$, so

$$E = -\frac{Q_1}{2\pi\epsilon_0 Lr} = -\frac{(8.99\times10^9\,N\cdot\text{m}^2/\text{C}^2)(3.40\times10^{-12}\,\text{C})}{(11.00\,\text{m})(26.0\times10^{-3}\,\text{m})} = -0.214\,\text{N/C}\,.$$

The magnitude of the field is $0.214\,\text{N/C}$. The negative sign indicates that the field points inward.

(c) and (d) Take the Gaussian surface to be a cylinder of length $L$ and radius $r$, concentric with the conducting rod and shell and with its curved surface between the conducting rod and the shell. As in (a), the flux through the curved portion of the surface is $\Phi = 2\pi rLE$, where $E$ is the magnitude of the field at the Gaussian surface, and the flux through the ends is zero. The charge enclosed by the Gaussian surface is only the charge $Q_1$ on the conducting rod. Gauss' law yields $2\pi\epsilon_0 rLE = Q_1$, so

$$E = \frac{Q_1}{2\pi\epsilon_0 Lr} = \frac{2(8.99\times10^9\,\text{N}\cdot\text{m}^2/\text{C}^2)(3.40\times10^{-12}\,\text{C})}{(11.00\,\text{m})(6.50\times10^{-3}\,\text{m})} = +0.855\,\text{N/C}\,.$$

The positive sign indicates that the field points outward.

(e) Consider a Gaussian surface in the form of a cylinder of length $L$ with the curved portion of its surface completely within the shell. The electric field is zero at all points on the curved surface and is parallel to the ends, so the total electric flux through the Gaussian surface is zero and the net charge within it is zero. Since the conducting rod, which is inside the Gaussian cylinder, has charge $Q_1$, the inner surface of the shell must have charge $-Q_1 = -3.40\times10^{-12}\,\text{C}$.

(f) Since the shell has total charge $-2.00Q_1$ and has charge $-Q_1$ on its inner surface, it must have charge $-Q_1 = -3.40\times10^{-12}\,\text{C}$ on its outer surface.

**35**

(a) To calculate the electric field at a point very close to the center of a large, uniformly charged conducting plate, we may replace the finite plate with an infinite plate with the same area charge density and take the magnitude of the field to be $E = \sigma/\epsilon_0$, where $\sigma$ is the area charge density for the surface just under the point. The charge is distributed uniformly over both sides of the original plate, with half being on the side near the field point. Thus

$$\sigma = \frac{q}{2A} = \frac{6.0\times10^{-6}\,\text{C}}{2(0.080\,\text{m})^2} = 4.69\times10^{-4}\,\text{C/m}^2\,.$$

The magnitude of the field is

$$E = \frac{\sigma}{\epsilon_0} = \frac{4.69 \times 10^{-4}\,\mathrm{C/m^2}}{8.85 \times 10^{-12}\,\mathrm{C^2/N \cdot m^2}} = 5.3 \times 10^{7}\,\mathrm{N/C}\,.$$

The field is normal to the plate and since the charge on the plate is positive, it points away from the plate.

(b) At a point far away from the plate, the electric field is nearly that of a point particle with charge equal to the total charge on the plate. The magnitude of the field is $E = q/4\pi\epsilon_0 r^2$, where $r$ is the distance from the plate. Thus

$$E = \frac{(8.99 \times 10^{9}\,\mathrm{N \cdot m^2/C^2})(6.0 \times 10^{-6}\,\mathrm{C})}{(30\,\mathrm{m})^2} = 60\,\mathrm{N/C}\,.$$

**41**

The forces on the ball are shown in the diagram to the right. The gravitational force has magnitude $mg$, where $m$ is the mass of the ball; the electrical force has magnitude $qE$, where $q$ is the charge on the ball and $E$ is the electric field at the position of the ball; and the tension in the thread is denoted by $T$. The electric field produced by the plate is normal to the plate and points to the right. Since the ball is positively charged, the electric force on it also points to the right. The tension in the thread makes the angle $\theta$ (= 30°) with the vertical.

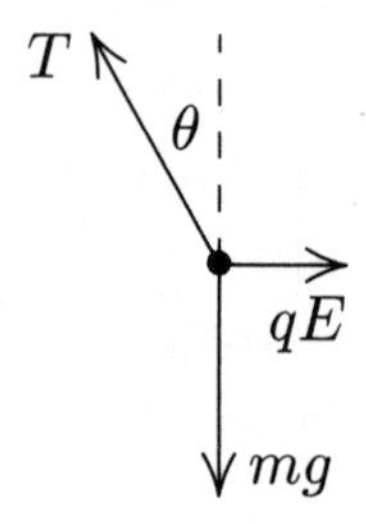


Since the ball is in equilibrium the net force on it vanishes. The sum of the horizontal components yields $qE - T\sin\theta = 0$ and the sum of the vertical components yields $T\cos\theta - mg = 0$. The expression $T = qE/\sin\theta$, from the first equation, is substituted into the second to obtain $qE = mg\tan\theta$.

The electric field produced by a large uniform plane of charge is given by $E = \sigma/2\epsilon_0$, where $\sigma$ is the surface charge density. Thus

$$\frac{q\sigma}{2\epsilon_0} = mg\tan\theta$$

and

$$\begin{aligned}\sigma &= \frac{2\epsilon_0 mg\tan\theta}{q} \\ &= \frac{2(8.85 \times 10^{-12}\,\mathrm{C^2/N \cdot m^2})(1.0 \times 10^{-6}\,\mathrm{kg})(9.8\,\mathrm{m/s^2})\tan 30^\circ}{2.0 \times 10^{-8}\,\mathrm{C}} \\ &= 5.0 \times 10^{-9}\,\mathrm{C/m^2}\,.\end{aligned}$$

**45**

Charge is distributed uniformly over the surface of the sphere and the electric field it produces at points outside the sphere is like the field of a point particle with charge equal to the net charge on the sphere. That is, the magnitude of the field is given by $E = q/4\pi\epsilon_0 r^2$, where $q$ is the magnitude of the charge on the sphere and $r$ is the distance from the center of the sphere to the point where the field is measured. Thus

$$q = 4\pi\epsilon_0 r^2 E = \frac{(0.15\,\text{m})^2(3.0 \times 10^3\,\text{N/C})}{8.99 \times 10^9\,\text{N}\cdot\text{m}^2/\text{C}^2} = 7.5 \times 10^{-9}\,\text{C}.$$

The field points inward, toward the sphere center, so the charge is negative: $-7.5 \times 10^{-9}\,\text{C}$.

**49**

To find an expression for the electric field inside the shell in terms of $A$ and the distance from the center of the shell, select $A$ so the field does not depend on the distance.

Use a Gaussian surface in the form of a sphere with radius $r_g$, concentric with the spherical shell and within it ($a < r_g < b$). Gauss' law is used to find the magnitude of the electric field a distance $r_g$ from the shell center.

The charge that is both in the shell and within the Gaussian sphere is given by the integral $q_{\text{enc}} = \int \rho\,dV$ over the portion of the shell within the Gaussian surface. Since the charge distribution has spherical symmetry, we may take $dV$ to be the volume of a spherical shell with radius $r$ and infinitesimal thickness $dr$: $dV = 4\pi r^2\,dr$. Thus

$$q_{\text{enc}} = 4\pi \int_a^{r_g} \rho r^2\,dr = 4\pi \int_a^{r_g} \frac{A}{r} r^2\,dr = 4\pi A \int_a^{r_g} r\,dr = 2\pi A(r_g^2 - a^2).$$

The total charge inside the Gaussian surface is $q + q_{\text{enc}} = q + 2\pi A(r_g^2 - a^2)$.

The electric field is radial, so the flux through the Gaussian surface is $\Phi = 4\pi r_g^2 E$, where $E$ is the magnitude of the field. Gauss' law yields

$$4\pi\epsilon_0 E r_g^2 = q + 2\pi A(r_g^2 - a^2).$$

Solve for $E$:

$$E = \frac{1}{4\pi\epsilon_0}\left[\frac{q}{r_g^2} + 2\pi A - \frac{2\pi A a^2}{r_g^2}\right].$$

For the field to be uniform, the first and last terms in the brackets must cancel. They do if $q - 2\pi A a^2 = 0$ or $A = q/2\pi a^2 = (45.0 \times 10^{-15}\,\text{C})/2\pi(2.00 \times 10^{-2}\,\text{m})^2 = 1.79 \times 10^{-11}\,\text{C/m}^2$.

**59**

(a) The magnitude $E_1$ of the electric field produced by the charge $q$ on the spherical shell is $E_1 = q/4\pi\epsilon_0 R_o^2$, where $R_o$ is the radius of the outer surface of the shell. Thus

$$q = 4\pi\epsilon_0 E_1 R_o^2 = \frac{(450\,\text{N/C})(0.20\,\text{m})^2}{8.99 \times 10^9\,\text{N}\cdot\text{m}^2/\text{C}^2} = 2.0 \times 10^{-9}\,\text{C}.$$

(b) Since the field at P is outward and is reduced in magnitude the field of $Q$ must be inward. $Q$ is a negative charge and the magnitude of its field at P is $E_2 = 450\,\text{N/C} - 180\,\text{N/C} = 270\,\text{N/C}$. The value of $Q$ is

$$Q = 4\pi\epsilon_0 E_2 R_o^2 = -\frac{(270\,\text{N/C})(0.20\,\text{m})^2}{8.99 \times 10^9\,\text{N}\cdot\text{m}^2/\text{C})^2} = -1.2 \times 10^{-9}\,\text{C}.$$

(c) Gauss' law tells us that since the electric field is zero inside a conductor the net charge inside a spherical surface with a radius that is slightly larger than the inside radius of the shell must be zero. Thus the charge on the inside surface of the shell is $+1.2 \times 10^{-9}$ C.

(d) The remaining charge on the shell must be on its outer surface and this is $2.0 \times 10^{-9}\,\text{C} - 1.2 \times 10^{-9}\,\text{C} = +0.8 \times 10^{-9}$ C.

**69**

(a) Draw a spherical Gaussian surface with radius $r$, concentric with the shells. The electric field, if it exists, is radial and so is normal to the surface. The integral in Gauss' law is $\oint \vec{E} \cdot d\vec{A} = 4\pi r^2 E$, where $E$ is the radial component of the field. For $r < a$ then charge enclosed is zero. Gauss' law gives $4\pi r^2 E = 0$, so $E = 0$.

(b) For $a < r < b$ the charge enclosed by the Gaussian surface is $q_a$, so the law gives $4\pi r^2 E = q_a/\epsilon_0$ and $E = q_a/4\pi\epsilon_0 r^2$.

(c) For $r > b$ the charge enclosed by the Gaussian surface is $q_a + q_b$, so $4\pi\epsilon_0 E = (q_a + q_b)/\epsilon_0$ and $E = (q_a + q_b)/4\pi\epsilon_0 r^2$.

(d) Consider first a spherical Gaussian with radius just slightly greater than $a$. The electric field is zero everywhere on this surface, so according to Gauss' law it encloses zero net charge. Since there is no charge in the cavity the charge on the inner surface of the smaller shell is zero. The total charge on the smaller shell is $q_a$ and this must reside on the outer surface. Now consider a spherical Gaussian surface with radius slight larger than the inner radius of the larger shell. This surface also encloses zero net charge, which is the sum of the charge on the outer surface of the smaller shell and the charge on the inner surface of the larger shell. Thus the charge on the inner surface of the larger shell is $-q_a$. The net charge on the larger shell is $q_b$, with $-q_a$ on its inner surface, so the charge on its outer surface must be $q_b - (-q_a) = q_b + q_a$.

**76**

(a) The magnitude of the electric field due to a large uniformly charged plate is given by $\sigma/2\epsilon_0$, where $\sigma$ is the surface charge density. In the region between the oppositely charged plates the fields of the plates are in the same direction, so the net field has magnitude $E = \sigma/\epsilon_0$. The electrical force on an electron has magnitude $eE = e\sigma/\epsilon_0$ and the gravitational force on it is $mg$, where $m$ is it mass. If these forces are to balance, they must have the same magnitude, so $mg = e\sigma/\epsilon_0$ and

$$\sigma = \frac{mg\epsilon_0}{e} = \frac{(9.11 \times 10^{-31}\,\text{kg})(9.8\,\text{m/s}^2)(8.85 \times 10^{-12}\,\text{C}^2/\text{N}\cdot\text{m}^2)}{1.60 \times 10^{-19}\,\text{C}} = 4.9 \times 10^{-22}\,\text{C/m}^2 .$$

(b) The gravitational force is downward, so the electrical force must be upward. Since an electron is negatively charged the electrical force on it is opposite to the electric field, so the electric field must be downward.

**79**

(a) Let $Q$ be the net charge on the shell, $q_i$ be the charge on its inner surface and $q_o$ be the charge on its outer surface. Then $Q = q_i + q_o$ and $q_i = Q - q_o = (-10\,\mu\text{C}) - (-14\,\mu\text{C}) = +4\,\mu\text{C}$.

(b) Let $q$ be the charge on the particle. Gauss' law tells us that since the electric field is zero inside the conducting shell the net charge inside any spherical surface that entirely within the shell is zero. Thus the sum of the charge on the particle and on the inner surface of the shell is zero, so $q + q_i = 0$ and $q = -q_i = -4\,\mu\text{C}$.

# Chapter 24

<u>3</u>

(a) An ampere is a coulomb per second, so

$$84\,\mathrm{A\cdot h} = \left(84\,\frac{\mathrm{C\cdot h}}{\mathrm{s}}\right)\left(3600\,\frac{\mathrm{s}}{\mathrm{h}}\right) = 3.0\times10^{5}\,\mathrm{C}\,.$$

(b) The change in potential energy is $\Delta U = q\,\Delta V = (3.0\times10^{5}\,\mathrm{C})(12\,\mathrm{V}) = 3.6\times10^{6}\,\mathrm{J}$.

<u>5</u>

The electric field produced by an infinite sheet of charge has magnitude $E = \sigma/2\epsilon_0$, where $\sigma$ is the surface charge density. The field is normal to the sheet and is uniform. Place the origin of a coordinate system at the sheet and take the $x$ axis to be parallel to the field and positive in the direction of the field. Then the electric potential is

$$V = V_s - \int_0^x E\,dx = V_s - Ex\,,$$

where $V_s$ is the potential at the sheet. The equipotential surfaces are surfaces of constant $x$; that is, they are planes that are parallel to the plane of charge. If two surfaces are separated by $\Delta x$ then their potentials differ in magnitude by $\Delta V = E\Delta x = (\sigma/2\epsilon_0)\Delta x$. Thus

$$\Delta x = \frac{2\epsilon_0\,\Delta V}{\sigma} = \frac{2(8.85\times10^{-12}\,\mathrm{C^2/N\cdot m^2})(50\,\mathrm{V})}{0.10\times10^{-6}\,\mathrm{C/m^2}} = 8.8\times10^{-3}\,\mathrm{m}\,.$$

<u>19</u>

(a) The electric potential $V$ at the surface of the drop, the charge $q$ on the drop, and the radius $R$ of the drop are related by $V = q/4\pi\epsilon_0 R$. Thus

$$R = \frac{q}{4\pi\epsilon_0 V} = \frac{(8.99\times10^{9}\,\mathrm{N\cdot m^2/C^2})(30\times10^{-12}\,\mathrm{C})}{500\,\mathrm{V}} = 5.4\times10^{-4}\,\mathrm{m}\,.$$

(b) After the drops combine the total volume is twice the volume of an original drop, so the radius $R'$ of the combined drop is given by $(R')^3 = 2R^3$ and $R' = 2^{1/3}R$. The charge is twice the charge of original drop: $q' = 2q$. Thus

$$V' = \frac{1}{4\pi\epsilon_0}\frac{q'}{R'} = \frac{1}{4\pi\epsilon_0}\frac{2q}{2^{1/3}R} = 2^{2/3}V = 2^{2/3}(500\,\mathrm{V}) = 790\,\mathrm{V}\,.$$

<u>29</u>

The disk is uniformly charged. This means that when the full disk is present each quadrant contributes equally to the electric potential at $P$, so the potential at $P$ due to a single quadrant is one-fourth the potential due to the entire disk. First find an expression for the potential at $P$ due to the entire disk.

Consider a ring of charge with radius $r$ and width $dr$. Its area is $2\pi r\,dr$ and it contains charge $dq = 2\pi\sigma r\,dr$. All the charge in it is a distance $\sqrt{r^2 + D^2}$ from $P$, so the potential it produces at $P$ is

$$dV = \frac{1}{4\pi\epsilon_0}\frac{2\pi\sigma r\,dr}{\sqrt{r^2+D^2}} = \frac{\sigma r\,dr}{2\epsilon_0\sqrt{r^2+D^2}}\,.$$

The total potential at $P$ is

$$V = \frac{\sigma}{2\epsilon_0}\int_0^R \frac{r\,dr}{\sqrt{r^2+D^2}} = \frac{\sigma}{2\epsilon_0}\sqrt{r^2+D^2}\bigg|_0^R = \frac{\sigma}{2\epsilon_0}\left[\sqrt{R^2+D^2} - D\right].$$

The potential $V_{sq}$ at $P$ due to a single quadrant is

$$\begin{aligned} V_{\text{sq}} &= \frac{V}{4} = \frac{\sigma}{8\epsilon_0}\left[\sqrt{R^2+D^2} - D\right] \\ &= \frac{7.73\times10^{-15}\,\text{C/m}^2}{8(8.85\times10^{-12}\,\text{C}^2/\text{N}\cdot\text{m}^2)}\left[\sqrt{(0.640\,\text{m})^2 + (0.259\,\text{m})^2} - 0.259\,\text{m}\right] \\ &= 4.71\times10^{-5}\,\text{V}\,. \end{aligned}$$

## 39

Take the negatives of the partial derivatives of the electric potential with respect to the coordinates and evaluate the results for $x = 3.00$ m, $y = -2.00$ m, and $z = 4.00$ m. This yields

$$\begin{aligned} E_x &= -\frac{\partial V}{\partial x} = -(2.00\,\text{V/m}^4)yz^2 = -(2.00\,\text{V/m}^4)((-2.00\,\text{m})(4.00\,\text{m})^2 = 64.0\,\text{V/m}\,, \\ E_y &= -\frac{\partial V}{\partial y} = -(2.00\,\text{V/m}^4)xz^2 = -(2.00\,\text{V/m}^4)(3.00\,\text{m})(4.00\,\text{m})^2 = -96.0\,\text{V/m}\,, \\ E_z &= -\frac{\partial V}{\partial z} = -2(2.00\,\text{V/m}^4)xyz = -2(2.00\,\text{V/m}^4)(3.00\,\text{m})(-2.00\,\text{m})(4.00\,\text{m}) = 96.0\,\text{V/m}\,. \end{aligned}$$

The magnitude of the electric field is

$$E = \sqrt{E_x^2 + E_y^2 + E_z^2} = \sqrt{(64.0\,\text{V/m})^2 + (-96.0\,\text{V/m})^2 + (96.0\,\text{V/m})^2} = 1.50\times10^2\,\text{V/m}\,.$$

## 41

The work required is equal to the potential energy of the system, relative to a potential energy of zero for infinite separation. Number the particles 1, 2, 3, and 4, in clockwise order starting with the particle in the upper left corner of the arrangement. The potential energy of the interaction of particles 1 and 2 is

$$\begin{aligned} U_{12} &= \frac{q_1 q_2}{4\pi\epsilon_0 a} = \frac{(8.99\times10^9\,\text{N}\cdot\text{m}^2/\text{C}^2)(2.30\times10^{-12}\,\text{C})(-2.30\times10^{-12}\,\text{C})}{0.640\,\text{m}} \\ &= -7.43\times10^{-14}\,\text{J}\,. \end{aligned}$$

The distance between particles 1 and 3 is $\sqrt{2}a$ and both these particles are positively charged, so the potential energy of the interaction between particles 1 and 3 is $U_{13} = -U_{12}/\sqrt{2} =$

$+5.25 \times 10^{-14}$ J. The potential energy of the interaction between particles 1 and 4 is $U_{14} = U_{12} = -7.43 \times 10^{-14}$ J. The potential energy of the interaction between particles 2 and 3 is $U_{23} = U_{12} = -7.43 \times 10^{-14}$ J. The potential energy of the interaction between particles 2 and 4 is $U_{24} = U_{13} = 5.25 \times 10^{-14}$ J. The potential energy of the interaction between particles 3 and 4 is $U_{34} = U_{12} = -7.43 \times 10^{-14}$ J.

The total potential energy of the system is

$$\begin{aligned} U &= U_{12} + U_{13} + U_{14} + U_{23} + U_{24} + U_{34} \\ &= -7.43 \times 10^{-14}\,\text{J} + 5.25 \times 10^{-14}\,\text{J} - 7.43 \times 10^{-14}\,\text{J} - 7.43 \times 10^{-14}\,\text{J} \\ &\quad - 7.43 \times 10^{-14}\,\text{J} + 5.25 \times 10^{-14}\,\text{J} = -1.92 \times 10^{-13}\,\text{J}. \end{aligned}$$

This is equal to the work that must be done to assemble the system from infinite separation.

**59**

(a) Use conservation of mechanical energy. The potential energy when the moving particle is at any coordinate $y$ is $qV$, where $V$ is the electric potential produced at that place by the two fixed particles. That is,

$$U = q\frac{2Q}{4\pi\epsilon_0\sqrt{x^2 + y^2}},$$

where $x$ is the coordinate and $Q$ is the charge of either one of the fixed particles. The factor 2 appears since the two fixed particles produce the same potential at points on the $y$ axis. Conservation of mechanical energy yields

$$K_f = K_i + q\frac{2Q}{4\pi\epsilon_0\sqrt{x^2 + y_i^2}} - q\frac{2Q}{4\pi\epsilon_0\sqrt{x^2 + y_f^2}} = K_i + \frac{2qQ}{4\pi\epsilon_0}\left(\frac{1}{\sqrt{x^2 + y_i^2}} - \frac{1}{\sqrt{x^2 + y_f^2}}\right),$$

where $K$ is the kinetic energy of the moving particle, the subscript $i$ refers to the initial position of the moving particle, and the subscript $f$ refers to the final position. Numerically

$$K_f = 1.2\,\text{J} + \frac{2(-15 \times 10^{-6}\,\text{C})(50 \times 10^{-6}\,\text{C})}{4\pi(8.85 \times 10^{-12}\,\text{C}^2/\text{N}\cdot\text{m}^2)}\left[\frac{1}{\sqrt{(3.0\,\text{m})^2 + (4.0\,\text{m})^2}} - \frac{1}{\sqrt{(3.0\,\text{m})^2}}\right] = 3.0\,\text{J}.$$

(b) Now $K_f = 0$ and we solve the energy conservation equation for $y_f$. Conservation of energy first yields $U_f = K_i + U_i$. The initial potential energy is

$$U_i = \frac{2qQ}{4\pi\epsilon_0\sqrt{x^2 + y_i^2}} = \frac{2(-15 \times 10^{-6}\,\text{C})(50 \times 10^{-6}\,\text{C})}{4\pi(8.85 \times 10^{-12}\,\text{C}^2/\text{N}\cdot\text{m}^2)\sqrt{(3.0\,\text{m})^2 + (4.0\,\text{m})^2}} = -2.7\,\text{J}.$$

Thus $K_f = 1.2\,\text{J} - 2.7\,\text{J} = -1.5\,\text{J}$.

Now

$$U_f = \frac{2qQ}{4\pi\epsilon_0\sqrt{x^2 + y_f^2}}$$

so

$$y = -\sqrt{\left(\frac{2qQ}{4\pi\epsilon_0 U_f}\right)^2 - x^2} = \sqrt{\left(\frac{2(-15\times10^{-6}\,\mathrm{C})(50\times10^{-6}\,\mathrm{C})}{4\pi(8.85\times10^{-12}\,\mathrm{C^2/N\cdot m^2})(-1.5\,\mathrm{J})}\right)^2 - (3.0\,\mathrm{m})^2} = -8.5\,\mathrm{m}\,.$$

**63**

If the electric potential is zero at infinity, then the electric potential at the surface of the sphere is given by $V = q/4\pi\epsilon_0 r$, where $q$ is the charge on the sphere and $r$ is its radius. Thus

$$q = 4\pi\epsilon_0 rV = \frac{(0.15\,\mathrm{m})(1500\,\mathrm{V})}{8.99\times10^9\,\mathrm{N\cdot m^2/C^2}} = 2.5\times10^{-8}\,\mathrm{C}\,.$$

**65**

(a) The electric potential is the sum of the contributions of the individual spheres. Let $q_1$ be the charge on one, $q_2$ be the charge on the other, and $d$ be their separation. The point halfway between them is the same distance $d/2$ (= 1.0 m) from the center of each sphere, so the potential at the halfway point is

$$V = \frac{q_1+q_2}{4\pi\epsilon_0 d/2} = \frac{(8.99\times10^9\,\mathrm{N\cdot m^2/C^2})(1.0\times10^{-8}\,\mathrm{C} - 3.0\times10^{-8}\,\mathrm{C})}{1.0\,\mathrm{m}} = -1.80\times10^2\,\mathrm{V}\,.$$

(b) The distance from the center of one sphere to the surface of the other is $d - R$, where $R$ is the radius of either sphere. The potential of either one of the spheres is due to the charge on that sphere and the charge on the other sphere. The potential at the surface of sphere 1 is

$$\begin{aligned} V_1 &= \frac{1}{4\pi\epsilon_0}\left[\frac{q_1}{R} + \frac{q_2}{d-R}\right] \\ &= (8.99\times10^9\,\mathrm{N\cdot m^2/C^2})\left[\frac{1.0\times10^{-8}\,\mathrm{C}}{0.030\,\mathrm{m}} - \frac{3.0\times10^{-8}\,\mathrm{C}}{2.0\,\mathrm{m} - 0.030\,\mathrm{m}}\right] \\ &= 2.9\times10^3\,\mathrm{V}\,. \end{aligned}$$

(c) The potential at the surface of sphere 2 is

$$\begin{aligned} V_2 &= \frac{1}{4\pi\epsilon_0}\left[\frac{q_1}{d-R} + \frac{q_2}{R}\right] \\ &= (8.99\times10^9\,\mathrm{N\cdot m^2/C^2})\left[\frac{1.0\times10^{-8}\,\mathrm{C}}{2.0\,\mathrm{m} - 0.030\,\mathrm{m}} - \frac{3.0\times10^{-8}\,\mathrm{C}}{0.030\,\mathrm{m}}\right] \\ &= -8.9\times10^3\,\mathrm{V}\,. \end{aligned}$$

**75**

The initial potential energy of the three-particle system is $U_i = 2(q^2/4\pi\epsilon_0 L) + U_{\text{fixed}}$, where $q$ is the charge on each particle, $L$ is the length of a triangle side, and $U_{\text{fixed}}$ is the potential energy associated with the interaction of the two fixed particles. The factor 2 appears since the potential energy is the same for the interaction of the movable particle and each of the fixed particles. The final potential energy is $U_f = 2[q^2/4\pi\epsilon_0(L/2)] + U_{\text{fixed}}$ and the change in the potential energy is

$$\Delta U = U_f - U_i = \frac{2q^2}{4\pi\epsilon_0}\left(\frac{2}{L} - \frac{1}{L}\right) = \frac{2q^2}{4\pi\epsilon_0 L}.$$

This is the work that is done by the external agent. If $P$ is the rate with energy is supplied by the agent and $t$ is the time for the move, then $Pt = \Delta U$, and

$$t = \frac{\Delta U}{P} = \frac{2q^2}{4\pi\epsilon_0 LP} = \frac{2(8.99\times10^9\,\text{N}\cdot\text{m}^2/\text{C}^2)(0.12\,\text{C})^2}{(1.7\,\text{m})(0.83\times10^3\,\text{W})} = 1.83\times10^5\,\text{s}.$$

This is 2.1 d.

77

(a) Use Gauss' law to find an expression for the electric field. The Gaussian surface is a cylindrical surface that is concentric with the cylinder and has a radius $r$ that is greater than the radius of the cylinder. The electric field is normal to the Gaussian surface and has uniform magnitude on it, so the integral in Gauss' law is $\oint \vec{E}\cdot d\vec{A} = 2\pi rEL$, where $L$ is the length of the Gaussian surface. The charge enclosed is $\lambda L$, where $\lambda$ is the charge per unit length on the cylinder. Thus $2\pi rRLE = \lambda L/\epsilon_0$ and $E = \lambda/2\pi\epsilon_0 r$.

Let $E_B$ be the magnitude of the field at B and $r_B$ be the distance from the central axis to B. Let $E_C$ be the magnitude of the field at C and $r_C$ be the distance from the central axis to C. Since $E$ is inversely proportional to the distance from the central axis,

$$E_C = \frac{r_B}{r_C}E_B = \frac{2.0\,\text{cm}}{5.0\,\text{cm}}(160\,\text{N/C}) = 64\,\text{N/C}.$$

(b) The magnitude of the field a distance $r$ from the central axis is $E = (r_B/r)E_B$, so the potential difference of points B and C is

$$\begin{aligned} V_B - V_C &= -\int_{r_C}^{r_B} \frac{r_B}{r}E_B\,dr = -r_BE_B\ln\left(\frac{r_B}{r_C}\right) \\ &= -(0.020\,\text{m})(160\,\text{N/C})\ln\left(\frac{0.020\,\text{m}}{0.050\,\text{m}}\right) = 2.9\,\text{V}. \end{aligned}$$

(c) The cylinder is conducting, so all points inside have the same potential, namely $V_B$, so $V_A - V_B = 0$.

85

Consider a point on the $z$ axis that has coordinate $z$. All points on the ring are the same distance from the point. The distance is $r = \sqrt{R^2 + z^2}$, where $R$ is the radius of the ring. If the electric potential is taken to be zero at points that are infinitely far from the ring, then the potential at the point is

$$V = \frac{Q}{4\pi\epsilon_0\sqrt{R^2+z^2}},$$

where $Q$ is the charge on the ring. Thus

$$V_B - V_A = \frac{Q}{4\pi\epsilon_0}\left[\frac{1}{\sqrt{R^2+z^2}} - \frac{1}{R}\right]$$

$$= (8.99 \times 10^9\,\mathrm{N\cdot m^2/C^2})(16.0 \times 10^{-6}\,\mathrm{C})\left[\frac{1}{\sqrt{(0.0300\,\mathrm{m})^2 + (0.0400\,\mathrm{m})^2}} - 0.300\,\mathrm{m}\right]$$

$$= -1.92 \times 10^6\,\mathrm{V}.$$

**93**

(a) For $r > r_2$ the field is like that of a point charge and

$$V = \frac{1}{4\pi\epsilon_0}\frac{Q}{r},$$

where the zero of potential was taken to be at infinity.

(b) To find the potential in the region $r_1 < r < r_2$, first use Gauss's law to find an expression for the electric field, then integrate along a radial path from $r_2$ to $r$. The Gaussian surface is a sphere of radius $r$, concentric with the shell. The field is radial and therefore normal to the surface. Its magnitude is uniform over the surface, so the flux through the surface is $\Phi = 4\pi r^2 E$. The volume of the shell is $(4\pi/3)(r_2^3 - r_1^3)$, so the charge density is

$$\rho = \frac{3Q}{4\pi(r_2^3 - r_1^3)}$$

and the charge enclosed by the Gaussian surface is

$$q = \left(\frac{4\pi}{3}\right)(r^3 - r_1^3)\rho = Q\left(\frac{r^3 - r_1^3}{r_2^3 - r_1^3}\right).$$

Gauss' law yields

$$4\pi\epsilon_0 r^2 E = Q\left(\frac{r^3 - r_1^3}{r_2^3 - r_1^3}\right)$$

and the magnitude of the electric field is

$$E = \frac{Q}{4\pi\epsilon_0}\frac{r^3 - r_1^3}{r^2(r_2^3 - r_1^3)}.$$

If $V_s$ is the electric potential at the outer surface of the shell ($r = r_2$) then the potential a distance $r$ from the center is given by

$$V = V_s - \int_{r_2}^{r} E\,dr = V_s - \frac{Q}{4\pi\epsilon_0}\frac{1}{r_2^3 - r_1^3}\int_{r_2}^{r}\left(r - \frac{r_1^3}{r^2}\right)dr$$

$$= V_s - \frac{Q}{4\pi\epsilon_0}\frac{1}{r_2^3 - r_1^3}\left(\frac{r^2}{2} - \frac{r_2^2}{2} + \frac{r_1^3}{r} - \frac{r_1^3}{r_2}\right).$$

The potential at the outer surface is found by placing $r = r_2$ in the expression found in part (a). It is $V_s = Q/4\pi\epsilon_0 r_2$. Make this substitution and collect like terms to find

$$V = \frac{Q}{4\pi\epsilon_0}\frac{1}{r_2^3 - r_1^3}\left(\frac{3r_2^2}{2} - \frac{r^2}{2} - \frac{r_1^3}{r}\right).$$

Since $\rho = 3Q/4\pi(r_2^3 - r_1^3)$ this can also be written

$$V = \frac{\rho}{3\epsilon_0}\left(\frac{3r_2^2}{2} - \frac{r^2}{2} - \frac{r_1^3}{r}\right).$$

(c) The electric field vanishes in the cavity, so the potential is everywhere the same inside and has the same value as at a point on the inside surface of the shell. Put $r = r_1$ in the result of part (b). After collecting terms the result is

$$V = \frac{Q}{4\pi\epsilon_0}\frac{3(r_2^2 - r_1^2)}{2(r_2^3 - r_1^3)},$$

or in terms of the charge density

$$V = \frac{\rho}{2\epsilon_0}(r_2^2 - r_1^2).$$

(d) The solutions agree at $r = r_1$ and at $r = r_2$.

**95**

The electric potential of a dipole at a point a distance $r$ away is given by Eq. 24–30:

$$V = \frac{1}{4\pi\epsilon_0}\frac{p\cos\theta}{r^2},$$

where $p$ is the magnitude of the dipole moment and $\theta$ is the angle between the dipole moment and the position vector of the point. The potential at infinity was taken to be zero. Take the $z$ axis to be the dipole axis and consider a point with $z$ positive (on the positive side of the dipole). For this point $r = z$ and $\theta = 0$. The $z$ component of the electric field is

$$E_z = -\frac{\partial V}{\partial x} = -\frac{\partial}{\partial z}\left(\frac{p}{4\pi\epsilon_0 z^2}\right) = \frac{p}{2\pi\epsilon_0 z^3}.$$

This is the only nonvanishing component at a point on the dipole axis.

For a point with a negative value of $z$, $r = -z$ and $\cos\theta = -1$, so

$$E_z = -\frac{\partial}{\partial z}\left(\frac{-p}{4\pi\epsilon_0 z^2}\right) = -\frac{p}{2\pi\epsilon_0 z^3}.$$

**103**

(a) The electric potential at the surface of the sphere is given by $V = q/4\pi\epsilon_0 R$, where $q$ is the charge on the sphere and $R$ is the sphere radius. The charge on the sphere when the potential reaches 1000 V is

$$q = 4\pi\epsilon_0 rV = \frac{(0.010\,\text{m})(1000\,\text{V})}{8.99\times10^9\,\text{N}\cdot\text{m}^2/\text{C}^2} = 1.11\times10^{-9}\,\text{C}.$$

The number of electrons that enter the sphere is $N = q/e = (1.11 \times 10^{-9}\,\text{C})/(1.60 \times 10^{-19}\,\text{C}) = 6.95 \times 10^9$. Let $R$ be the decay rate and $t$ be the time for the potential to reach it final value. Since half the resulting electrons enter the sphere $N = (P/2)t$ and $t = 2N/P = 2(6.95 \times 10^9)/(3.70 \times 10^8\,\text{s}^{-1}) = 38\,\text{s}$.

(b) The increase in temperature is $\Delta T = N\Delta E/C$, where $E$ is the energy deposited by a single electron and $C$ is the heat capacity of the sphere. Since $N = (P/2)t$, this is $\Delta T = (P/2)t\Delta E/C$ and

$$t = \frac{2C\,\Delta T}{P\,\Delta E} = \frac{2(14\,\text{J/K})(5.0\,\text{K})}{(3.70 \times 10^8\,\text{s}^{-1})(100 \times 10^3\,\text{eV})(1.60 \times 10^{-19}\,\text{J/eV})} = 2.4 \times 10^7\,\text{s}\,.$$

This is about 280 d.

# Chapter 25

5

(a) The capacitance of a parallel-plate capacitor is given by $C = \epsilon_0 A/d$, where $A$ is the area of each plate and $d$ is the plate separation. Since the plates are circular, the plate area is $A = \pi R^2$, where $R$ is the radius of a plate. Thus

$$C = \frac{\epsilon_0 \pi R^2}{d} = \frac{(8.85 \times 10^{-12}\,\mathrm{F/m})\pi(8.20 \times 10^{-2}\,\mathrm{m})^2}{1.30 \times 10^{-3}\,\mathrm{m}} = 1.44 \times 10^{-10}\,\mathrm{F} = 144\,\mathrm{pF}\,.$$

(b) The charge on the positive plate is given by $q = CV$, where $V$ is the potential difference across the plates. Thus $q = (1.44 \times 10^{-10}\,\mathrm{F})(120\,\mathrm{V}) = 1.73 \times 10^{-8}\,\mathrm{C} = 17.3\,\mathrm{nC}$.

15

The charge initially on the charged capacitor is given by $q = C_1 V_0$, where $C_1$ (= 100 pF) is the capacitance and $V_0$ (= 50 V) is the initial potential difference. After the battery is disconnected and the second capacitor wired in parallel to the first, the charge on the first capacitor is $q_1 = C_1 V$, where $v$ (= 35 V) is the new potential difference. Since charge is conserved in the process, the charge on the second capacitor is $q_2 = q - q_1$, where $C_2$ is the capacitance of the second capacitor. Substitute $C_1 V_0$ for $q$ and $C_1 V$ for $q_1$ to obtain $q_2 = C_1(V_0 - V)$. The potential difference across the second capacitor is also $V$, so the capacitance is

$$C_2 = \frac{q_2}{V} = \frac{V_0 - V}{V} C_1 = \frac{50\,\mathrm{V} - 35\,\mathrm{V}}{35\,\mathrm{V}}(100\,\mathrm{pF}) = 43\,\mathrm{pF}\,.$$

19

(a) After the switches are closed, the potential differences across the capacitors are the same and the two capacitors are in parallel. The potential difference from $a$ to $b$ is given by $V_{ab} = Q/C_{eq}$, where $Q$ is the net charge on the combination and $C_{eq}$ is the equivalent capacitance.
The equivalent capacitance is $C_{eq} = C_1 + C_2 = 4.0 \times 10^{-6}\,\mathrm{F}$. The total charge on the combination is the net charge on either pair of connected plates. The charge on capacitor 1 is

$$q_1 = C_1 V = (1.0 \times 10^{-6}\,\mathrm{F})(100\,\mathrm{V}) = 1.0 \times 10^{-4}\,\mathrm{C}$$

and the charge on capacitor 2 is

$$q_2 = C_2 V = (3.0 \times 10^{-6}\,\mathrm{F})(100\,\mathrm{V}) = 3.0 \times 10^{-4}\,\mathrm{C}\,,$$

so the net charge on the combination is $3.0 \times 10^{-4}\,\mathrm{C} - 1.0 \times 10^{-4}\,\mathrm{C} = 2.0 \times 10^{-4}\,\mathrm{C}$. The potential difference is

$$V_{ab} = \frac{2.0 \times 10^{-4}\,\mathrm{C}}{4.0 \times 10^{-6}\,\mathrm{F}} = 50\,\mathrm{V}\,.$$

(b) The charge on capacitor 1 is now $q_1 = C_1 V_{ab} = (1.0 \times 10^{-6}\,\mathrm{F})(50\,\mathrm{V}) = 5.0 \times 10^{-5}\,\mathrm{C}$.

(c) The charge on capacitor 2 is now $q_2 = C_2 V_{ab} = (3.0 \times 10^{-6}\,\text{F})(50\,\text{V}) = 1.5 \times 10^{-4}\,\text{C}$.

**29**

The total energy is the sum of the energies stored in the individual capacitors. Since they are connected in parallel, the potential difference $V$ across the capacitors is the same and the total energy is $U = \frac{1}{2}(C_1 + C_2)V^2 = \frac{1}{2}(2.0 \times 10^{-6}\,\text{F} + 4.0 \times 10^{-6}\,\text{F})(300\,\text{V})^2 = 0.27\,\text{J}$.

**35**

(a) Let $q$ be the charge on the positive plate. Since the capacitance of a parallel-plate capacitor is given by $\epsilon_0 A/d$, the charge is $q = CV = \epsilon_0 AV/d$. After the plates are pulled apart, their separation is $d'$ and the potential difference is $V'$. Then $q = \epsilon_0 AV'/d'$ and

$$V' = \frac{d'}{\epsilon_0 A} q = \frac{d'}{\epsilon_0 A} \frac{\epsilon_0 A}{d} V = \frac{d'}{d} V = \frac{8.00\,\text{mm}}{3.00\,\text{mm}}(6.00\,\text{V}) = 16.0\,\text{V}\,.$$

(b) The initial energy stored in the capacitor is

$$U_i = \frac{1}{2} CV^2 = \frac{\epsilon_0 AV^2}{2d} = \frac{(8.85 \times 10^{-12}\,\text{F/m})(8.50 \times 10^{-4}\,\text{m}^2)(6.00\,\text{V})}{2(3.00 \times 10^{-3}\,\text{mm})} = 4.51 \times 10^{-11}\,\text{J}$$

and the final energy stored is

$$U_f = \frac{1}{2} C'(V')^2 = \frac{1}{2} \frac{\epsilon_0 A}{d'} (V')^2 = \frac{(8.85 \times 10^{-12}\,\text{F/m})(8.50 \times 10^{-4}\,\text{m}^2)(16.0\,\text{V})}{2(8.00 \times 10^{-3}\,\text{mm})} = 1.20 \times 10^{-10}\,\text{J}\,.$$

(c) The work done to pull the plates apart is the difference in the energy: $W = U_f - U_i = 1.20 \times 10^{-10}\,\text{J} - 4.51 \times 10^{-11}\,\text{J} = 7.49 \times 10^{-11}\,\text{J}$.

**43**

The capacitance of a cylindrical capacitor is given by

$$C = \kappa C_0 = \frac{2\pi\kappa\epsilon_0 L}{\ln(b/a)}\,,$$

where $C_0$ is the capacitance without the dielectric, $\kappa$ is the dielectric constant, $L$ is the length, $a$ is the inner radius, and $b$ is the outer radius. See Eq. 25–14. The capacitance per unit length of the cable is

$$\frac{C}{L} = \frac{2\pi\kappa\epsilon_0}{\ln(b/a)} = \frac{2\pi(2.6)(8.85 \times 10^{-12}\,\text{F/m})}{\ln\left[(0.60\,\text{mm})/(0.10\,\text{mm})\right]} = 8.1 \times 10^{-11}\,\text{F/m} = 81\,\text{pF/m}\,.$$

**45**

The capacitance is given by $C = \kappa C_0 = \kappa\epsilon_0 A/d$, where $C_0$ is the capacitance without the dielectric, $\kappa$ is the dielectric constant, $A$ is the plate area, and $d$ is the plate separation. The electric field between the plates is given by $E = V/d$, where $V$ is the potential difference between the plates. Thus $d = V/E$ and $C = \kappa\epsilon_0 AE/V$. Solve for $A$:

$$A = \frac{CV}{\kappa\epsilon_0 E}.$$

For the area to be a minimum, the electric field must be the greatest it can be without breakdown occurring. That is,

$$A = \frac{(7.0\times10^{-8}\,\text{F})(4.0\times10^{3}\,\text{V})}{2.8(8.85\times10^{-12}\,\text{F/m})(18\times10^{6}\,\text{V/m})} = 0.63\,\text{m}^2.$$

**51**

(a) The electric field in the region between the plates is given by $E = V/d$, where $V$ is the potential difference between the plates and $d$ is the plate separation. The capacitance is given by $C = \kappa\epsilon_0 A/d$, where $A$ is the plate area and $\kappa$ is the dielectric constant, so $d = \kappa\epsilon_0 A/C$ and

$$E = \frac{VC}{\kappa\epsilon_0 A} = \frac{(50\,\text{V})(100\times10^{-12}\,\text{F})}{5.4(8.85\times10^{-12}\,\text{F/m})(100\times10^{-4}\,\text{m}^2)} = 1.0\times10^{4}\,\text{V/m}.$$

(b) The free charge on the plates is $q_f = CV = (100\times10^{-12}\,\text{F})(50\,\text{V}) = 5.0\times10^{-9}\,\text{C}$.

(c) The electric field is produced by both the free and induced charge. Since the field of a large uniform layer of charge is $q/2\epsilon_0 A$, the field between the plates is

$$E = \frac{q_f}{2\epsilon_0 A} + \frac{q_f}{2\epsilon_0 A} - \frac{q_i}{2\epsilon_0 A} - \frac{q_i}{2\epsilon_0 A},$$

where the first term is due to the positive free charge on one plate, the second is due to the negative free charge on the other plate, the third is due to the positive induced charge on one dielectric surface, and the fourth is due to the negative induced charge on the other dielectric surface. Note that the field due to the induced charge is opposite the field due to the free charge, so the fields tend to cancel. The induced charge is therefore

$$\begin{aligned} q_i &= q_f - \epsilon_0 AE \\ &= 5.0\times10^{-9}\,\text{C} - (8.85\times10^{-12}\,\text{F/m})(100\times10^{-4}\,\text{m}^2)(1.0\times10^{4}\,\text{V/m}) \\ &= 4.1\times10^{-9}\,\text{C} = 4.1\,\text{nC}. \end{aligned}$$

**61**

Capacitors 3 and 4 are in parallel and may be replaced by a capacitor with capacitance $C_{34} = C_3 + C_4 = 30\,\mu\text{F}$. Capacitors 1, 2, and the equivalent capacitor that replaced 3 and 4 are all in series, so the sum of their potential differences must equal the potential difference across the battery. Since all of these capacitors have the same capacitance the potential difference across each of them is one-third the battery potential difference or 3.0 V. The potential difference across capacitor 4 is the same as the potential difference across the equivalent capacitor that replaced 3 and 4, so the charge on capacitor 4 is $q_4 = C_4V_4 = (15\times10^{-6}\,\text{F})(3.0\,\text{V}) = 45\times10^{-6}\,\text{C}$.

**69**

(a) and (b) The capacitors have the same plate separation $d$ and the same potential difference $V$ across their plates, so the electric field are the same within them. The magnitude of the field in either one is $E = V/d = (600\text{ V})/(3.00 \times 10^{-3}\text{ m}) = 2.00 \times 10^5\text{ V/m}$.

(c) Let $A$ be the area of a plate. Then the surface charge density on the positive plate is $\sigma_A = q_A/A = C_A V/A = (\epsilon_0 A/d)V/A = \epsilon_0 V/d = \epsilon_0 E = (8.85 \times 10^{-12}\text{ N}\cdot\text{m}^2/\text{C}^2)(2.00 \times 10^5\text{ V/m}) = 1.77 \times 10^{-6}\text{ C/m}^2$, where $CV$ was substituted for $q$ and the expression $\epsilon_0 A/d$ for the capacitance of a parallel-plate capacitor was substituted for $C$.

(d) Now the capacitance is $\kappa\epsilon_0 A/d$, where $\kappa$ is the dielectric constant. The surface charge density on the positive plate is $\sigma_B = \kappa\epsilon_0 E = \kappa\sigma_A = (2.60)(1.77 \times 10^{-6}\text{ C/m}^2) = 4.60 \times 10^{-6}\text{ C/m}^2$.

(e) The electric field in B is produced by the charge on the plates and the induced charge together while the field in A is produced by the charge on the plates alone. since the fields are the same $\sigma_B + \sigma_{\text{induced}} = \sigma_A$, so $\sigma_{\text{induced}} = \sigma_A - \sigma_B = 1.77 \times 10^{-6}\text{ C/m}^2 - 4.60 \times 10^{-6}\text{ C/m}^2 = -2.83 \times 10^{-6}\text{ C/m}^2$.

**73**

The electric field in the lower region is due to the charge on both plates and the charge induced on the upper and lower surfaces of the dielectric in the region. The charge induced on the dielectric surfaces of the upper region has the same magnitude but opposite sign on the two surfaces and so produces a net field of zero in the lower region. Similarly, the electric field in the upper region is due to the charge on the plates and the charge induced on the upper and lower surfaces of dielectric in that region. Thus the electric field in the upper region has magnitude $E_{\text{upper}} = q\kappa_{\text{upper}}\epsilon_0 A$ and the potential difference across that region is $V_{\text{upper}} = E_{\text{upper}}d$, where $d$ is the thickness of the region. The electric field in the lower region is $E_{\text{lower}} = q\kappa_{\text{lower}}\epsilon_0 A$ and the potential difference across that region is $V_{\text{lower}} = E_{\text{lower}}d$. The sum of the potential differences must equal the potential difference $V$ across the entire capacitor, so

$$V = E_{\text{upper}}d + E_{\text{lower}}d = \frac{qd}{\epsilon_0 A}\left[\frac{1}{\kappa_{\text{upper}}} + \frac{1}{\kappa_{\text{lower}}}\right].$$

The solution for $q$ is

$$q = \frac{\kappa_{\text{upper}}\kappa_{\text{lower}}}{\kappa_{\text{upper}} + \kappa_{\text{lower}}}\frac{\epsilon_0 A}{d}V = \frac{(3.00)(4.00)}{3.00 + 4.00}\frac{(8.85 \times 10^{-12}\text{ N}\cdot m^2/\text{C}^2)(2.00 \times 10^{-2}\text{ m}^2)}{2.00 \times 10^{-3}\text{ m}}(7.00\text{ V})$$
$$= 1.06 \times 10^{-9}\text{ C}.$$

# Chapter 26

<u>7</u>

(a) The magnitude of the current density is given by $J = nqv_d$, where $n$ is the number of particles per unit volume, $q$ is the charge on each particle, and $v_d$ is the drift speed of the particles. The particle concentration is $n = 2.0 \times 10^8\,\text{cm}^{-3} = 2.0 \times 10^{14}\,\text{m}^{-3}$, the charge is $q = 2e = 2(1.60 \times 10^{-19}\,\text{C}) = 3.20 \times 10^{-19}\,\text{C}$, and the drift speed is $1.0 \times 10^5\,\text{m/s}$. Thus

$$J = (2 \times 10^{14}\,\text{m}^{-3})(3.2 \times 10^{-19}\,\text{C})(1.0 \times 10^5\,\text{m/s}) = 6.4\,\text{A/m}^2 .$$

(b) Since the particles are positively charged, the current density is in the same direction as their motion, to the north.

(c) The current cannot be calculated unless the cross-sectional area of the beam is known. Then $i = JA$ can be used.

<u>17</u>

The resistance of the wire is given by $R = \rho L/A$, where $\rho$ is the resistivity of the material, $L$ is the length of the wire, and $A$ is the cross-sectional area of the wire. The cross-sectional area is $A = \pi r^2 = \pi(0.50 \times 10^{-3}\,\text{m})^2 = 7.85 \times 10^{-7}\,\text{m}^2$. Here $r = 0.50\,\text{mm} = 0.50 \times 10^{-3}\,\text{m}$ is the radius of the wire. Thus

$$\rho = \frac{RA}{L} = \frac{(50 \times 10^{-3}\,\Omega)(7.85 \times 10^{-7}\,\text{m}^2)}{2.0\,\text{m}} = 2.0 \times 10^{-8}\,\Omega \cdot \text{m} .$$

<u>19</u>

The resistance of the coil is given by $R = \rho L/A$, where $L$ is the length of the wire, $\rho$ is the resistivity of copper, and $A$ is the cross-sectional area of the wire. Since each turn of wire has length $2\pi r$, where $r$ is the radius of the coil, $L = (250)2\pi r = (250)(2\pi)(0.12\,\text{m}) = 188.5\,\text{m}$. If $r_w$ is the radius of the wire, its cross-sectional area is $A = \pi r_w^2 = \pi(0.65 \times 10^{-3}\,\text{m})^2 = 1.33 \times 10^{-6}\,\text{m}^2$. According to Table 26—1, the resistivity of copper is $1.69 \times 10^{-8}\,\Omega \cdot \text{m}$. Thus

$$R = \frac{\rho L}{A} = \frac{(1.69 \times 10^{-8}\,\Omega \cdot \text{m})(188.5\,\text{m})}{1.33 \times 10^{-6}\,\text{m}^2} = 2.4\,\Omega .$$

<u>21</u>

Since the mass and density of the material do not change, the volume remains the same. If $L_0$ is the original length, $L$ is the new length, $A_0$ is the original cross-sectional area, and $A$ is the new cross-sectional area, then $L_0 A_0 = LA$ and $A = L_0 A_0/L = L_0 A_0/3L_0 = A_0/3$. The new resistance is

$$R = \frac{\rho L}{A} = \frac{\rho 3 L_0}{A_0/3} = 9\frac{\rho L_0}{A_0} = 9R_0 ,$$

where $R_0$ is the original resistance. Thus $R = 9(6.0\,\Omega) = 54\,\Omega$.

**23**

The resistance of conductor $A$ is given by

$$R_A = \frac{\rho L}{\pi r_A^2},$$

where $r_A$ is the radius of the conductor. If $r_o$ is the outside radius of conductor $B$ and $r_i$ is its inside radius, then its cross-sectional area is $\pi(r_o^2 - r_i^2)$ and its resistance is

$$R_B = \frac{\rho L}{\pi(r_o^2 - r_i^2)}.$$

The ratio is

$$\frac{R_A}{R_B} = \frac{r_o^2 - r_i^2}{r_A^2} = \frac{(1.0\,\text{mm})^2 - (0.50\,\text{mm})^2}{(0.50\,\text{mm})^2} = 3.0\,.$$

**39**

(a) Electrical energy is transferred to internal energy at a rate given by

$$P = \frac{V^2}{R},$$

where $V$ is the potential difference across the heater and $R$ is the resistance of the heater. Thus

$$P = \frac{(120\,\text{V})^2}{14\,\Omega} = 1.0 \times 10^3\,\text{W} = 1.0\,\text{kW}\,.$$

(b) The cost is given by

$$C = (1.0\,\text{kW})(5.0\,\text{h})(\$0.050\,/\text{kW}\cdot\text{h}) = \$0.25\,.$$

**43**

(a) Let $P$ be the rate of energy dissipation, $i$ be the current in the heater, and $V$ be the potential difference across the heater. They are related by $P = iV$. Solve for $i$:

$$i = \frac{P}{V} = \frac{1250\,\text{W}}{115\,\text{V}} = 10.9\,\text{A}\,.$$

(b) According to the definition of resistance $V = iR$, where $R$ is the resistance of the heater. Solve for $R$:

$$R = \frac{V}{i} = \frac{115\,\text{V}}{10.9\,\text{A}} = 10.6\,\Omega\,.$$

(c) The thermal energy $E$ produced by the heater in time $t$ (= 1.0 h = 3600 s) is

$$E = Pt = (1250\,\text{W})(3600\,\text{s}) = 4.5 \times 10^6\,\text{J}\,.$$

**53**

(a) and (b) Calculate the electrical resistances of the wires. Let $\rho_C$ be the resistivity of wire C, $r_C$ be its radius, and $L_C$ be its length. Then the resistance of this wire is

$$R_C = \rho_C \frac{L_C}{\pi r_C^2} = (2.0 \times 10^{-6}\,\Omega \cdot \text{m})\frac{1.0\,\text{m}}{\pi(0.50 \times 10^{-3}\,\text{m})^2} = 2.54\,\Omega\,.$$

Let $\rho_D$ be the resistivity of wire D, $r_D$ be its radius, and $L_D$ be its length. Then the resistance of this wire is

$$R_D = \rho_D \frac{L_D}{\pi r_D^2} = (1.0 \times 10^{-6}\,\Omega \cdot \text{m})\frac{1.0\,\text{m}}{\pi(0.25 \times 10^{-3}\,\text{m})^2} = 5.09\,\Omega\,.$$

If $i$ is the current in the wire, the potential difference between points 1 and 2 is

$$\Delta V_{12} = iR_C = (2.0\,\text{A})(2.54\,\Omega) = 5.1\,\text{V}$$

and the potential difference between points 2 and 3 is

$$\Delta V_{23} = iR_D = (2.0\,\text{A})(5.09\,\Omega) = 10\,\text{V}\,.$$

(c) and (d) The rate of energy dissipation between points 1 and 2 is

$$P_{12} = i^2 R_C = (2.0\,\text{A})^2(2.54\,\Omega) = 10\,\text{W}$$

and the rate of energy dissipation between points 2 and 3 is

$$P_{23} = i^2 R_D = (2.0\,\text{A})^2(5.09\,\Omega) = 20\,\text{W}\,.$$

**55**

(a) The charge that strikes the surface in time $\Delta t$ is given by $\Delta q = i\,\Delta t$, where $i$ is the current. Since each particle carries charge $2e$, the number of particles that strike the surface is

$$N = \frac{\Delta q}{2e} = \frac{i\,\Delta t}{2e} = \frac{(0.25 \times 10^{-6}\,\text{A})(3.0\,\text{s})}{2(1.6 \times 10^{-19}\,\text{C})} = 2.3 \times 10^{12}\,.$$

(b) Now let $N$ be the number of particles in a length $L$ of the beam. They will all pass through the beam cross section at one end in time $t = L/v$, where $v$ is the particle speed. The current is the charge that moves through the cross section per unit time. That is, $i = 2eN/t = 2eNv/L$. Thus, $N = iL/2ev$.

Now find the particle speed. The kinetic energy of a particle is

$$K = 20\,\text{MeV} = (20 \times 10^6\,\text{eV})(1.60 \times 10^{-19}\,\text{J/eV}) = 3.2 \times 10^{-12}\,\text{J}\,.$$

Since $K = \frac{1}{2}mv^2$, $v = \sqrt{2K/m}$. The mass of an alpha particle is four times the mass of a proton or $m = 4(1.67 \times 10^{-27}\,\text{kg}) = 6.68 \times 10^{-27}\,\text{kg}$, so

$$v = \sqrt{\frac{2(3.2 \times 10^{-12}\,\text{J})}{6.68 \times 10^{-27}\,\text{kg}}} = 3.1 \times 10^7\,\text{m/s}$$

and

$$N = \frac{iL}{2ev} = \frac{(0.25 \times 10^{-6}\,\text{A})(20 \times 10^{-2}\,\text{m})}{2(1.60 \times 10^{-19}\,\text{C})(3.1 \times 10^{7}\,\text{m/s})} = 5.0 \times 10^{3}\,.$$

(c) Use conservation of energy. The initial kinetic energy is zero, the final kinetic energy is $20\,\text{MeV} = 3.2 \times 10^{-12}\,\text{J}$, the initial potential energy is $qV = 2eV$, and the final potential energy is zero. Here $V$ is the electric potential through which the particles are accelerated. Conservation of energy leads to $K_f = U_i = 2eV$, so

$$V = \frac{K_f}{2e} = \frac{3.2 \times 10^{-12}\,\text{J}}{2(1.60 \times 10^{-19}\,\text{C})} = 10 \times 10^{6}\,\text{V}\,.$$

## 59

Let $R_H$ be the resistance at the higher temperature (800° C) and let $R_L$ be the resistance at the lower temperature (200° C). Since the potential difference is the same for the two temperatures, the rate of energy dissipation at the lower temperature is $P_L = V^2/R_L$, and the rate of energy dissipation at the higher temperature is $P_H = V^2/R_H$, so $P_L = (R_H/R_L)P_H$. Now $R_L = R_H + \alpha R_H\,\Delta T$, where $\Delta T$ is the temperature difference $T_L - T_H = -600°\,\text{C}$. Thus,

$$P_L = \frac{R_H}{R_H + \alpha R_H\,\Delta T} P_H = \frac{P_H}{1 + \alpha\,\Delta T} = \frac{500\,\text{W}}{1 + (4.0 \times 10^{-4}\,/°\text{C})(-600°\,\text{C})} = 660\,\text{W}\,.$$

## 75

If the resistivity is $\rho_0$ at temperature $T_0$, then the resistivity at temperature $T$ is $\rho = \rho_0 + \alpha\rho_0(T - T_0)$, where $\alpha$ is the temperature coefficient of resistivity. The solution for $T$ is

$$T = \frac{\rho - \rho_0 + \alpha\rho_0 T_0}{\alpha\rho_0}\,.$$

Substitute $\rho = 2\rho_0$ to obtain

$$T = T_0 + \frac{1}{\alpha} = 20.0°\text{C} + \frac{1}{4.3 \times 10^{-3}\,\text{K}^{-1}} = 250°\text{C}\,.$$

The value of $\alpha$ was obtained from Table 26–1.

# Chapter 27

7

(a) Let $i$ be the current in the circuit and take it to be positive if it is to the left in $R_1$. Use Kirchhoff's loop rule: $\mathcal{E}_1 - iR_2 - iR_1 - \mathcal{E}_2 = 0$. Solve for $i$:

$$i = \frac{\mathcal{E}_1 - \mathcal{E}_2}{R_1 + R_2} = \frac{12\text{ V} - 6.0\text{ V}}{4.0\,\Omega + 8.0\,\Omega} = 0.50\text{ A}.$$

A positive value was obtained, so the current is counterclockwise around the circuit.

(b) and (c) If $i$ is the current in a resistor with resistance $R$, then the power dissipated by that resistor is given by $P = i^2R$. For $R_1$ the power dissipated is

$$P_1 = (0.50\text{ A})^2(4.0\,\Omega) = 1.0\text{ W}$$

and for $R_2$ the power dissipated is

$$P_2 = (0.50\text{ A})^2(8.0\,\Omega) = 2.0\text{ W}.$$

(d) and (e) If $i$ is the current in a battery with emf $\mathcal{E}$, then the battery supplies energy at the rate $P = i\mathcal{E}$ provided the current and emf are in the same direction. The battery absorbs energy at the rate $P = i\mathcal{E}$ if the current and emf are in opposite directions. For battery 1 the power is

$$P_1 = (0.50\text{ A})(12\text{ V}) = 6.0\text{ W}$$

and for battery 2 it is

$$P_2 = (0.50\text{ A})(6.0\text{ V}) = 3.0\text{ W}.$$

(f) and (g) In battery 1, the current is in the same direction as the emf so this battery supplies energy to the circuit. The battery is discharging. The current in battery 2 is opposite the direction of the emf, so this battery absorbs energy from the circuit. It is charging.

13

(a) If $i$ is the current and $\Delta V$ is the potential difference, then the power absorbed is given by $P = i\,\Delta V$. Thus

$$\Delta V = \frac{P}{i} = \frac{50\text{ W}}{1.0\text{ A}} = 50\text{ V}.$$

Since energy is absorbed, point A is at a higher potential than point B; that is, $V_A - V_B = 50\text{ V}$.

(b) The end-to-end potential difference is given by $V_A - V_B = +iR + \mathcal{E}$, where $\mathcal{E}$ is the emf of element C and is taken to be positive if it is to the left in the diagram. Thus $\mathcal{E} = V_A - V_B - iR = 50\text{ V} - (1.0\text{ A})(2.0\,\Omega) = 48\text{ V}$.

(c) A positive value was obtained for $\mathcal{E}$, so it is toward the left. The negative terminal is at B.

**21**

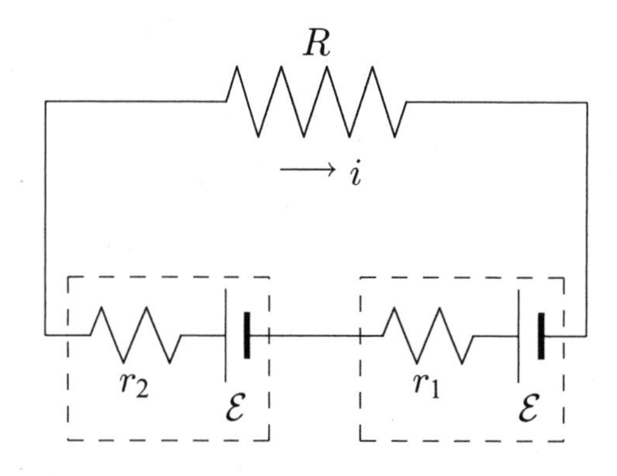


(a) and (b) The circuit is shown in the diagram to the right. The current is taken to be positive if it is clockwise. The potential difference across battery 1 is given by $V_1 = \mathcal{E} - ir_1$ and for this to be zero, the current must be $i = \mathcal{E}/r_1$. Kirchhoff's loop rule gives $2\mathcal{E} - ir_1 - ir_2 - iR = 0$. Substitute $i = \mathcal{E}/r_1$ and solve for $R$. You should get $R = r_1 - r_2 = 0.016\,\Omega - 0.012\,\Omega = 0.004\,\Omega$.

Now assume that the potential difference across battery 2 is zero and carry out the same analysis. You should find $R = r_2 - r_1$. Since $r_1 > r_2$ and $R$ must be positive, this situation is not possible. Only the potential difference across the battery with the larger internal resistance can be made to vanish with the proper choice of $R$.

**29**

Let $r$ be the resistance of each of the thin wires. Since they are in parallel, the resistance $R$ of the composite can be determined from

$$\frac{1}{R} = \frac{9}{r},$$

or $R = r/9$. Now

$$r = \frac{4\rho\ell}{\pi d^2}$$

and

$$R = \frac{4\rho\ell}{\pi D^2},$$

where $\rho$ is the resistivity of copper. Here $\pi d^2/4$ was used for the cross-sectional area of any one of the original wires and $\pi D^2/4$ was used for the cross-sectional area of the replacement wire. Here $d$ and $D$ are diameters. Since the replacement wire is to have the same resistance as the composite,

$$\frac{4\rho\ell}{\pi D^2} = \frac{4\rho\ell}{9\pi d^2}.$$

Solve for $D$ and obtain $D = 3d$.

**33**

Replace the two resistors on the left with their equivalent resistor. They are in parallel, so the equivalent resistance is $R_{eq} = 1.0\,\Omega$. The circuit now consists of the two emf devices and four resistors. Take the current to be upward in the right-hand emf device. Then the loop rule gives $\mathcal{E}_2 - iR_{eq} - 3iR - \mathcal{E}_2$, where $R = 2.0\,\Omega$. The current is

$$i = \frac{\mathcal{E}_2 - \mathcal{E}_1}{R_{eq} + 3R} = \frac{12\,\text{V} - 5.0\,\text{V}}{1.0\,\Omega + 3(2.0\,\Omega)} = 1.0\,\text{A}.$$

To find the potential at point 1 take a path from ground, through the equivalent resistor and $\mathcal{E}_2$, to the point. The result is $V_1 = iR_{eq} - \mathcal{E}_1 = (1.0\,\text{A})(1.0\,\Omega) - 12\,\text{V} = -11\,\text{V}$. To find the potential at point 2 continue the path through the lowest resistor on the digram. It is $V_2 = V_1 + iR = -11\,\text{V} + (1.0\,\text{A})(2.0\,\Omega) = -9.0\,\text{V}$.

**47**

(a) and (b) The copper wire and the aluminum jacket are connected in parallel, so the potential difference is the same for them. Since the potential difference is the product of the current and the resistance, $i_C R_C = i_A R_A$, where $i_C$ is the current in the copper, $i_A$ is the current in the aluminum, $R_C$ is the resistance of the copper, and $R_A$ is the resistance of the aluminum. The resistance of either component is given by $R = \rho L/A$, where $\rho$ is the resistivity, $L$ is the length, and $A$ is the cross-sectional area. The resistance of the copper wire is

$$R_C = \frac{\rho_C L}{\pi a^2}$$

and the resistance of the aluminum jacket is

$$R_A = \frac{\rho_A L}{\pi(b^2 - a^2)}\,.$$

Substitute these expressions into $i_C R_C = i_A R_A$ and cancel the common factors $L$ and $\pi$ to obtain

$$\frac{i_C \rho_C}{a^2} = \frac{i_A \rho_A}{b^2 - a^2}\,.$$

Solve this equation simultaneously with $i = i_C + i_A$, where $i$ is the total current. You should get

$$i_C = \frac{a^2 \rho_C i}{(b^2 - a^2)\rho_C + a^2 \rho_A}$$

and

$$i_A = \frac{(b^2 - a^2)\rho_C i}{(b^2 - a^2)\rho_C + a^2 \rho_A}\,.$$

The denominators are the same and each has the value

$$\begin{aligned}(b^2 - a^2)\rho_C + a^2\rho_A &= \left[(0.380 \times 10^{-3}\,\text{m})^2 - (0.250 \times 10^{-3}\,\text{m})^2\right](1.69 \times 10^{-8}\,\Omega\cdot\text{m}) \\ &\quad + (0.250 \times 10^{-3}\,\text{m})^2(2.75 \times 10^{-8}\,\Omega\cdot\text{m}) \\ &= 3.10 \times 10^{-15}\,\Omega\cdot\text{m}^3\,.\end{aligned}$$

Thus

$$i_C = \frac{(0.250 \times 10^{-3}\,\text{m})^2(2.75 \times 10^{-8}\Omega\cdot\text{m})(2.00\,\text{A})}{3.10 \times 10^{-15}\,\Omega\cdot\text{m}^3} = 1.11\,\text{A}$$

and

$$\begin{aligned}i_A &= \frac{\left[(0.380 \times 10^{-3}\,\text{m})^2 - (0.250 \times 10^{-3}\,\text{m})^2\right](1.69 \times 10^{-8}\,\Omega\cdot\text{m})(2.00\,\text{A})}{3.10 \times 10^{-15}\,\Omega\cdot\text{m}^3} \\ &= 0.893\,\text{A}\,.\end{aligned}$$

(c) Consider the copper wire. If $V$ is the potential difference, then the current is given by $V = i_C R_C = i_C \rho_C L/\pi a^2$, so

$$L = \frac{\pi a^2 V}{i_C \rho_C} = \frac{\pi(0.250 \times 10^{-3}\,\mathrm{m})^2(12.0\,\mathrm{V})}{(1.11\,\mathrm{A})(1.69 \times 10^{-8}\,\Omega\cdot\mathrm{m})} = 126\,\mathrm{m}\,.$$

**57**

During charging the charge on the positive plate of the capacitor is given by Eq. 27–33, with $RC = \tau$. That is,

$$q = C\mathcal{E}\left[1 - e^{-t/\tau}\right],$$

where $C$ is the capacitance, $\mathcal{E}$ is applied emf, and $\tau$ is the time constant. You want the time for which $q = 0.990C\mathcal{E}$, so

$$0.990 = 1 - e^{-t/\tau}\,.$$

Thus

$$e^{-t/\tau} = 0.010\,.$$

Take the natural logarithm of both sides to obtain $t/\tau = -\ln 0.010 = 4.61$ and $t = 4.61\tau$.

**65**

(a), (b), and (c) At $t = 0$, the capacitor is completely uncharged and the current in the capacitor branch is as it would be if the capacitor were replaced by a wire. Let $i_1$ be the current in $R_1$ and take it to be positive if it is to the right. Let $i_2$ be the current in $R_2$ and take it to be positive if it is downward. Let $i_3$ be the current in $R_3$ and take it to be positive if it is downward. The junction rule produces $i_1 = i_2 + i_3$, the loop rule applied to the left-hand loop produces

$$\mathcal{E} - i_1 R_1 - i_2 R_2 = 0\,,$$

and the loop rule applied to the right-hand loop produces

$$i_2 R_2 - i_3 R_3 = 0\,.$$

Since the resistances are all the same, you can simplify the mathematics by replacing $R_1$, $R_2$, and $R_3$ with $R$. The solution to the three simultaneous equations is

$$i_1 = \frac{2\mathcal{E}}{3R} = \frac{2(1.2 \times 10^3\,\mathrm{V})}{3(0.73 \times 10^6\,\Omega)} = 1.1 \times 10^{-3}\,\mathrm{A}$$

and

$$i_2 = i_3 = \frac{\mathcal{E}}{3R} = \frac{1.2 \times 10^3\,\mathrm{V}}{3(0.73 \times 10^6\,\Omega)} = 5.5 \times 10^{-4}\,\mathrm{A}\,.$$

(d), (e), and (f) At $t = \infty$, the capacitor is fully charged and the current in the capacitor branch is zero. Then $i_1 = i_2$ and the loop rule yields

$$\mathcal{E} - i_1 R_1 - i_1 R_2 = 0\,.$$

The solution is

$$i_1 = i_2 = \frac{\mathcal{E}}{2R} = \frac{1.2 \times 10^3\,\text{V}}{2(0.73 \times 10^6\,\Omega)} = 8.2 \times 10^{-4}\,\text{A}\,.$$

(g) and (h) The potential difference across resistor 2 is $V_2 = i_2 R_2$. At $t = 0$ it is

$$V_2 = (5.5 \times 10^{-4}\,\text{A})(0.73 \times 10^6\,\Omega) = 4.0 \times 10^2\,\text{V}$$

and at $t = \infty$ it is

$$V_2 = (8.2 \times 10^{-4}\,\text{A})(0.73 \times 10^6\,\Omega) = 6.0 \times 10^2\,\text{V}\,.$$

(i) The graph of $V_2$ versus $t$ is shown to the right.

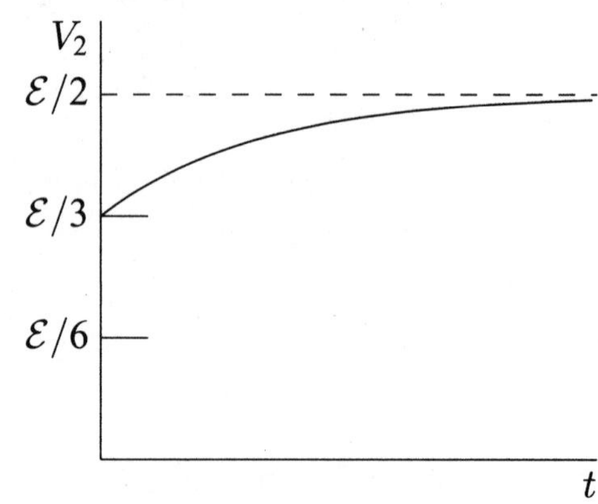


**73**

As the capacitor discharges the potential difference across its plates at time $t$ is given by $V = V_0 e^{-t/\tau}$, where $V_0$ is the potential difference at time $t = 0$ and $\tau$ is the capacitive time constant. This equation is solved for the time constant, with result

$$\tau = -\frac{t}{\ln(V/V_0)}\,.$$

Since the time constant is $\tau = RC$, where $RR$ is the resistance and $C$ is the capacitance,

$$R = -\frac{t}{C\ln(V/V_0)}\,.$$

For the smaller time interval

$$R = -\frac{10.0 \times 10^{-6}\,\text{s}}{(0.220 \times 10^{-6}\,\text{F})\ln\left(\dfrac{0.800\,\text{V}}{5.00\,\text{V}}\right)} = 24.8\,\Omega\,.$$

and for the larger time interval

$$R = -\frac{6.00 \times 10^{-3}\,\text{s}}{(0.220 \times 10^{-6}\,\text{F})\ln\left(\dfrac{0.800\,\text{V}}{5.00\,\text{V}}\right)} = 1.49 \times 10^4\,\Omega\,.$$

**75**

(a) Let $i$ be the current, which is the same in both wires, and $\mathcal{E}$ be the applied potential difference. Then the loop equation gives $\mathcal{E} - iR_A - iR_B = 0$ and the current is

$$i = \frac{\mathcal{E}}{R_A + R_B} = \frac{60.0\,\text{V}}{0.127\,\Omega + 0.729\,\Omega} = 70.1\,\text{A}\,.$$

The current density in wire A is

$$J_A = \frac{i}{\pi r_A^2} = \frac{70.1\,\text{A}}{\pi(1.30\times10^{-3}\,\text{m})^2} = 1.32\times10^7\,\text{A/m}^2\,.$$

(b) The potential difference across wire A is $V_A = iR_A = (70.1\,\text{A})(0.127\,\Omega) = 8.90\,\text{V}$.

(c) The resistance is $R_A = \rho_A L/A$, where $\rho$ is the resistivity, $A$ is the cross-sectional area, and $L$ is the length. The resistivity of wire A is

$$\rho_A = \frac{R_A A}{L} = \frac{(0.127\,\Omega)\pi(1.30\times10^{-3}\,\text{m})^2}{40.0\,\text{m}} = 1.69\times10^{-8}\,\Omega\cdot\text{m}\,.$$

According to Table 26–1 the material is copper.

(d) Since wire B has the same diameter and length as wire A and carries the same current, the current density in it is the same, $1.32\times10^7\,\text{A/m}^2$.

(e) The potential difference across wire B is $V_B = iR_B = (70.1\,\text{A})(0.729\,\Omega) = 51.1\,\text{V}$.

(f) The resistivity of wire B is

$$\rho_B = \frac{R_B A}{L} = \frac{(0.729\,\Omega)\pi(1.30\times10^{-3}\,\text{m})^2}{40.0\,\text{m}} = 9.68\times10^{-8}\,\Omega\cdot\text{m}\,.$$

According to Table 26–1 the material is iron.

**77**

The three circuit elements are in series, so the current is the same in all of them. Since the battery is discharging, the potential difference across its terminals is $V_{\text{batt}} = \mathcal{E} - ir$, where $\mathcal{E}$ is its emf and $r$ is its internal resistance. Thus

$$r = \frac{\mathcal{E} - V}{i} = \frac{12\,\text{V} - 11.4\,\text{V}}{50\,\text{A}} = 0.012\,\Omega\,.$$

This is less than $0.0200\,\Omega$, so the battery is not defective.

The resistance of the cable is $R_{\text{cable}} = V_{\text{cable}}/i = (3.0\,\text{V})/(50\,\text{A} = 0.060\,\Omega$, which is greater than $0.040\,\Omega$. The cable is defective.

The potential difference across the motor is $V_{\text{motor}} = 11.4\,\text{V} - 3.0\,\text{V} = 8.4\,\text{V}$ and its resistance is $R_{\text{motor}} = V_{\text{motor}}/i = (8.4\,\text{V})/(50\,\text{A}) = 0.17\,\Omega$, which is less than $0.200\,\Omega$. The motor is not defective.

**85**

Let $R_{S0}$ be the resistance of the silicon resistor at $20^\circ$ and $R_{I0}$ be the resistance of the iron resistor at that temperature. At some other temperature $T$ the resistance of the silicon resistor is $R_S = R_{S0} + \alpha_S R_{S0}(T - 20^\circ\text{C})$ and the resistance of the iron resistor is $R_I = R_{I0} + \alpha_I R_{I0}(T - 20^\circ\text{C})$. Here $\alpha_S$ and $\alpha_I$ are the temperature coefficients of resistivity. The resistors are series so the resistance of the combination is

$$R = R_{S0} + R_{I0} + (\alpha_S R_{S0} + \alpha_I R_{I0})(T - 20^\circ\text{C})\,.$$

We want $R_{S0} + R_{I0}$ to be $1000\,\Omega$ and $\alpha_S R_{S0} + \alpha_I R_{I0}$ to be zero. Then the resistance of the combination will be independent of the temperature.

The second equation gives $R_{I0} = -(\alpha_S/\alpha_I)R_{S0}$ and when this is used to substitute for $R_{I0}$ in the first equation the result is $R_{S0} - (\alpha_S/\alpha_I)R_{S0} = 1000\,\Omega$. The solution for $R_{S0}$ is

$$R_{S0} = \frac{1000\,\Omega}{\dfrac{\alpha_S}{\alpha_I} - 1} = \frac{1000\,\Omega}{\dfrac{-70 \times 10^{-3}\,\text{K}^{-1}}{6.5 \times 10^{-3}\,\text{K}^{-1}} - 1} = 85\,\Omega\,,$$

where values for the temperature coefficients of resistivity were obtained from Table 26−1. The resistance of the iron resistor is $R_{I0} = 1000\,\Omega - 85\,\Omega = 915\,\Omega$.

## 95

When the capacitor is fully charged the potential difference across its plates is $\mathcal{E}$ and the energy stored in it is $U = \frac{1}{2}C\mathcal{E}^2$.

(a) The current is given as a function of time by $i = (\mathcal{E}/R)e^{-t/\tau}$, where $\tau$ (= $RC$) is the capacitive time constant. The rate with which the emf device supplies energy is $P_\mathcal{E} = i\mathcal{E}$ and the energy supplied in fully charging the capacitor is

$$E_\mathcal{E} = \int_0^\infty P_\mathcal{E}\,dt = \frac{\mathcal{E}^2}{R}\int_0^\infty e^{-t/\tau}\,dt = \frac{\mathcal{E}^2\tau}{R} = \frac{\mathcal{E}^2 RC}{R} = C\mathcal{E}^2\,.$$

This is twice the energy stored in the capacitor.

(b) The rate with which energy is dissipated in the resistor is $P_R = i^2R$ and the energy dissipated as the capacitor is fully charged is

$$E_R = \int_0^\infty P_R\,dt = \frac{\mathcal{E}^2}{R}\int_0^\infty e^{-2t/\tau}\,dt = \frac{\mathcal{E}^2\tau}{2R} = \frac{\mathcal{E}^2 RC}{2R} = \tfrac{1}{2}C\mathcal{E}^2\,.$$

## 97

(a) Immediately after the switch is closed the capacitor is uncharged and since the charge on the capacitor is given by $q = CV_C$, the potential difference across its plates is zero. Apply the loop rule to the right-hand loop to find that the potential difference across $R_2$ must also be zero. Now apply the loop rule to the left-hand loop to find that $\mathcal{E} - i_1R_1 = 0$ and $i_1 = \mathcal{E}/R_1 = (30\,\text{V})/(20 \times 10^3\,\Omega) = 1.5 \times 10^{-3}\,\text{A}$.

(b) Since the potential difference across $R_2$ is zero and this potential difference is given by $V_{R2} = i_2R_2$, $i_2 = 0$.

(c) A long time later, when the capacitor is fully charged, the current is zero in the capacitor branch and the current is the same in the two resistors. The loop rule applied to the left-hand loop gives $\mathcal{E} - iR_1 - iR_2 = 0$, so $i = \mathcal{E}/(R_1 + R_2) = (30\,\text{V})/(20 \times 10^3\,\Omega + 10 \times 10^3\,\Omega) = 1.0 \times 10^{-3}\,\text{A}$.

## 99

(a) $R_2$ and $R_3$ are in parallel, with an equivalent resistance of $R_2R_3/(R_2 + R_3)$, and this combination is in series with $R_1$, so the circuit can be reduced to a single loop with an emf $\mathcal{E}$ and a resistance $R_{\text{eq}} = R_1 + R_2R_3/(R_2 + R_3) = (R_1R_2 + R_1R_3 + R_2R_3)/(R_1 + R_2)$. The current is

$$i = \frac{\mathcal{E}}{R_{\text{eq}}} = \frac{(R_2 + R_3)\mathcal{E}}{R_1R_2 + R_1R_3 + R_2R_3}\,.$$

The rate with which the battery supplies energy is

$$P = i\mathcal{E} = \frac{(R_2 + R_3)\mathcal{E}^2}{R_1R_2 + R_1R_3 + R_2R_3} .$$

The derivative with respect to $R_3$ is

$$\frac{dP}{dR_3} = \frac{\mathcal{E}^2}{R_1R_2 + R_1R_3 + R_2R_3} - \frac{(R_2 + R_3)(R_1 + R_2)\mathcal{E}^2}{(R_1R_2 + R_1R_3 + R_2R_3)^2} = -\frac{\mathcal{E}^2R_2^2}{(R_1R_2 + R_1R_3 + R_2R_3)^2} ,$$

where the last form was obtained with a little algebra. The derivative is negative for all (positive) values of the resistances, so $P$ has its maximum value for $R_3 = 0$.

(b) Substitute $R_3 = 0$ in the expression for $P$ to obtain

$$P = \frac{R_1\mathcal{E}^2}{R_1R_2} = \frac{\mathcal{E}^2}{R_1} = \frac{12.0\,\text{V}}{10.0\,\Omega} = 14.4\,\text{W} .$$

**101**

If the batteries are connected in series the total emf in the circuit is $N\mathcal{E}$ and the equivalent resistance is $R + nr$, so the current is $i = N\mathcal{E}/(R + Nr)$. If $R = r$, then $i = N\mathcal{E}/(N + 1)r$.

If the batteries are connected in parallel then the emf in the circuit is $\mathcal{E}$ and the equivalent resistance is $R + r/N$, so the current is $i = \mathcal{E}/(R + r/N) = N\mathcal{E}/(NR + r)$. If $R = r$, $i = N\mathcal{E}/(N + 1)r$, the same as when they are connected in series.

# Chapter 28

**3**

(a) The magnitude of the magnetic force on the proton is given by $F_B = evB\sin\phi$, where $v$ is the speed of the proton, $B$ is the magnitude of the magnetic field, and $\phi$ is the angle between the particle velocity and the field when they are drawn with their tails at the same point. Thus

$$v = \frac{F_B}{eB\sin\phi} = \frac{6.50\times10^{-17}\,\mathrm{N}}{(1.60\times10^{-19}\,\mathrm{C})(2.60\times10^{-3}\,\mathrm{T})\sin 23.0^\circ} = 4.00\times10^{5}\,\mathrm{m/s}\,.$$

(b) The kinetic energy of the proton is

$$K = \tfrac{1}{2}mv^2 = \tfrac{1}{2}(1.67\times10^{-27}\,\mathrm{kg})(4.00\times10^{5}\,\mathrm{m/s})^2 = 1.34\times10^{-16}\,\mathrm{J}\,.$$

This is $(1.34\times10^{-16}\,\mathrm{J})/(1.60\times10^{-19}\,\mathrm{J/eV}) = 835\,\mathrm{eV}$.

**17**

(a) Since the kinetic energy is given by $K = \frac{1}{2}mv^2$, where $m$ is the mass of the electron and $v$ is its speed,

$$v = \sqrt{\frac{2K}{m}} = \sqrt{\frac{2(1.20\times10^{3}\,\mathrm{eV})(1.60\times10^{-19}\,\mathrm{J/eV})}{9.11\times10^{-31}\,\mathrm{kg}}} = 2.05\times10^{7}\,\mathrm{m/s}\,.$$

(b) The magnitude of the magnetic force is given by $evB$ and the acceleration of the electron is given by $v^2/r$, where $r$ is the radius of the orbit. Newton's second law is $evB = mv^2/r$, so

$$B = \frac{mv}{er} = \frac{(9.11\times10^{-31}\,\mathrm{kg})(2.05\times10^{7}\,\mathrm{m/s})}{(1.60\times10^{-19}\,\mathrm{C})(25.0\times10^{-2}\,\mathrm{m})} = 4.68\times10^{-4}\,\mathrm{T} = 468\,\mu\mathrm{T}\,.$$

(c) The frequency $f$ is the number of times the electron goes around per unit time, so

$$f = \frac{v}{2\pi r} = \frac{2.05\times10^{7}\,\mathrm{m/s}}{2\pi(25.0\times10^{-2}\,\mathrm{m})} = 1.31\times10^{7}\,\mathrm{Hz} = 13.1\,\mathrm{MHz}\,.$$

(d) The period is the reciprocal of the frequency:

$$T = \frac{1}{f} = \frac{1}{1.31\times10^{7}\,\mathrm{Hz}} = 7.63\times10^{-8}\,\mathrm{s} = 76.3\,\mathrm{ns}\,.$$

**29**

(a) If $v$ is the speed of the positron, then $v\sin\phi$ is the component of its velocity in the plane that is perpendicular to the magnetic field. Here $\phi$ is the angle between the velocity and the field (89°). Newton's second law yields $eBv\sin\phi = m(v\sin\phi)^2/r$, where $r$ is the radius of the orbit. Thus $r = (mv/eB)\sin\phi$. The period is given by

$$T = \frac{2\pi r}{v\sin\phi} = \frac{2\pi m}{eB} = \frac{2\pi(9.11\times10^{-31}\,\mathrm{kg})}{(1.60\times10^{-19}\,\mathrm{C})(0.100\,\mathrm{T})} = 3.58\times10^{-10}\,\mathrm{s}\,.$$

The expression for $r$ was substituted to obtain the second expression for $T$.

(b) The pitch $p$ is the distance traveled along the line of the magnetic field in a time interval of one period. Thus $p = vT\cos\phi$. Use the kinetic energy to find the speed: $K = \frac{1}{2}mv^2$ yields

$$v = \sqrt{\frac{2K}{m}} = \sqrt{\frac{2(2.0 \times 10^3\,\mathrm{eV})(1.60 \times 10^{-19}\,\mathrm{J/eV})}{9.11 \times 10^{-31}\,\mathrm{kg}}} = 2.651 \times 10^7\,\mathrm{m/s}\,.$$

Thus

$$p = (2.651 \times 10^7\,\mathrm{m/s})(3.58 \times 10^{-10}\,\mathrm{s})\cos 89.0° = 1.66 \times 10^{-4}\,\mathrm{m}\,.$$

(c) The orbit radius is

$$r = \frac{mv\sin\phi}{eB} = \frac{(9.11 \times 10^{-31}\,\mathrm{kg})(2.651 \times 10^7\,\mathrm{m/s})\sin 89.0°}{(1.60 \times 10^{-19}\,\mathrm{C})(0.100\,\mathrm{T})} = 1.51 \times 10^{-3}\,\mathrm{m}\,.$$

**41**

(a) The magnitude of the magnetic force on the wire is given by $F_B = iLB\sin\phi$, where $i$ is the current in the wire, $L$ is the length of the wire, $B$ is the magnitude of the magnetic field, and $\phi$ is the angle between the current and the field. In this case $\phi = 70°$. Thus

$$F_B = (5000\,\mathrm{A})(100\,\mathrm{m})(60.0 \times 10^{-6}\,\mathrm{T})\sin 70° = 28.2\,\mathrm{N}\,.$$

(b) Apply the right-hand rule to the vector product $\vec{F}_B = i\vec{L} \times \vec{B}$ to show that the force is to the west.

**47**

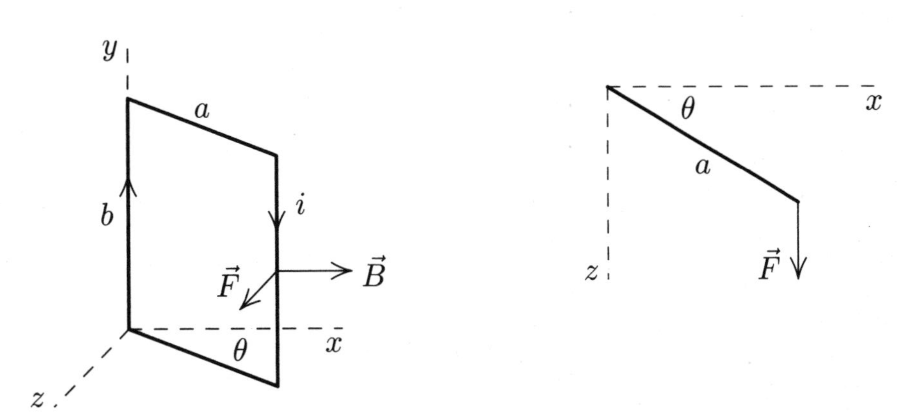


The situation is shown in the left diagram above. The $y$ axis is along the hinge and the magnetic field is in the positive $x$ direction. A torque around the hinge is associated with the wire opposite the hinge and not with the other wires. The force on this wire is in the positive $z$ direction and has magnitude $F = NibB$, where $N$ is the number of turns.

The right diagram shows the view from above. The magnitude of the torque is given by

$$\begin{aligned}\tau &= Fa\cos\theta = NibBa\cos\theta \\ &= 20(0.10\,\mathrm{A})(0.10\,\mathrm{m})(0.50 \times 10^{-3}\,\mathrm{T})(0.050\,\mathrm{m})\cos 30° \\ &= 4.3 \times 10^{-3}\,\mathrm{N\cdot m}\,.\end{aligned}$$

Use the right-hand rule to show that the torque is directed downward, in the negative $y$ direction. Thus $\vec{\tau} = -(4.3 \times 10^{-3}\,\text{N}\cdot\text{m})\hat{\text{j}}$.

**55**

(a) The magnitude of the magnetic dipole moment is given by $\mu = NiA$, where $N$ is the number of turns, $i$ is the current in each turn, and $A$ is the area of a loop. In this case the loops are circular, so $A = \pi r^2$, where $r$ is the radius of a turn. Thus

$$i = \frac{\mu}{N\pi r^2} = \frac{2.30\,\text{A}\cdot\text{m}^2}{(160)(\pi)(0.0190\,\text{m})^2} = 12.7\,\text{A}\,.$$

(b) The maximum torque occurs when the dipole moment is perpendicular to the field (or the plane of the loop is parallel to the field). It is given by $\tau = \mu B = (2.30\,\text{A}\cdot\text{m}^2)(35.0 \times 10^{-3}\,\text{T}) = 8.05 \times 10^{-2}\,\text{N}\cdot\text{m}$.

**59**

The magnitude of a magnetic dipole moment of a current loop is given by $\mu = iA$, where $i$ is the current in the loop and $A$ is the area of the loop. Each of these loops is a circle and its area is given by $A = \pi R^2$, where $R$ is the radius. Thus the dipole moment of the inner loop has a magnitude of $\mu_i = i\pi r_1^2 = (7.00\,\text{A})\pi(0.200\,\text{m})^2 = 0.880\,\text{A}\cdot\text{m}^2$ and the dipole moment of the outer loop has a magnitude of $\mu_o = i\pi r_2^2 = (7.00\,\text{A}\pi(0.300\,\text{m})^2 = 1.979\,\text{A}\cdot\text{m}^2$.

(a) Both currents are clockwise in Fig. 28–51 so, according to the right-hand rule, both dipole moments are directed into the page. The magnitude of the net dipole moment is the sum of the magnitudes of the individual moments: $\mu_{\text{net}} = \mu_i + \mu_o = 0.880\,\text{A}\cdot\text{m}^2 + 1.979\,\text{A}\cdot\text{m}^2 = 2.86\,\text{A}\cdot\text{m}^2$. The net dipole moment is directed into the page.

(b) Now the dipole moment of the inner loop is directed out of the page. The moments are in opposite directions, so the magnitude of the net moment is $\mu_{\text{net}} = \mu_o - \mu_i = 1.979\,\text{A}\cdot\text{m}^2 - 0.880\,\text{A}\cdot\text{m}^2 = 1.10\,\text{A}\cdot\text{m}^2$. The net dipole moment is again into the page.

**63**

If $N$ closed loops are formed from the wire of length $L$, the circumference of each loop is $L/N$, the radius of each loop is $R = L/2\pi N$, and the area of each loop is $A = \pi R^2 = \pi(L/2\pi N)^2 = L^2/4\pi N^2$. For maximum torque, orient the plane of the loops parallel to the magnetic field, so the dipole moment is perpendicular to the field. The magnitude of the torque is then

$$\tau = NiAB = (Ni)\left(\frac{L^2}{4\pi N^2}\right)B = \frac{iL^2B}{4\pi N}\,.$$

To maximize the torque, take $N$ to have the smallest possible value, 1. Then

$$\tau = \frac{iL^2B}{4\pi} = \frac{(4.51 \times 10^{-3}\,\text{A})(0.250\,\text{m})^2(5.71 \times 10^{-3}\,\text{T})}{4\pi} = 1.28 \times 10^{-7}\,\text{N}\cdot\text{m}\,.$$

**65**

(a) the magnetic potential energy is given by $U = -\vec{\mu}\cdot\vec{B}$, where $\vec{\mu}$ is the magnetic dipole moment of the coil and $\vec{B}$ is the magnetic field. The magnitude of the magnetic moment is $\mu = NiA$, where $i$ is the current in the coil, $A$ is the area of the coil, and $N$ is the number of turns. The moment is in the negative $y$ direction, as you can tell by wrapping the fingers of your right hand around the coil in the direction of the current. Your thumb is then in the negative $y$ direction. Thus $\vec{\mu} = -(3.00)(2.00\,\text{A})(4.00 \times 10^{-3}\,\text{m}^2)\hat{\jmath} = -(2.40 \times 10^{-2}\,\text{A}\cdot\text{m}^2)\hat{\jmath}$. The magnetic potential energy is

$$\begin{aligned} U &= -(\mu_y \hat{\jmath}) \cdot (B_x \hat{\imath} + B_y \hat{\jmath} + B_z \hat{k}) = -\mu_y By \\ &= -(-2.40 \times 10^{-2}\,\text{A}\cdot\text{m}^2)(-3.00 \times 10^{-3}\,\text{T}) = -7.20 \times 10^{-5}\,\text{J}, \end{aligned}$$

where $\hat{\jmath} \cdot \hat{\imath} = 0$, $\hat{\jmath} \cdot \hat{\jmath} = 1$, and $\hat{\jmath} \cdot \hat{k} = 0$ were used.

(b) The magnetic torque on the coil is

$$\begin{aligned} \vec{\tau} &= \vec{\mu} \times \vec{B} = (\mu_y \hat{\jmath}) \times (B_x \hat{\imath} + B_y \hat{\jmath} + B_z \hat{k}) = \mu_y B_z \hat{\imath} - \mu_y B_x \hat{k} \\ &= (-2.40 \times 10^{-2}\,\text{A}\cdot\text{m}^2)(-4.00 \times 10^{-3}\,\text{T})\hat{\imath} - (-2.40 \times 10^{-2}\,\text{A}\cdot\text{m}^2)(2.00 \times 10^{-3}\,\text{T})\hat{k} \\ &= (9.6 \times 10^{-5}\,\text{N}\cdot\text{m})\hat{\imath} + (4.80 \times 10^{-5}\,\text{N}\cdot\text{m})\hat{k}, \end{aligned}$$

where $\hat{\jmath} \times \hat{\imath} = -\hat{k}$, $\hat{\jmath} \times \hat{\jmath} = 0$, and $\hat{\jmath} \times \hat{k} = \hat{\imath}$ were used.

### 73

The net force on the electron is given by $\vec{F} = -e(\vec{E} + \vec{v} \times \vec{B})$, where $\vec{E}$ is the electric field, $\vec{B}$ is the magnetic field, and $\vec{v}$ is the electron's velocity. Since the electron moves with constant velocity you know that the net force must vanish. Thus

$$\vec{E} = -\vec{v} \times \vec{B} = -(v\hat{\imath}) \times (B\hat{k}) = -vB\hat{\jmath} = -(100\,\text{m/s})(5.00\,\text{T})\hat{\jmath} = (500\,\text{V/m})\hat{\jmath}\,.$$

### 75

(a) and (b) Suppose the particles are accelerated from rest through an electric potential difference $V$. Since energy is conserved the kinetic energy of a particle is $K = qV$, where $q$ is the particle's charge. The ratio of the proton's kinetic energy to the alpha particle's kinetic energy is $K_p/K_\alpha = e/2e = 0.50$. The ratio of the deuteron's kinetic energy to the alpha particle's kinetic energy is $K_d/K_\alpha = e/2e = 0.50$.

(c) The magnitude of the magnetic force on a particle is $qvB$ and, according to Newton's second law, this must equal $mv^2/R$, where $v$ is its speed and $R$ is the radius of its orbit. Since $v = \sqrt{2K/m} = \sqrt{2qV/m}$,

$$R = \frac{mv}{qB} = \frac{m}{qB}\sqrt{\frac{2K}{m}} = \frac{m}{qB}\sqrt{\frac{2qV}{m}} = \sqrt{\frac{2m}{qB^2}}\,.$$

The ratio of the radius of the deuteron's path to the radius of the proton's path is

$$\frac{R_d}{R_p} = \sqrt{\frac{2.0\,\text{u}}{1.0\,\text{u}}}\sqrt{\frac{e}{e}} = 1.4\,.$$

Since the radius of the proton's path is 10 cm, the radius of the deuteron's path is $(1.4)(10\,\text{cm}) =$ 14 cm.

(d) The ratio of the radius of the alpha particle's path to the radius of the proton's path is

$$\frac{R_\alpha}{R_p} = \sqrt{\frac{4.0\,\mathrm{u}}{1.0\,\mathrm{u}}}\sqrt{\frac{e}{2e}} = 1.4\,.$$

Since the radius of the proton's path is 10 cm, the radius of the deuteron's path is (1.4)(10 cm) = 14 cm.

## 77

Take the velocity of the particle to be $\vec{v} = v_x\,\hat{\imath} + v_y\,\hat{\jmath}$ and the magnetic field to be $B\,\hat{\imath}$. The magnetic force on the particle is then

$$\vec{F} = q\vec{v}\times\vec{B} = q(v_x\,\hat{\imath} + v_y\,\hat{\jmath})\times(B\,\hat{\imath}) = -qv_yB\,\hat{k}\,,$$

where $q$ is the charge of the particle. We used $\hat{\imath}\times\hat{\imath} = 0$ and $\hat{\jmath}\times\hat{\imath} = -\hat{k}$. The charge is

$$q = \frac{F}{-v_yB} = \frac{0.48\,\mathrm{N}}{-(4.0\times10^3\,\mathrm{m/s})(\sin 37^\circ)(5.0\times10^{-3}\,\mathrm{T})} = -4.0\times10^{-2}\,\mathrm{C}\,.$$

## 81

(a) If $K$ is the kinetic energy of the electron and $m$ is its mass, then its speed is

$$v = \sqrt{\frac{2K}{m}} = \sqrt{\frac{2(12\times10^3\,\mathrm{eV})(1.60\times10^{-19}\,\mathrm{J/eV})}{9.11\times10^{-31}}} = 6.49\times10^7\,\mathrm{m/s}\,.$$

Since the electron is traveling along a line that is parallel to the horizontal component of Earth's magnetic field, that component does not enter into the calculation of the magnetic force on the electron. The magnitude of the force on the electron is $evB$ and since $F = ma$, where $a$ is the magnitude of its acceleration, $evB = ma$ and

$$a = \frac{evB}{m} = \frac{(1.60\times10^{-19}\,\mathrm{C})(6.49\times10^7\,\mathrm{m/s})(55.0\times10^{-6}\,\mathrm{T})}{9.11\times10^{-31}\,\mathrm{kg}} = 6.3\times10^{14}\,\mathrm{m/s^2}\,.$$

(b) If the electron does not get far from the $x$ axis we may neglect the influence of the horizontal component of Earth's field and assume the electron follows a circular path. Its acceleration is given by $a = v^2/R$, where $R$ is the radius of the path. Thus

$$R = \frac{v^2}{a} = \frac{(6.49\times10^7\,\mathrm{m/s})^2}{6.27\times10^{14}\,\mathrm{m/s^2}} = 6.72\,\mathrm{m}\,.$$

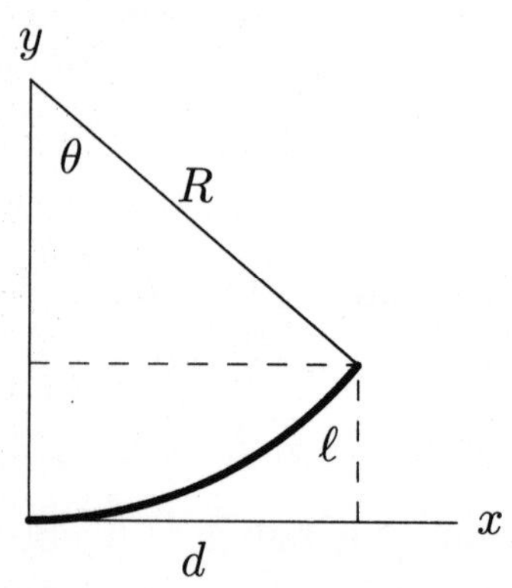


The solid curve on the diagram is the path. Suppose it subtends the angle $\theta$ at its center. $d$ (= 0.200 m) is the distance traveled along the $x$ axis and $\ell$ is the deflection. The right triangle yields $d = R\sin\theta$, so $\sin\theta = d/R$ and $\cos\theta = \sqrt{1-\sin^2\theta} = \sqrt{1-(x/R)^2}$. The triangle also gives $\ell = R - R\cos\theta$, so $\ell = R - R\sqrt{1-(x/R)^2}$. Substitute $R = 6.72$ m and $d = 0.2$ m to obtain $\ell = 0.0030$ m.

# Chapter 29

**1**

(a) The field due to the wire, at a point 8.0 cm from the wire, must be 39 $\mu$T and must be directed toward due south. Since $B = \mu_0 i/2\pi r$,

$$i = \frac{2\pi r B}{\mu_0} = \frac{2\pi(0.080\,\text{m})(39 \times 10^{-6}\,\text{T})}{4\pi \times 10^{-7}\,\text{T}\cdot\text{m/A}} = 16\,\text{A}\,.$$

(b) The current must be from west to east to produce a field to the south at points above it.

**7**

(a) If the currents are parallel, the two magnetic fields are in opposite directions in the region between the wires. Since the currents are the same, the net field is zero along the line that runs halfway between the wires. There is no possible current for which the field does not vanish. If there is to be a field on the bisecting line the currents must be in opposite directions. Then the fields are in the same direction in the region between the wires.

(b) At a point halfway between the wires, the fields have the same magnitude, $\mu_0 i/2\pi r$. Thus the net field at the midpoint has magnitude $B = \mu_0 i/\pi r$ and

$$i = \frac{\pi r B}{\mu_0} = \frac{\pi(0.040\,\text{m})(300 \times 10^{-6}\,\text{T})}{4\pi \times 10^{-7}\,\text{T}\cdot\text{m/A}} = 30\,\text{A}\,.$$

**15**

Sum the fields of the two straight wires and the circular arc. Look at the derivation of the expression for the field of a long straight wire, leading to Eq. 29–6. Since the wires we are considering are infinite in only one direction, the field of either of them is half the field of an infinite wire. That is, the magnitude is $\mu_0 i/4\pi R$, where $R$ is the distance from the end of the wire to the center of the arc. It is the radius of the arc. The fields of both wires are out of the page at the center of the arc.

Now find an expression for the field of the arc at its center. Divide the arc into infinitesimal segments. Each segment produces a field in the same direction. If $ds$ is the length of a segment, the magnitude of the field it produces at the arc center is $(\mu_0 i/4\pi R^2)\,ds$. If $\theta$ is the angle subtended by the arc in radians, then $R\theta$ is the length of the arc and the net field of the arc is $\mu_0 i\theta/4\pi R$. For the arc of the diagram, the field is into the page. The net field at the center, due to the wires and arc together, is

$$B = \frac{\mu_0 i}{4\pi R} + \frac{\mu_0 i}{4\pi R} - \frac{\mu_0 i\theta}{4\pi R} = \frac{\mu_0 i}{4\pi R}(2 - \theta)\,.$$

For this to vanish, $\theta$ must be exactly 2 radians.

**19**

Each wire produces a field with magnitude given by $B = \mu_0 i/2\pi r$, where $r$ is the distance from the corner of the square to the center. According to the Pythagorean theorem, the diagonal of the square has length $\sqrt{2}a$, so $r = a/\sqrt{2}$ and $B = \mu_0 i/\sqrt{2}\pi a$. The fields due to the wires at the upper left and lower right corners both point toward the upper right corner of the square. The fields due to the wires at the upper right and lower left corners both point toward the upper left corner. The horizontal components cancel and the vertical components sum to

$$\begin{aligned} B_{\text{net}} &= 4\frac{\mu_0 i}{\sqrt{2}\pi a}\cos 45^\circ = \frac{2\mu_0 i}{\pi a} \\ &= \frac{2(4\pi \times 10^{-7}\,\text{T}\cdot\text{m/A})(20\,\text{A})}{\pi(0.20\,\text{m})} = 8.0 \times 10^{-5}\,\text{T}\,. \end{aligned}$$

In the calculation $\cos 45^\circ$ was replaced with $1/\sqrt{2}$. In unit vector notation $\vec{B} = (8.0 \times 10^{-5}\,\text{T})\hat{\text{j}}$.

**21**

Follow the same steps as in the solution of Problem 17 above but change the lower limit of integration to $-L$, and the upper limit to 0. The magnitude of the net field is

$$\begin{aligned} B &= \frac{\mu_0 i R}{4\pi}\int_{-L}^{0}\frac{dx}{(x^2+R^2)^{3/2}} = \frac{\mu_0 i R}{4\pi}\frac{1}{R^2}\frac{x}{(x^2+R^2)^{1/2}}\bigg|_{-L}^{0} = \frac{\mu_0 i}{4\pi R}\frac{L}{\sqrt{L^2+R^2}} \\ &= \frac{4\pi \times 10^{-7}\,\text{T}\cdot\text{m/A})(0.693\,\text{A})}{4\pi(0.251\,\text{m})}\frac{0.136\,\text{m}}{\sqrt{(0.136\,\text{m})^2+(0.251\,\text{m})^2}} = 1.32 \times 10^{-7}\,\text{T}\,. \end{aligned}$$

**31**

The current per unit width of the strip is $i/w$ and the current through a width $dx$ is $(i/w)\,dx$. Treat this as a long straight wire. The magnitude of the field it produces at a point that is a distance $d$ from the edge of the strip is $dB = (\mu_0/2\pi)(i/w)\,dx/x$ and the net field is

$$\begin{aligned} B &= \frac{\mu_0 i}{2\pi w}\int_{d}^{d+w}\frac{dx}{x} = \frac{\mu_0}{2\pi w}\ln\frac{d+w}{d} \\ &= \frac{(4\pi \times 10^{-7}\,\text{T}\cdot\text{m/A})(4.61 \times 10^{-6}\,\text{A})}{2\pi(0.0491\,\text{m})}\ln\frac{0.0216\,\text{m}+0.0491\,\text{m}}{0.0216\,\text{m}} \\ &= 2.23 \times 10^{-11}\,\text{T}\,. \end{aligned}$$

**35**

The magnitude of the force of wire 1 on wire 2 is given by $\mu_0 i_1 i_2/2\pi r$, where $i_1$ is the current in wire 1, $i_2$ is the current in wire 2, and $r$ is the separation of the wires. The distance between the wires is $r = \sqrt{d_1^2 + d_2^2}$. Since the currents are in opposite directions the wires repel each other so the force on wire 2 is along the line that joins the wires and is away from wire 1.

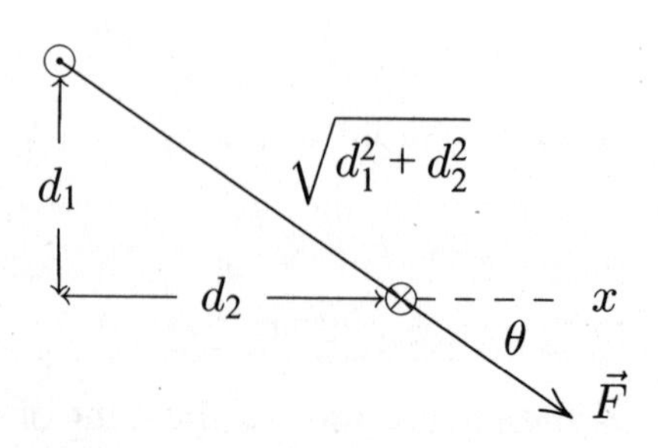

To find the $x$ component of the force, multiply the magnitude of the force by the cosine of the angle $\theta$ that the force makes with the $x$ axis. This is $\cos\theta = d_2/\sqrt{d_1^2 + d_2^2}$. Thus the $x$ component of the force is

$$\begin{aligned} F_x &= \frac{\mu_0 i_1 i_2}{2\pi} \frac{d_2}{d_1^2 + d_2^2} \\ &= \frac{(2\pi \times 10^{-7}\,\mathrm{T\cdot m/A})(4.00 \times 10^{-3}\,\mathrm{A})(6.80 \times 10^{-3}\,\mathrm{A})}{2\pi} \frac{0.0500\,\mathrm{m}}{(0.024\,\mathrm{m})^2 + (5.00\,\mathrm{m})^2} \\ &= 8.84 \times 10^{-11}\,\mathrm{T}\,. \end{aligned}$$

**43**

(a) Two of the currents are out of the page and one is into the page, so the net current enclosed by the path is 2.0 A, out of the page. Since the path is traversed in the clockwise sense, a current into the page is positive and a current out of the page is negative, as indicated by the right-hand rule associated with Ampere's law. Thus $i_{\text{enc}} = -i$ and

$$\oint \vec{B} \cdot d\vec{s} = -\mu_0 i = -(4\pi \times 10^{-7}\,\mathrm{T\cdot m/A})(2.0\,\mathrm{A}) = -2.5 \times 10^{-6}\,\mathrm{T\cdot m}\,.$$

(b) The net current enclosed by the path is zero (two currents are out of the page and two are into the page), so $\oint \vec{B} \cdot d\vec{s} = \mu_0 i_{\text{enc}} = 0$.

**53**

(a) Assume that the point is inside the solenoid. The field of the solenoid at the point is parallel to the solenoid axis and the field of the wire is perpendicular to the solenoid axis. The net field makes an angle of 45° with the axis if these two fields have equal magnitudes.

The magnitude of the magnetic field produced by a solenoid at a point inside is given by $B_{\text{sol}} = \mu_0 i_{\text{sol}} n$, where $n$ is the number of turns per unit length and $i_{\text{sol}}$ is the current in the solenoid. The magnitude of the magnetic field produced by a long straight wire at a point a distance $r$ away is given by $B_{\text{wire}} = \mu_0 i_{\text{wire}}/2\pi r$, where $i_{\text{wire}}$ is the current in the wire. We want $\mu_0 n i_{\text{sol}} = \mu_0 i_{\text{wire}}/2\pi r$. The solution for $r$ is

$$r = \frac{i_{\text{wire}}}{2\pi n i_{\text{sol}}} = \frac{6.00\,\mathrm{A}}{2\pi(10.0 \times 10^2\,\mathrm{m^{-1}})(20.0 \times 10^{-3}\,\mathrm{A})} = 4.77 \times 10^{-2}\,\mathrm{m} = 4.77\,\mathrm{cm}\,.$$

This distance is less than the radius of the solenoid, so the point is indeed inside as we assumed.

(b) The magnitude of the either field at the point is

$$B_{\text{sol}} = B_{\text{wire}} = \mu_0 n i_{\text{sol}} = (4\pi \times 10^{-7}\,\mathrm{T\cdot m/A})(10.0 \times 10^2\,\mathrm{m^{-1}})(20.0 \times 10^{-3}\,\mathrm{A}) = 2.51 \times 10^{-5}\,\mathrm{T}\,.$$

Each of the two fields is a vector component of the net field, so the magnitude of the net field is the square root of the sum of the squares of the individual fields: $B = \sqrt{2(2.51 \times 10^{-5}\,\mathrm{T})^2} = 3.55 \times 10^{-5}\,\mathrm{T}$.

**57**

The magnitude of the dipole moment is given by $\mu = NiA$, where $N$ is the number of turns, $i$ is the current, and $A$ is the area. Use $A = \pi R^2$, where $R$ is the radius. Thus

$$\mu = Ni\pi R^2 = (200)(0.30\,\text{A})\pi(0.050\,\text{m})^2 = 0.47\,\text{A}\cdot\text{m}^2\,.$$

**59**

(a) The magnitude of the dipole moment is given by $\mu = NiA$, where $N$ is the number of turns, $i$ is the current, and $A$ is the area. Use $A = \pi R^2$, where $R$ is the radius. Thus

$$\mu = Ni\pi R^2 = (300)(4.0\,\text{A})\pi(0.025\,\text{m})^2 = 2.4\,\text{A}\cdot\text{m}^2\,.$$

(b) The magnetic field on the axis of a magnetic dipole, a distance $z$ away, is given by Eq. 29–27:

$$B = \frac{\mu_0}{2\pi}\frac{\mu}{z^3}\,.$$

Solve for $z$:

$$z = \left[\frac{\mu_0}{2\pi}\frac{\mu}{B}\right]^{1/3} = \left[\frac{4\pi\times10^{-7}\,\text{T}\cdot\text{m/A}}{2\pi}\frac{2.36\,\text{A}\cdot\text{m}^2}{5.0\times10^{-6}\,\text{T}}\right]^{1/3} = 46\,\text{cm}\,.$$

**71**

Use the Biot-Savart law in the form

$$\vec{B} = \frac{\mu_0}{4\pi}\frac{i\Delta\vec{s}\times\vec{r}}{r^3}\,.$$

Take $\Delta\vec{s}$ to be $\Delta s\,\hat{\text{j}}$, and $\vec{r}$ to be $x\,\hat{\text{i}}+y\,\hat{\text{j}}+z\,\hat{\text{k}}$. Then $\Delta\vec{s}\times\vec{r} = \Delta s\,\hat{\text{j}}\times(x\,\hat{\text{i}}+y\,\hat{\text{j}}+z\,\hat{\text{k}}) = \Delta s(z\,\hat{\text{i}}-x\,\hat{\text{k}})$, where $\hat{\text{j}}\times\hat{\text{i}} = -\hat{\text{k}}$, $\hat{\text{j}}\times\hat{\text{j}} = 0$, and $\hat{\text{j}}\times\hat{\text{k}} = \hat{\text{i}}$ were used. In addition, $r = \sqrt{x^2+y^2+z^2}$. The Biot-Savart equation becomes

$$\vec{B} = \frac{\mu_0}{4\pi}\frac{i\,\Delta s(z\,\hat{\text{i}} - z\,\hat{\text{k}})}{(x^2+y^2+z^2)^{3/2}}\,.$$

(a) For $x = 0$, $y = 0$, and $z = 5.0\,\text{m}$,

$$\vec{B} = \frac{4\pi\times10^{-7}\,\text{T}\cdot\text{m/A}}{4\pi}\frac{(2.0\,\text{A})(0.030\,\text{m})(5.0\,\text{m})\,\hat{\text{i}}}{(5.0\,\text{m})^3} = (2.4\times10^{-10}\,\text{T})\,\hat{\text{i}}\,.$$

(b) For $x = 0$, $y = 6.0\,\text{m}$, and $z = 0$, $\vec{B} = 0$.

(c) For $x = 7.0\,\text{m}$, $y = 7.0\,\text{m}$, and $z = 0$,

$$\vec{B} = \frac{4\pi\times10^{-7}\,\text{T}\cdot\text{m/A}}{4\pi}\frac{(2.0\,\text{A})(0.030\,\text{m})(-7.0\,\text{m})\,\hat{\text{k}}}{[(7.0\,\text{m})^2+(7.0\,\text{m})^2]^{3/2}} = (4.3\times10^{-11}\,\text{T})\,\hat{\text{k}}\,.$$

(d) For $x = -3.0\,\text{m}$, $y = -4.0\,\text{m}$, and $z = 0$,

$$\vec{B} = \frac{4\pi \times 10^{-7}\,\text{T}\cdot\text{m/A}}{4\pi} \frac{(2.0\,\text{A})(0.030\,\text{m})(3.0\,\text{m})\,\hat{\text{k}}}{[(-3.0\,\text{m})^2 + (-4.0\,\text{m})^2]^{3/2}} = (1.4 \times 10^{-10}\,\text{T})\,\hat{\text{k}}\,.$$

77

First consider the finite wire segment shown on the right. It extends from $y = -d$ to $y = a - d$, where $a$ is the length of the segment, and it carries current $i$ in the positive $y$ direction. Let $dy$ be an infinitesimal length of wire at coordinate $y$. According to the Biot-Savart law the magnitude of the magnetic field at P due to this infinitesimal length is $dB = (\mu_0/4\pi)(i \sin\theta/r^2)\,dy$. Now $r^2 = y^2 + R^2$ and $\sin\theta = R/r = R/\sqrt{y^2 + R^2}$, so

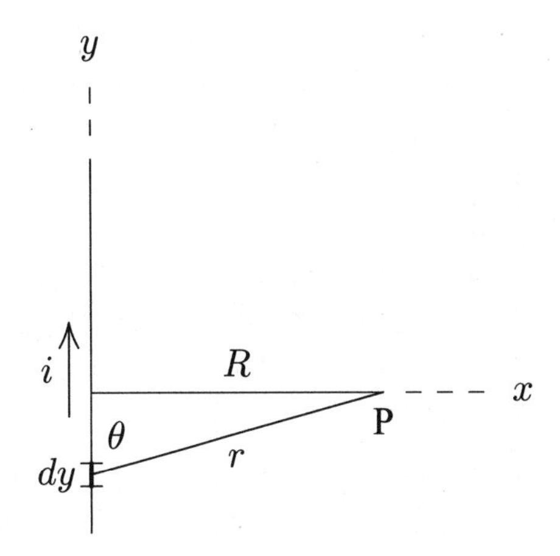


$$dB = \frac{\mu_0}{4\pi}\frac{iR}{(y^2 + R^2)^{3/2}}dy$$

and the field of the entire segment is

$$B = \frac{\mu_0}{4\pi}iR\int_{-d}^{a-d}\frac{y}{(y^2 + R^2)^{3/2}}\,dy = \mu_0/4\pi\,\frac{i}{R}\left[\frac{a-d}{\sqrt{R^2 + (a-d)^2}} + \frac{d}{\sqrt{R^2 + d^2}}\right],$$

where integral 19 of Appendix E was used.

All four sides of the square produce magnetic fields that are into the page at P, so we sum their magnitudes. To calculate the field of the left side of the square put $d = 3a/4$ and $R = a/4$. The result is

$$B_{\text{left}} = \frac{\mu_0}{4\pi}\frac{4i}{a}\left[\frac{1}{\sqrt{2}} + \frac{3}{\sqrt{10}}\right] = \frac{\mu_O}{3\pi}\frac{4i}{a}(1.66)\,.$$

The field of the upper side of the square is the same. To calculate the field of the right side of the square put $d = a/4$ and $R = 3a/4$. The result is

$$B_{\text{right}} = \frac{\mu_0}{4\pi}\frac{4i}{3a}\left[\frac{3}{\sqrt{18}} + \frac{1}{\sqrt{10}}\right] = \frac{\mu_O}{3\pi}\frac{4i}{a}(0.341)\,.$$

The field of the bottom side is the same. The total field at P is

$$\begin{aligned} B = B_{\text{left}} + B_{\text{upper}} + B_{\text{right}} + B_{\text{lower}} &= \frac{\mu_0}{4\pi}\frac{4i}{a}(1.66 + 1.66 + 0.341 + 0.341) \\ &= \frac{4\pi \times 10^{-7}\,\text{T}\cdot\text{m/A}}{4\pi}\frac{4(10\,\text{A})}{0.080\,\text{m}}(4.00) = 2.0 \times 10^{-4}\,\text{T}\,. \end{aligned}$$

## 79

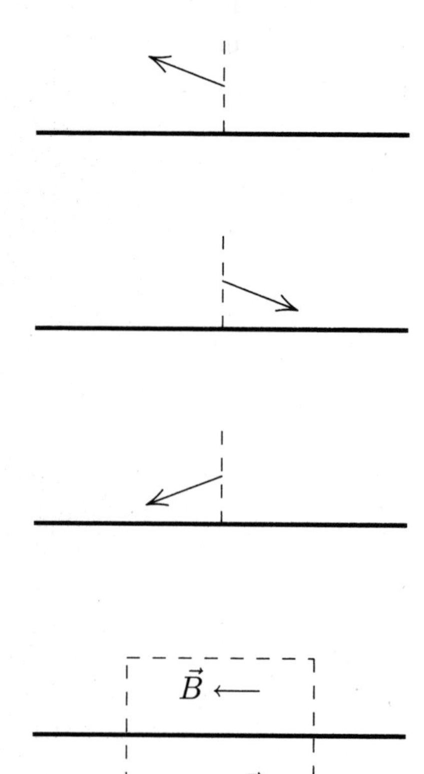

(a) Suppose the field is not parallel to the sheet, as shown in the upper diagram. Reverse the direction of the current. According to the Biot-Savart law, the field reverses, so it will be as in the second diagram. Now rotate the sheet by 180° about a line that is perpendicular to the sheet. The field, of course, will rotate with it and end up in the direction shown in the third diagram. The current distribution is now exactly as it was originally, so the field must also be as it was originally. But it is not. Only if the field is parallel to the sheet will be final direction of the field be the same as the original direction. If the current is out of the page, any infinitesimal portion of the sheet in the form of a long straight wire produces a field that is to the left above the sheet and to the right below the sheet. The field must be as drawn in Fig. 29–85.

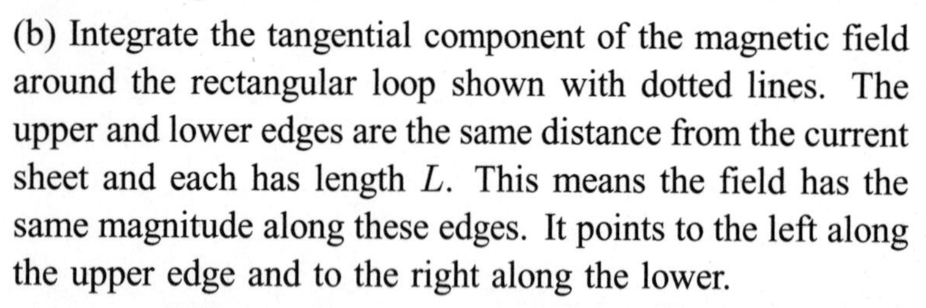

(b) Integrate the tangential component of the magnetic field around the rectangular loop shown with dotted lines. The upper and lower edges are the same distance from the current sheet and each has length $L$. This means the field has the same magnitude along these edges. It points to the left along the upper edge and to the right along the lower.

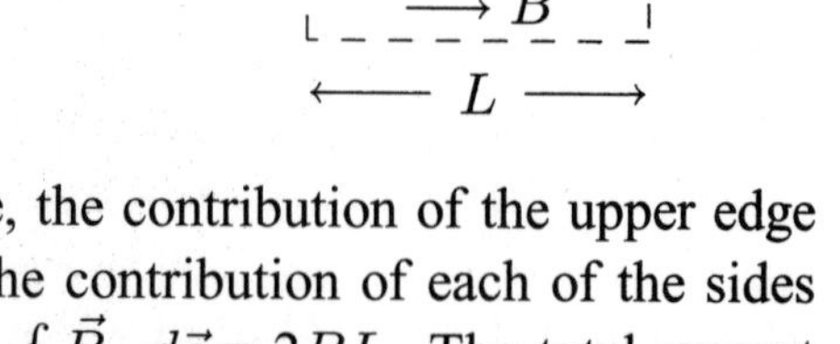


If the integration is carried out in the counterclockwise sense, the contribution of the upper edge is $BL$, the contribution of the lower edge is also $BL$, and the contribution of each of the sides is zero because the field is perpendicular to the sides. Thus $\oint \vec{B} \cdot d\vec{s} = 2BL$. The total current through the loop is $\lambda L$. Ampere's law yields $2BL = \mu_0 \lambda L$, so $B = \mu_0 \lambda / 2$.

## 81

(a) Use a circular Amperian path that has radius $r$ and is concentric with the cylindrical shell as shown by the dotted circle on Fig. 29–86. The magnetic field is tangent to the path and has uniform magnitude on it, so the integral on the left side of the Ampere's law equation is $\oint \vec{B} \cdot d\vec{s} = 2\pi r B$. The current through the Amperian path is the current through the region outside the circle of radius $b$ and inside the circle of radius $r$. Since the current is uniformly distributed through a cross section of the shell, the enclosed current is $i(r^2 - b^2)/(a^2 - b^2)$. Thus

$$2\pi r B = \frac{r^2 - b^2}{a^2 - b^2}\, i$$

and

$$B = \frac{\mu_0 i}{2\pi(a^2 - b^2)} \frac{r^2 - b^2}{r} .$$

(b) When $r = a$ this expression reduce to $B = \mu_0 i / 2\pi r$, which is the correct expression for the field of a long straight wire. When $r = b$ it reduces to $B = 0$, which is correct since there is no field inside the shell. When $b = 0$ it reduces to $B = \mu_0 i r / 2\pi a^2$, which is correct for the field inside a cylindrical conductor.

(c) The graph is shown below.

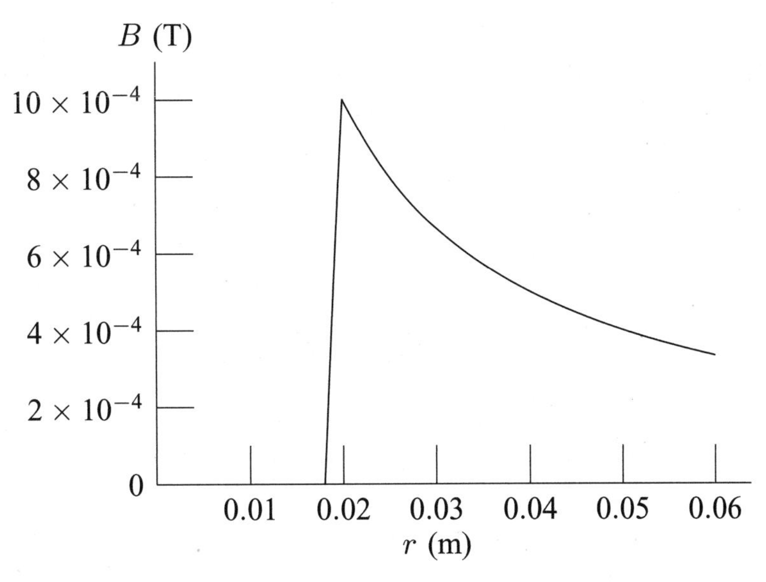


**89**

The result of Problem 11 is used four times, once for each of the sides of the square loop. A point on the axis of the loop is also on a perpendicular bisector of each of the loop sides. The diagram shows the field due to one of the loop sides, the one on the left. In the expression found in Problem 11, replace $L$ with $a$ and $R$ with $\sqrt{x^2 + a^2/4} = \frac{1}{2}\sqrt{4x^2 + a^2}$. The field due to the side is therefore

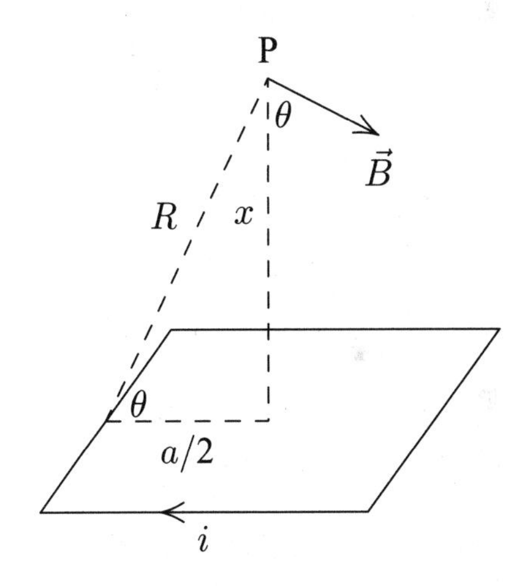


$$B = \frac{\mu_0 i a}{\pi \sqrt{4x^2 + a^2}\sqrt{4x^2 + 2a^2}}.$$

The field is in the plane of the dotted triangle shown and is perpendicular to the line from the midpoint of the loop side to the point P. Therefore it makes the angle $\theta$ with the vertical.

When the fields of the four sides are summed vectorially the horizontal components add to zero. The vertical components are all the same, so the total field is given by

$$B_{\text{total}} = 4B\cos\theta = \frac{4Ba}{2R} = \frac{4Ba}{\sqrt{4x^2 + a^2}}.$$

Thus

$$B_{\text{total}} = \frac{4\mu_0 i a^2}{\pi(4x^2 + a^2)\sqrt{4x^2 + 2a^2}}.$$

For $x = 0$, the expression reduces to

$$B_{\text{total}} = \frac{4\mu_0 i a^2}{\pi a^2 \sqrt{2} a} = \frac{2\sqrt{2}\mu_0 i}{\pi a},$$

in agreement with the result of Problem 12.

**91**

Use Ampere's law: $\oint \vec{B} \cdot d\vec{s} = \mu_0 i_{\text{enc}}$, where the integral is around a closed loop and $i_{\text{enc}}$ is the net current through the loop. For the dashed loop shown on the diagram $i = 0$. Assume the integral $\int \vec{B} \cdot d\vec{s}$ is zero along the bottom, right, and top sides of the loop as it would be if the field lines are as shown on the diagram. Along the right side the field is zero and along the top and bottom sides the field is perpendicular to $d\vec{s}$. If $\ell$ is the length of the left edge, then direct integration yields $\oint \vec{B} \cdot d\vec{s} = B\ell$, where $B$ is the magnitude of the field at the left side of the loop. Since neither $B$ nor $\ell$ is zero, Ampere's law is contradicted. We conclude that the geometry shown for the magnetic field lines is in error. The lines actually bulge outward and their density decreases gradually, not precipitously as shown.

# Chapter 30

## 5

The magnitude of the magnetic field inside the solenoid is $B = \mu_0 n i_s$, where $n$ is the number of turns per unit length and $i_s$ is the current. The field is parallel to the solenoid axis, so the flux through a cross section of the solenoid is $\Phi_B = A_s B = \mu_0 \pi r_s^2 n i_s$, where $A_s$ $(= \pi r_s^2)$ is the cross-sectional area of the solenoid. Since the magnetic field is zero outside the solenoid, this is also the flux through the coil. The emf in the coil has magnitude

$$\mathcal{E} = \frac{N d\Phi_B}{dt} = \mu_0 \pi r_s^2 N n \frac{di_s}{dt}$$

and the current in the coil is

$$i_c = \frac{\mathcal{E}}{R} = \frac{\mu_0 \pi r_s^2 N n}{R} \frac{di_s}{dt},$$

where $N$ is the number of turns in the coil and $R$ is the resistance of the coil. The current changes linearly by 3.0 A in 50 ms, so $di_s/dt = (3.0\,\text{A})/(50 \times 10^{-3}\,\text{s}) = 60\,\text{A/s}$. Thus

$$i_c = \frac{(4\pi \times 10^{-7}\,\text{T}\cdot\text{m/A})\pi(0.016\,\text{m})^2(120)(220 \times 10^2\,\text{m}^{-1})}{5.3\,\Omega}(60\,\text{A/s}) = 3.0 \times 10^{-2}\,\text{A}.$$

## 21

(a) In the region of the smaller loop, the magnetic field produced by the larger loop may be taken to be uniform and equal to its value at the center of the smaller loop, on the axis. Eq. 29–26, with $z = x$ and much greater than $R$, gives

$$B = \frac{\mu_0 i R^2}{2x^3}$$

for the magnitude. The field is upward in the diagram. The magnetic flux through the smaller loop is the product of this field and the area $(\pi r^2)$ of the smaller loop:

$$\Phi_B = \frac{\pi \mu_0 i r^2 R^2}{2x^3}.$$

(b) The emf is given by Faraday's law:

$$\mathcal{E} = -\frac{d\Phi_B}{dt} = -\left(\frac{\pi \mu_0 i r^2 R^2}{2}\right)\frac{d}{dt}\left(\frac{1}{x^3}\right) = -\left(\frac{\pi \mu_0 i r^2 R^2}{2}\right)\left(-\frac{3}{x^4}\frac{dx}{dt}\right) = \frac{3\pi \mu_0 i r^2 R^2 v}{2x^4}.$$

(c) The field of the larger loop is upward and decreases with distance away from the loop. As the smaller loop moves away, the flux through it decreases. The induced current is directed so as to produce a magnetic field that is upward through the smaller loop, in the same direction as the field of the larger loop. It is counterclockwise as viewed from above, in the same direction as the current in the larger loop.

<u>29</u>

Thermal energy is generated at the rate $\mathcal{E}^2/R$, where $\mathcal{E}$ is the emf in the wire and $R$ is the resistance of the wire. The resistance is given by $R = \rho L/A$, where $\rho$ is the resistivity of copper, $L$ is the length of the wire, and $A$ is the cross-sectional area of the wire. The resistivity can be found in Table 26–1. Thus

$$R = \frac{\rho L}{A} = \frac{(1.69 \times 10^{-8}\,\Omega \cdot \text{m})(0.500\,\text{m})}{\pi(0.500 \times 10^{-3}\,\text{m})^2} = 1.076 \times 10^{-2}\,\Omega\,.$$

Faraday's law is used to find the emf. If $B$ is the magnitude of the magnetic field through the loop, then $\mathcal{E} = A\,dB/dt$, where $A$ is the area of the loop. The radius $r$ of the loop is $r = L/2\pi$ and its area is $\pi r^2 = \pi L^2/4\pi^2 = L^2/4\pi$. Thus

$$\mathcal{E} = \frac{L^2}{4\pi}\frac{dB}{dt} = \frac{(0.500\,\text{m})^2}{4\pi}(10.0 \times 10^{-3}\,\text{T/s}) = 1.989 \times 10^{-4}\,\text{V}\,.$$

The rate of thermal energy generation is

$$P = \frac{\mathcal{E}^2}{R} = \frac{(1.989 \times 10^{-4}\,\text{V})^2}{1.076 \times 10^{-2}\,\Omega} = 3.68 \times 10^{-6}\,\text{W}\,.$$

<u>37</u>

(a) The field point is inside the solenoid, so Eq. 30–25 applies. The magnitude of the induced electric field is

$$E = \frac{1}{2}\frac{dB}{dt}r = \frac{1}{2}(6.5 \times 10^{-3}\,\text{T/s})(0.0220\,\text{m}) = 7.15 \times 10^{-5}\,\text{V/m}\,.$$

(b) Now the field point is outside the solenoid and Eq. 30–27 applies. The magnitude of the induced field is

$$E = \frac{1}{2}\frac{dB}{dt}\frac{R^2}{r} = \frac{1}{2}(6.5 \times 10^{-3}\,\text{T/s})\frac{(0.0600\,\text{m})^2}{(0.0820\,\text{m})} = 1.43 \times 10^{-4}\,\text{V/m}\,.$$

<u>51</u>

Starting with zero current when the switch is closed, at time $t = 0$, the current in an $RL$ series circuit at a later time $t$ is given by

$$i = \frac{\mathcal{E}}{R}\left(1 - e^{-t/\tau_L}\right)\,,$$

where $\tau_L$ is the inductive time constant, $\mathcal{E}$ is the emf, and $R$ is the resistance. You want to calculate the time $t$ for which $i = 0.9990\mathcal{E}/R$. This means

$$0.9990\frac{\mathcal{E}}{R} = \frac{\mathcal{E}}{R}\left(1 - e^{-t/\tau_L}\right)\,,$$

so

$$0.9990 = 1 - e^{-t/\tau_L}$$

or

$$e^{-t/\tau_L} = 0.0010 .$$

Take the natural logarithm of both sides to obtain $-(t/\tau_L) = \ln(0.0010) = -6.91$. That is, 6.91 inductive time constants must elapse.

**55**

(a) If the battery is switched into the circuit at time $t = 0$, then the current at a later time $t$ is given by

$$i = \frac{\mathcal{E}}{R}\left(1 - e^{-t/\tau_L}\right) ,$$

where $\tau_L = L/R$. You want to find the time for which $i = 0.800\mathcal{E}/R$. This means

$$0.800 = 1 - e^{-t/\tau_L}$$

or

$$e^{-t/\tau_L} = 0.200 .$$

Take the natural logarithm of both sides to obtain $-(t/\tau_L) = \ln(0.200) = -1.609$. Thus

$$t = 1.609\tau_L = \frac{1.609L}{R} = \frac{1.609(6.30 \times 10^{-6}\,\text{H})}{1.20 \times 10^{3}\,\Omega} = 8.45 \times 10^{-9}\,\text{s} .$$

(b) At $t = 1.0\tau_L$ the current in the circuit is

$$i = \frac{\mathcal{E}}{R}\left(1 - e^{-1.0}\right) = \left(\frac{14.0\,\text{V}}{1.20 \times 10^{3}\,\Omega}\right)\left(1 - e^{-1.0}\right) = 7.37 \times 10^{-3}\,\text{A} .$$

**59**

(a) Assume $i$ is from left to right through the closed switch. Let $i_1$ be the current in the resistor and take it to be downward. Let $i_2$ be the current in the inductor and also take it to be downward. The junction rule gives $i = i_1 + i_2$ and the loop rule gives $i_1 R - L(di_2/dt) = 0$. Since $di/dt = 0$, the junction rule yields $(di_1/dt) = -(di_2/dt)$. Substitute into the loop equation to obtain

$$L\frac{di_1}{dt} + i_1 R = 0 .$$

This equation is similar to Eq. 30—44, and its solution is the function given as Eq. 30—45:

$$i_1 = i_0 e^{-Rt/L} ,$$

where $i_0$ is the current through the resistor at $t = 0$, just after the switch is closed. Now, just after the switch is closed, the inductor prevents the rapid build-up of current in its branch, so at that time, $i_2 = 0$ and $i_1 = i$. Thus $i_0 = i$, so

$$i_1 = ie^{-Rt/L}$$

and

$$i_2 = i - i_1 = i\left[1 - e^{-Rt/L}\right].$$

(b) When $i_2 = i_1$,

$$e^{-Rt/L} = 1 - e^{-Rt/L},$$

so

$$e^{-Rt/L} = \frac{1}{2}.$$

Take the natural logarithm of both sides and use $\ln(1/2) = -\ln 2$ to obtain $(Rt/L) = \ln 2$ or

$$t = \frac{L}{R}\ln 2.$$

**63**

(a) If the battery is applied at time $t = 0$, the current is given by

$$i = \frac{\mathcal{E}}{R}\left(1 - e^{-t/\tau_L}\right),$$

where $\mathcal{E}$ is the emf of the battery, $R$ is the resistance, and $\tau_L$ is the inductive time constant. In terms of $R$ and the inductance $L$, $\tau_L = L/R$. Solve the current equation for the time constant. First obtain

$$e^{-t/\tau_L} = 1 - \frac{iR}{\mathcal{E}},$$

then take the natural logarithm of both sides to obtain

$$-\frac{t}{\tau_L} = \ln\left[1 - \frac{iR}{\mathcal{E}}\right].$$

Since

$$\ln\left[1 - \frac{iR}{\mathcal{E}}\right] = \ln\left[1 - \frac{(2.00 \times 10^{-3}\,\text{A})(10.0 \times 10^{3}\,\Omega)}{50.0\,\text{V}}\right] = -0.5108,$$

the inductive time constant is $\tau_L = t/0.5108 = (5.00 \times 10^{-3}\,\text{s})/(0.5108) = 9.79 \times 10^{-3}\,\text{s}$ and the inductance is

$$L = \tau_L R = (9.79 \times 10^{-3}\,\text{s})(10.0 \times 10^{3}\,\Omega) = 97.9\,\text{H}.$$

(b) The energy stored in the coil is

$$U_B = \frac{1}{2}Li^2 = \frac{1}{2}(97.9\,\text{H})(2.00 \times 10^{-3}\,\text{A})^2 = 1.96 \times 10^{-4}\,\text{J}.$$

**69**

(a) At any point, the magnetic energy density is given by $u_B = B^2/2\mu_0$, where $B$ is the magnitude of the magnetic field at that point. Inside a solenoid, $B = \mu_0 n i$, where $n$ is the number of turns per unit length and $i$ is the current. For the solenoid of this problem, $n = (950)/(0.850\,\text{m}) = 1.118 \times 10^3\,\text{m}^{-1}$. The magnetic energy density is

$$u_B = \frac{1}{2}\mu_0 n^2 i^2 = \frac{1}{2}(4\pi \times 10^{-7}\,\text{T}\cdot\text{m/A})(1.118 \times 10^3\,\text{m}^{-1})^2(6.60\,\text{A})^2 = 34.2\,\text{J/m}^3 .$$

(b) Since the magnetic field is uniform inside an ideal solenoid, the total energy stored in the field is $U_B = u_B V$, where $V$ is the volume of the solenoid. $V$ is calculated as the product of the cross-sectional area and the length. Thus

$$U_B = (34.2\,\text{J/m}^3)(17.0 \times 10^{-4}\,\text{m}^2)(0.850\,\text{m}) = 4.94 \times 10^{-2}\,\text{J} .$$

**73**

(a) The mutual inductance $M$ is given by

$$\mathcal{E}_1 = M\,\frac{di_2}{dt} ,$$

where $\mathcal{E}_1$ is the emf in coil 1 due to the changing current $i_2$ in coil 2. Thus

$$M = \frac{\mathcal{E}_1}{di_2/dt} = \frac{25.0 \times 10^{-3}\,\text{V}}{15.0\,\text{A/s}} = 1.67 \times 10^{-3}\,\text{H} .$$

(b) The flux linkage in coil 2 is

$$N_2\Phi_{21} = Mi_1 = (1.67 \times 10^{-3}\,\text{H})(3.60\,\text{A}) = 6.01 \times 10^{-3}\,\text{Wb} .$$

**75**

(a) Assume the current is changing at the rate $di/dt$ and calculate the total emf across both coils. First consider the left-hand coil. The magnetic field due to the current in that coil points to the left. So does the magnetic field due to the current in coil 2. When the current increases, both fields increase and both changes in flux contribute emf's in the same direction. Thus the emf in coil 1 is

$$\mathcal{E}_1 = -(L_1 + M)\,\frac{di}{dt} .$$

The magnetic field in coil 2 due to the current in that coil points to the left, as does the field in coil 2 due to the current in coil 1. The two sources of emf are again in the same direction and the emf in coil 2 is

$$\mathcal{E}_2 = -(L_2 + M)\,\frac{di}{dt} .$$

The total emf across both coils is

$$\mathcal{E} = \mathcal{E}_1 + \mathcal{E}_2 = -(L_1 + L_2 + 2M)\,\frac{di}{dt} .$$

This is exactly the emf that would be produced if the coils were replaced by a single coil with inductance $L_{eq} = L_1 + L_2 + 2M$.

(b) Reverse the leads of coil 2 so the current enters at the back of the coil rather than the front as pictured in the diagram. Then the field produced by coil 2 at the site of coil 1 is opposite the field produced by coil 1 itself. The fluxes have opposite signs. An increasing current in coil 1 tends to increase the flux in that coil but an increasing current in coil 2 tends to decrease it. The emf across coil 1 is

$$\mathcal{E}_1 = -(L_1 - M)\frac{di}{dt}.$$

Similarly the emf across coil 2 is

$$\mathcal{E}_2 = -(L_2 - M)\frac{di}{dt}.$$

The total emf across both coils is

$$\mathcal{E} = -(L_1 + L_2 - 2M)\frac{di}{dt}.$$

This the same as the emf that would be produced by a single coil with inductance $L_{eq} = L_1 + L_2 - 2M$.

**79**

(a) The electric field lines are circles that are concentric with the cylindrical region and the magnitude of the field is uniform around any circle. Thus the emf around a circle of radius $r$ is $\mathcal{E} = \oint \vec{E} \cdot d\vec{s} = 2\pi r E$. Here $r$ is inside the cylindrical region so the magnetic flux is $\pi r^2 B$. According to Faraday's law $2\pi r E = -\pi r^2(dB/dt)$ and

$$E = -\tfrac{1}{2} r \frac{dB}{dt} = -\tfrac{1}{2}(0.050\,\text{m})(-10 \times 10^{-3}\,\text{T/s}) = 2.5 \times 10^{-4}\,\text{V/m}.$$

Since the normal used to compute the flux was taken to be into the page, in the direction of the magnetic field, the positive direction for the electric is clockwise. The calculated value of $E$ is positive, so the electric field at point a is toward the left and $\vec{E} = -(2.5 \times 10^{-4}\,\text{V/m})\,\hat{\text{i}}$.
The force on the electron is $\vec{F} = -e\vec{E}$ and, according to Newton's second law, its acceleration is

$$\vec{a} = \frac{\vec{F}}{m} = -\frac{e\vec{E}}{m} = -\frac{(1.60 \times 10^{-19}\,\text{C})(-2.5 \times 10^{-4}\,\text{V/m})\,\hat{\text{i}}}{9.11 \times 10^{-31}\,\text{kg}} = (4.4 \times 10^{7}\,\text{m/s}^2)\,\hat{\text{i}}.$$

The mass and charge of an electron can be found in Appendix B.

(b) The electric field at $r = 0$ is zero, so the force and acceleration of an election placed at point b are zero.

(c) The electric field at point c has the same magnitude as the field at point a but now the field is to the right. That is $\vec{E} = (2.5 \times 10^{-4}\,\text{V/m})\,\hat{\text{i}}$ and $\vec{a} = -(4.4 \times 10^{7}\,\text{m/s}^2)\,\hat{\text{i}}$.

**81**

(a) The magnetic flux through the loop is $\Phi_B = BA$, where $B$ is the magnitude of the magnetic field and $A$ is the area of the loop. The magnitude of the average emf is given by Faraday's law : $\mathcal{E}_{avg} = B\Delta A/\Delta t$, where $\Delta A$ is the change in the area in time $\Delta t$. Since the final area is zero, the change in area is the initial area and $\mathcal{E}_{avg} = BA/\Delta t = (2.0\,\text{T})(0.20\,\text{m})^2/(0.20\,\text{s}) = 0.40\,\text{V}$.

(b) The average current in the loop is the emf divided by the resistance of the loop: $i_{\text{avg}} = \mathcal{E}_{\text{avg}}/R = (0.40\,\text{V})/(20 \times 10^{-3}\,\Omega) = 20\,\text{A}$.

**85**

(a), (b), (c), (d), and (e) Just after the switch is closed the current $i_2$ through the inductor is zero. The loop rule applied to the left loop gives $\mathcal{E} - I_1 R_1 = 0$, so $i_1 = \mathcal{E}/R_1 = (10\,\text{V})/(5.0\,\Omega) = 2.0\,\text{A}$. The junction rule gives $i_s = i_1 = 2.0\,\text{A}$. Since $i_2 = 0$, the potential difference across $R_2$ is $V_2 = i_2 R_2 = 0$. The potential differences across the inductor and resistor must sum to $\mathcal{E}$ and, since $V_2 = 0$, $V_L = \mathcal{E} = 10\,\text{V}$. The rate of change of $i_2$ is $di_2/dt = V_L/L = (10\,\text{V})/(5.0\,\text{H}) = 2.0\,\text{A/s}$.

(g), (h), (i), (j), (k), and (l) After the switch has been closed for a long time the current $i_2$ reaches a constant value. Since its derivative is zero the potential difference across the inductor is $V_L = 0$. The potential differences across both $R_1$ and $R_2$ are equal to the emf of the battery, so $i_1 = \mathcal{E}/R_1 = (10\,\text{V})/(5.0\,\Omega) = 2.0\,\text{A}$ and $i_2 = \mathcal{E}/R_2 = (10\,\text{V})/(10\,\Omega) = 1.0\,\text{A}$. The junction rule gives $i_s = i_1 + i_2 = 3.0\,\text{A}$.

**95**

(a) Because the inductor is in series with the battery the current in the circuit builds slowly and just after the switch is closed it is zero.

(b) Since all currents are zero just after the switch is closed the emf of the inductor must match the emf of the battery in magnitude. Thus $L(di_{\text{bat}}/dt) = \mathcal{E}$ and $di_{\text{bat}} = \mathcal{E}/L = (40\,\text{V})/(50 \times 10^{-3}\,\text{H}) = 8.0 \times 10^2\,\text{A/s}$.

(c) Replace the two resistors in parallel with their equivalent resistor. The equivalent resistance is

$$R_{\text{eq}} = \frac{R_1 R_2}{R_1 + R_2} = \frac{(20\,\text{k}\Omega)(20\,\text{k}\Omega)}{20\,\text{k}\Omega + 20\,\text{k}\Omega} = 10\,\text{k}\Omega\,.$$

The current as a function of time is given by

$$i_{\text{bat}} = \frac{\mathcal{E}}{R_{\text{eq}}}\left[1 - e^{-t/\tau_L}\right],$$

where $\tau_L$ is the inductive time constant. Its value is $\tau_L = L/R_{\text{eq}} = (50 \times 10^{-3}\,\text{H})/(10 \times 10^3\,\Omega) = 5.0 \times 10^{-6}\,\text{s}$. At $t = 3.0 \times 10^{-6}\,\text{s}$, $t/\tau_L = (3.0)/(5.0) = 0.60$ and

$$i_{\text{bat}} = \frac{40\,\text{V}}{10 \times 10^3\,\Omega}\left[1 - e^{-0.60}\right] = 1.8 \times 10^{-3}\,\text{A}\,.$$

(d) Differentiate the expression for $i_{\text{bat}}$ to obtain

$$\frac{di_{\text{bat}}}{dt} = \frac{\mathcal{E}}{R_{\text{eq}}}\frac{1}{\tau_L}e^{-t/\tau_L} = \frac{\mathcal{E}}{L}e^{-t/\tau_L},$$

where $\tau_L = L/R_{\text{eq}}$ was used to obtain the last form. At $t = 3.0 \times 10^{-6}\,\text{s}$

$$\frac{di_{\text{bat}}}{dt} = \frac{40\,\text{V}}{50 \times 10^{-3}\,\text{H}}e^{-0.60} = 4.4 \times 10^2\,\text{A/s}\,.$$

(e) A long time after the switch is closed the currents are constant and the emf of the inductor is zero. The current in the battery is $i_{\text{bat}} = \mathcal{E}/R_{\text{eq}} = (40\,\text{V})/(10 \times 10^3\,\Omega) = 4.0 \times 10^{-3}\,\text{A}$.

(f) The currents are constant and $di_{\text{bat}}/dt = 0$.

**97**

(a) and (b) Take clockwise current to be positive and counterclockwise current to be negative. Then according to the right-hand rule we must take the normal to the loop to be into the page, so the flux is negative if the magnetic field is out of the page and positive if it is into the page. Assume the field in region 1 is out of the page. We will obtain a negative result for the field if the assumption is incorrect. Let $x$ be the distance that the front edge of the loop is into region 1. Then while the loop is entering this region flux is $-B_1Hx$ and, according to Faraday's law, the emf induced around the loop is $\mathcal{E} = B_1H(dx/dt) = B_1Hv$. The current in the loop is $i = \mathcal{E}/R = B_1Hv/R$, so

$$B_1 = \frac{iR}{Hv} = \frac{(3.0\times10^{-6}\,\text{A})(0.020\,\Omega)}{(0.0150\,\text{m})(0.40\,\text{m/s})} = 1.0\times10^{-5}\,\text{T}\,.$$

The field is positive and therefore out of the page.

(c) and (d) Assume that the field $B_2$ of region 2 is out of the page. Let $x$ now be the distance the front end of the loop is into region 2 as the loop enters that region. The flux is $-B_1H(D-x) - B_2Hx$, the emf is $\mathcal{E} = -B_1Hv + B_2Hv = (B_2 - B_1)Hv$, and the current is $i = (B_2 - B_1)Hv/R$. The field of region 2 is

$$B_2 = B_1 + \frac{iR}{Hv} = 1.0\times10^{-5}\,\text{T} + \frac{(-2.0\times10^{-6}\,\text{A}(0.020\,\Omega)}{(0.015\,\text{m})(0.40\,\text{m/s})} = 3.3\times10^{-6}\,\text{T}\,.$$

The field is positive, indicating that it is out of the page.

# Chapter 31

7

(a) The mass $m$ corresponds to the inductance, so $m = 1.25\,\text{kg}$.

(b) The spring constant $k$ corresponds to the reciprocal of the capacitance. Since the total energy is given by $U = Q^2/2C$, where $Q$ is the maximum charge on the capacitor and $C$ is the capacitance,

$$C = \frac{Q^2}{2U} = \frac{(175 \times 10^{-6}\,\text{C})^2}{2(5.70 \times 10^{-6}\,\text{J})} = 2.69 \times 10^{-3}\,\text{F}$$

and

$$k = \frac{1}{2.69 \times 10^{-3}\,\text{m/N}} = 372\,\text{N/m}\,.$$

(c) The maximum displacement $x_m$ corresponds to the maximum charge, so

$$x_m = 1.75 \times 10^{-4}\,\text{m}\,.$$

(d) The maximum speed $v_m$ corresponds to the maximum current. The maximum current is

$$I = Q\omega = \frac{Q}{\sqrt{LC}} = \frac{175 \times 10^{-6}\,\text{C}}{\sqrt{(1.25\,\text{H})(2.69 \times 10^{-3}\,\text{F})}} = 3.02 \times 10^{-3}\,\text{A}\,.$$

Thus $v_m = 3.02 \times 10^{-3}\,\text{m/s}$.

15

(a) Since the frequency of oscillation $f$ is related to the inductance $L$ and capacitance $C$ by $f = 1/2\pi\sqrt{LC}$, the smaller value of $C$ gives the larger value of $f$. Hence, $f_{\text{max}} = 1/2\pi\sqrt{LC_{\text{min}}}$, $f_{\text{min}} = 1/2\pi\sqrt{LC_{\text{max}}}$, and

$$\frac{f_{\text{max}}}{f_{\text{min}}} = \frac{\sqrt{C_{\text{max}}}}{\sqrt{C_{\text{min}}}} = \frac{\sqrt{365\,\text{pF}}}{\sqrt{10\,\text{pF}}} = 6.0\,.$$

(b) You want to choose the additional capacitance $C$ so the ratio of the frequencies is

$$r = \frac{1.60\,\text{MHz}}{0.54\,\text{MHz}} = 2.96\,.$$

Since the additional capacitor is in parallel with the tuning capacitor, its capacitance adds to that of the tuning capacitor. If $C$ is in picofarads, then

$$\frac{\sqrt{C + 365\,\text{pF}}}{\sqrt{C + 10\,\text{pF}}} = 2.96\,.$$

The solution for $C$ is

$$C = \frac{(365\,\text{pF}) - (2.96)^2(10\,\text{pF})}{(2.96)^2 - 1} = 36\,\text{pF}\,.$$

(c) Solve $f = 1/2\pi\sqrt{LC}$ for $L$. For the minimum frequency, $C = 365\,\text{pF} + 36\,\text{pF} = 401\,\text{pF}$ and $f = 0.54\,\text{MHz}$. Thus

$$L = \frac{1}{(2\pi)^2 C f^2} = \frac{1}{(2\pi)^2(401 \times 10^{-12}\,\text{F})(0.54 \times 10^6\,\text{Hz})^2} = 2.2 \times 10^{-4}\,\text{H}\,.$$

**27**

Let $t$ be a time at which the capacitor is fully charged in some cycle and let $q_{\max 1}$ be the charge on the capacitor then. The energy in the capacitor at that time is

$$U(t) = \frac{q_{\max 1}^2}{2C} = \frac{Q^2}{2C}\,e^{-Rt/L}\,,$$

where

$$q_{\max 1} = Q\,e^{-Rt/2L}$$

was used. Here $Q$ is the charge at $t = 0$. One cycle later, the maximum charge is

$$q_{\max 2} = Q\,e^{-R(t+T)/2L}$$

and the energy is

$$U(t+T) = \frac{q_{\max 2}^2}{2C} = \frac{Q^2}{2C}\,e^{-R(t+T)/L}\,,$$

where $T$ is the period of oscillation. The fractional loss in energy is

$$\frac{\Delta U}{U} = \frac{U(t) - U(t+T)}{U(t)} = \frac{e^{-Rt/L} - e^{-R(t+T)/L}}{e^{-Rt/L}} = 1 - e^{-RT/L}\,.$$

Assume that $RT/L$ is small compared to 1 (the resistance is small) and use the Maclaurin series to expand the exponential. The first two terms are:

$$e^{-RT/L} \approx 1 - \frac{RT}{L}\,.$$

Replace $T$ with $2\pi/\omega$, where $\omega$ is the angular frequency of oscillation. Thus

$$\frac{\Delta U}{U} \approx 1 - \left(1 - \frac{RT}{L}\right) = \frac{RT}{L} = \frac{2\pi R}{\omega L}\,.$$

**33**

(a) The generator emf is a maximum when $\sin(\omega_d t - \pi/4) = 1$ or $\omega_d t - \pi/4 = (\pi/2) \pm 2n\pi$, where $n$ is an integer, including zero. The first time this occurs after $t = 0$ is when $\omega_d t - \pi/4 = \pi/2$ or

$$t = \frac{3\pi}{4\omega_d} = \frac{3\pi}{4(350\,\text{s}^{-1})} = 6.73 \times 10^{-3}\,\text{s}\,.$$

(b) The current is a maximum when $\sin(\omega_d t - 3\pi/4) = 1$, or $\omega_d t - 3\pi/4 = \pi/2 \pm 2n\pi$. The first time this occurs after $t = 0$ is when

$$t = \frac{5\pi}{4\omega_d} = \frac{5\pi}{4(350\,\text{s}^{-1})} = 1.12 \times 10^{-2}\,\text{s}\,.$$

(c) The current lags the inductor by $\pi/2$ rad, so the circuit element must be an inductor.

(d) The current amplitude $I$ is related to the voltage amplitude $V_L$ by $V_L = IX_L$, where $X_L$ is the inductive reactance, given by $X_L = \omega_d L$. Furthermore, since there is only one element in the circuit, the amplitude of the potential difference across the element must be the same as the amplitude of the generator emf: $V_L = \mathcal{E}_m$. Thus $\mathcal{E}_m = I\omega_d L$ and

$$L = \frac{\mathcal{E}_m}{I\omega_d} = \frac{30.0\,\text{V}}{(620 \times 10^{-3}\,\text{A})(350\,\text{rad/s})} = 0.138\,\text{H}\,.$$

**39**

(a) The capacitive reactance is

$$X_C = \frac{1}{\omega_d C} = \frac{1}{2\pi f_d C} = \frac{1}{2\pi(60.0\,\text{Hz})(70.0 \times 10^{-6}\,\text{F})} = 37.9\,\Omega\,.$$

The inductive reactance is

$$X_L = \omega_d L = 1\pi f_d L = 2\pi(60.0\,\text{Hz})(230 \times 10^{-3}\,\text{H}) = 86.7\,\Omega\,.$$

The impedance is

$$Z = \sqrt{R^2 + (X_L - X_C)^2} = \sqrt{(200\,\Omega)^2 + (37.9\,\Omega - 86.7\,\Omega)^2} = 206\,\Omega\,.$$

(b) The phase angle is

$$\phi = \tan^{-1}\left(\frac{X_L - X_C}{R}\right) = \tan^{-1}\left(\frac{86.7\,\Omega - 37.9\,\Omega}{200\,\Omega}\right) = 13.7^\circ\,.$$

(c) The current amplitude is

$$I = \frac{\mathcal{E}_m}{Z} = \frac{36.0\,\text{V}}{206\,\Omega} = 0.175\,\text{A}\,.$$

(d) The voltage amplitudes are

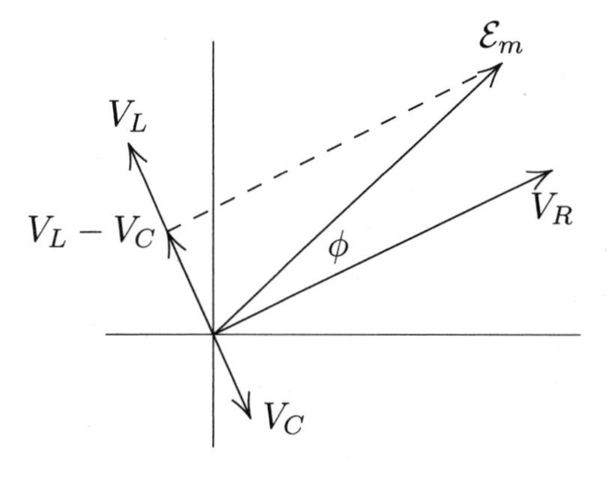


$$V_R = IR = (0.175\,\text{A})(200\,\Omega) = 35.0\,\text{V},$$

$$V_L = IX_L = (0.i75\,\text{A})(86.7\,\Omega) = 15.2\,\text{V},$$

and

$$V_C = IX_C = (0.175\,\text{A})(37.9\,\Omega) = 6,63\,\text{V}.$$

Note that $X_L > X_C$, so that $\mathcal{E}_m$ leads $I$. The phasor diagram is drawn to scale on the right.

**45**

(a) For a given amplitude $\mathcal{E}_m$ of the generator emf, the current amplitude is given by

$$I = \frac{\mathcal{E}_m}{Z} = \frac{\mathcal{E}_m}{\sqrt{R^2 + (\omega_d L - 1/\omega_d C)^2}}\,,$$

where $R$ is the resistance, $L$ is the inductance, $C$ is the capacitance, and $\omega_d$ is the angular frequency. To find the maximum, set the derivative with respect to $\omega_d$ equal to zero and solve for $\omega_d$. The derivative is

$$\frac{dI}{d\omega_d} = -\mathcal{E}_m\left[R^2 + (\omega_d L - 1/\omega_d C)^2\right]^{-3/2}\left[\omega_d L - \frac{1}{\omega_d C}\right]\left[L + \frac{1}{\omega_d^2 C}\right].$$

The only factor that can equal zero is $\omega_d L - (1/\omega_d C)$ and it does for $\omega_d = 1/\sqrt{LC}$. For the given circuit,

$$\omega_d = \frac{1}{\sqrt{LC}} = \frac{1}{\sqrt{(1.00\,\text{H})(20.0\times 10^{-6}\,\text{F})}} = 224\,\text{rad/s}.$$

(b) For this value of the angular frequency, the impedance is $Z = R$ and the current amplitude is

$$I = \frac{\mathcal{E}_m}{R} = \frac{30.0\,\text{V}}{5.00\,\Omega} = 6.00\,\text{A}.$$

(c) and (d) You want to find the values of $\omega_d$ for which $I = \mathcal{E}_m/2R$. This means

$$\frac{\mathcal{E}_m}{\sqrt{R^2 + (\omega_d L - 1/\omega_d C)^2}} = \frac{\mathcal{E}_m}{2R}.$$

Cancel the factors $\mathcal{E}_m$ that appear on both sides, square both sides, and set the reciprocals of the two sides equal to each other to obtain

$$R^2 + \left(\omega_d L - \frac{1}{\omega_d C}\right)^2 = 4R^2.$$

Thus

$$\left(\omega_d L - \frac{1}{\omega_d C}\right)^2 = 3R^2.$$

Now take the square root of both sides and multiply by $\omega_d C$ to obtain

$$\omega_d^2(LC) \pm \omega_d\left(\sqrt{3}CR\right) - 1 = 0,$$

where the symbol $\pm$ indicates the two possible signs for the square root. The last equation is a quadratic equation for $\omega_d$. Its solutions are

$$\omega_d = \frac{\pm\sqrt{3}CR \pm \sqrt{3C^2R^2 + 4LC}}{2LC}.$$

You want the two positive solutions. The smaller of these is

$$\begin{aligned}\omega_2 &= \frac{-\sqrt{3}CR + \sqrt{3C^2R^2 + 4LC}}{2LC}\\ &= \frac{-\sqrt{3}(20.0\times 10^{-6}\,\text{F})(5.00\,\Omega)}{2(1.00\,\text{H})(20.0\times 10^{-6}\,\text{F})}\\ &\quad + \frac{\sqrt{3(20.0\times 10^{-6}\,\text{F})^2(5.00\,\Omega)^2 + 4(1.00\,\text{H})(20.0\times 10^{-6}\,\text{F})}}{2(1.00\,\text{H})(20.0\times 10^{-6}\,\text{F})}\\ &= 219\,\text{rad/s}\end{aligned}$$

and the larger is

$$\omega_1 = \frac{+\sqrt{3}CR + \sqrt{3C^2R^2 + 4LC}}{2LC}$$
$$= \frac{+\sqrt{3}(20.0 \times 10^{-6}\,\text{F})(5.00\,\Omega)}{2(1.00\,\text{H})(20.0 \times 10^{-6}\,\text{F})}$$
$$+ \frac{\sqrt{3(20.0 \times 10^{-6}\,\text{F})^2(5.00\,\Omega)^2 + 4(1.00\,\text{H})(20.0 \times 10^{-6}\,\text{F})}}{2(1.00\,\text{H})(20.0 \times 10^{-6}\,\text{F})}$$
$$= 228\,\text{rad/s}\,.$$

(e) The fractional width is

$$\frac{\omega_1 - \omega_2}{\omega_0} = \frac{228\,\text{rad/s} - 219\,\text{rad/s}}{224\,\text{rad/s}} = 0.04\,.$$

**49**

Use the expressions found in Problem 31–45:

$$\omega_1 = \frac{+\sqrt{3}CR + \sqrt{3C^2R^2 + 4LC}}{2LC}$$

and

$$\omega_2 = \frac{-\sqrt{3}CR + \sqrt{3C^2R^2 + 4LC}}{2LC}\,.$$

Also use

$$\omega = \frac{1}{\sqrt{LC}}\,.$$

Thus

$$\frac{\Delta\omega_d}{\omega} = \frac{\omega_1 - \omega_2}{\omega} = \frac{2\sqrt{3}CR\sqrt{LC}}{2LC} = R\sqrt{\frac{3C}{L}}\,.$$

**55**

(a) The impedance is given by

$$Z = \sqrt{R^2 + (X_L - X_C)^2}\,,$$

where $R$ is the resistance, $X_L$ is the inductive reactance, and $X_C$ is the capacitive reactance. Thus

$$Z = \sqrt{(12.0\,\Omega)^2 + (1.30\,\Omega - 0)^2} = 12.1\,\Omega\,.$$

(b) The average rate at which energy is supplied to the air conditioner is given by

$$P_{\text{avg}} = \frac{\mathcal{E}_{\text{rms}}^2}{Z}\cos\phi\,,$$

where $\cos\phi$ is the power factor. Now

$$\cos\phi = \frac{R}{Z} = \frac{12\,\Omega}{12.1\,\Omega} = 0.992\,,$$

so

$$P_{\text{avg}} = \left[\frac{(120\text{ V})^2}{12.1\,\Omega}\right](0.992) = 1.18 \times 10^3\text{ W}\,.$$

<u>57</u>

(a) The power factor is $\cos\phi$, where $\phi$ is the phase angle when the current is written $i = I\sin(\omega_d t - \phi)$. Thus $\phi = -42.0°$ and $\cos\phi = \cos(-42.0°) = 0.743$.

(b) Since $\phi < 0$, $\omega_d t - \phi > \omega_d t$ and the current leads the emf.

(c) The phase angle is given by $\tan\phi = (X_L - X_C)/R$, where $X_L$ is the inductive reactance, $X_C$ is the capacitive reactance, and $R$ is the resistance. Now $\tan\phi = \tan(-42.0°) = -0.900$, a negative number. This means $X_L - X_C$ is negative, or $X_C > X_L$. The circuit in the box is predominantly capacitive.

(d) If the circuit is in resonance, $X_L$ is the same as $X_C$, $\tan\phi$ is zero, and $\phi$ would be zero. Since $\phi$ is not zero, we conclude the circuit is not in resonance.

(e), (f), and (g) Since $\tan\phi$ is negative and finite, neither the capacitive reactance nor the resistance is zero. This means the box must contain a capacitor and a resistor. The inductive reactance may be zero, so there need not be an inductor. If there is an inductor, its reactance must be less than that of the capacitor at the operating frequency.

(h) The average power is

$$P_{\text{avg}} = \frac{1}{2}\mathcal{E}_m I\cos\phi = \frac{1}{2}(75.0\text{ V})(1.20\text{ A})(0.743) = 33.4\text{ W}\,.$$

(i) The answers above depend on the frequency only through the phase angle $\phi$, which is given. If values are given for $R$, $L$, and $C$, then the value of the frequency would also be needed to compute the power factor.

<u>63</u>

(a) If $N_p$ is the number of primary turns and $N_s$ is the number of secondary turns, then

$$V_s = \frac{N_s}{N_p}V_p = \left(\frac{10}{500}\right)(120\text{ V}) = 2.4\text{ V}\,.$$

(b) and (c) The current in the secondary is given by Ohm's law:

$$I_s = \frac{V_s}{R_s} = \frac{2.4\text{ V}}{15\,\Omega} = 0.16\text{ A}\,.$$

The current in the primary is

$$I_p = \frac{N_s}{N_p}I_s = \left(\frac{10}{500}\right)(0.16\text{ A}) = 3.2 \times 10^{-3}\text{ A}\,.$$

**67**

Use the trigonometric identity, found in Appendix E,

$$\sin\alpha - \sin\beta = 2\sin\left(\frac{\alpha-\beta}{2}\right)\cos\left(\frac{\alpha+\beta}{2}\right),$$

where $\alpha$ and $\beta$ are any two angles. Thus

$$V_1 - V_2 = A\sin(\omega_d t) - A\sin(\omega_d t - 120°) = 2A\sin(120°)\cos(\omega_d t - 60°) = \sqrt{3}A\cos(\omega_d t - 60°),$$

where $\sin(120°) = \sqrt{3}/2$ was used. Similarly,

$$V_1 - V_3 = A\sin(\omega_d t) - A\sin(\omega_d t - 240°) = 2A\sin(240°)\cos(\omega_d t - 120°) = -\sqrt{3}A\cos(\omega_d t - 120°),$$

where $\sin(240°) = -\sqrt{3}/2$ was used, and

$$\begin{aligned} V_2 - V_3 &= A\sin(\omega_d t - 120°) - A\sin(\omega_d t - 240°) = 2A\sin(120°)\cos(\omega_d t - 180°) \\ &= \sqrt{3}A\cos(\omega_d t - 180°). \end{aligned}$$

All of these are sinusoidal functions of $\omega_d$ and all have amplitudes of $\sqrt{3}A$.

**71**

(a) Let $V_C$ be the maximum potential difference across the capacitor, $V_L$ be the maximum potential difference across the inductor, and $V_R$ be the maximum potential difference across the resistor. Then the phase constant $\phi$ is

$$\tan^{-1}\left(\frac{V_L - V_C}{V_R}\right) = \tan^{-1}\left(\frac{2.00V_R - V_R}{V_R}\right) = \tan^{-1}(1.00) = 45.0°.$$

(b) Since the maximum emf is related to the current amplitude by $\mathcal{E}_m = IZ$, where $Z$ is the impedance and $R = Z\cos\phi$,

$$R = \frac{\mathcal{E}_m\cos\phi}{I} = \frac{(30.0\text{ V})\cos 45°}{300\times10^{-3}\text{ A}} = 70.7\,\Omega.$$

**73**

(a) The frequency of oscillation of an LC circuit is $f = 1/2\pi\sqrt{LC}$, where $L$ is the inductance and $C$ is the capacitance. Thus

$$L = \frac{1}{4\pi^2 f^2 C} = \frac{1}{4\pi^2(10.4\times10^3\text{ Hz})^2(340\times10^{-6}\text{ F})} = 6.89\times10^{-7}\text{ H}.$$

(b) The total energy is $U = \frac{1}{2}LI^2$, where $I$ is the current amplitude. Thus $U = \frac{1}{2}(6.89\times10^{-7}\text{ H})(7.20\times10^{-3}\text{ A})^2 = 1.79\times10^{-11}$ J.

(c) The total energy is also given by $U = Q^2/2C$, where $Q$ is the charge amplitude. Thus $Q = \sqrt{2UC} = \sqrt{2(1.79 \times 10^{-11}\,\text{J})(340 \times 10^{-6}\,\text{F})} = 1.10 \times 10^{-7}\,\text{C}$.

**83**

(a) The total energy $U$ of the circuit is the sum of the energy $U_E$ stored in the capacitor and the energy $U_B$ stored in the inductor at the same time. Since $U_B = 2.00U_E$, the total energy is $U = 3.00U_E$. Now $U = Q^2/2C$ and $U_E = q^2/2C$, where $Q$ is the maximum charge, $q$ is the charge when the magnetic energy is twice the electrical energy, and $C$ is the capacitance. Thus $Q^2/2C = 3.00q^2/2C$ and $q = Q/\sqrt{3.00} = 0.577Q$.

(b) If the capacitor has maximum charge at time $t = 0$, then $q = Q\cos(\omega t)$, where $\omega$ is the angular frequency of oscillation. This means $\omega t = \cos^{-1}(0.577) = 0.964\,\text{rad}$. Since $\omega = 2\pi/T$, where $T$ is the period,

$$t = \frac{0.964}{2\pi}T = 0.153T\,.$$

**85**

(a) The energy stored in a capacitor is given by $U_E = q^2/2C$, where $q$ is the charge and $C$ is the capacitance. Now $q^2$ is periodic with a period of $T/2$, where $T$ is the period of the driving emf, so $U_E$ has the same value at the beginning and end of each cycle. Actually $U_E$ has the same value at the beginning and end of each half cycle.

(b) The energy stored in an inductor is given by $Li^2/2$, where $i$ is the current and $L$ is the inductance. The square of the current is periodic with a period of $T/2$, so it has the same value at the beginning and end of each cycle.

(c) The rate with which the driving emf device supplies energy is

$$P_\mathcal{E} = i\mathcal{E} = I\mathcal{E}_m \sin(\omega_d t)\sin(\omega_d t - \phi)\,,$$

where $I$ is the current amplitude, $\mathcal{E}_m$ is the emf amplitude, $\omega$ is the angular frequency, and $\phi$ is a phase constant. The energy supplied over a cycle is

$$\begin{aligned} E_\mathcal{E} &= \int_0^T P_\mathcal{E}\,dt = I\mathcal{E}_m \int_0^T \sin(\omega_d t)\sin(\omega_d t - \phi)\,dt \\ &= I\mathcal{E}_m \int_0^T \sin(\omega_d t)\,[\sin(\omega_d t)cos(\phi) - \cos(\omega_d t)\sin(\phi)]\,dt\,, \end{aligned}$$

where the trigonometric identity $\sin(\alpha - \beta) = \sin\alpha\cos\beta - \cos\alpha\sin\beta$ was used. Now the integral of $\sin^2(\omega_d t)$ over a cycle is $T/2$ and the integral of $\sin(\omega_d t)\cos(\omega_d t)$ over a cycle is zero, so $E_\mathcal{E} = \frac{1}{2}I\mathcal{E}_m\cos\phi$.

(d) The rate of energy dissipation in a resistor is given by

$$P_R = i^2R = I^2\sin^2(\omega_d t - \phi)$$

and the energy dissipated over a cycle is

$$E_R = I^2 \int_0^T \sin^2(\omega_d t - \phi)\,dt = \tfrac{1}{2}I^2RT\,.$$

(e) Now $\mathcal{E}_m = IZ$, where $Z$ is the impedance, and $R = Z\cos\phi$, so $E_\mathcal{E} = \frac{1}{2}I^2TZ\cos\phi = \frac{1}{2}I^2RT = E_R$.

# Chapter 32

3

(a) Use Gauss' law for magnetism: $\oint \vec{B} \cdot d\vec{A} = 0$. Write $\oint \vec{B} \cdot d\vec{A} = \Phi_1 + \Phi_2 + \Phi_C$, where $\Phi_1$ is the magnetic flux through the first end mentioned, $\Phi_2$ is the magnetic flux through the second end mentioned, and $\Phi_C$ is the magnetic flux through the curved surface. Over the first end, the magnetic field is inward, so the flux is $\Phi_1 = -25.0\,\mu\text{Wb}$. Over the second end, the magnetic field is uniform, normal to the surface, and outward, so the flux is $\Phi_2 = AB = \pi r^2 B$, where $A$ is the area of the end and $r$ is the radius of the cylinder. Its value is

$$\Phi_2 = \pi(0.120\,\text{m})^2(1.60 \times 10^{-3}\,\text{T}) = +7.24 \times 10^{-5}\,\text{Wb} = +72.4\,\mu\text{Wb}\,.$$

Since the three fluxes must sum to zero,

$$\Phi_C = -\Phi_1 - \Phi_2 = 25.0\,\mu\text{Wb} - 72.4\,\mu\text{Wb} = -47.4\,\mu\text{Wb}\,.$$

(b) The minus sign indicates that the flux is inward through the curved surface.

5

Consider a circle of radius $r$ (= 6.0 mm), between the plates and with its center on the axis of the capacitor. The current through this circle is zero, so the Ampere-Maxwell law becomes

$$\oint \vec{B} \cdot d\vec{s} = \mu_0\epsilon_0 \frac{d\Phi_E}{dt}\,,$$

where $\vec{B}$ is the magnetic field at points on the circle and $\Phi_E$ is the electric flux through the circle. The magnetic field is tangent to the circle at all points on it, so $\oint \vec{B} \cdot d\vec{s} = 2\pi rB$. The electric flux through the circle is $\Phi_E = \pi R^2 E$, where $R$ (= 3.0 mm) is the radius of a capacitor plate. When these substitutions are made, the Ampere-Maxwell law becomes

$$2\pi rB = \mu_0\epsilon_0\pi R^2 \frac{dE}{dt}\,.$$

Thus

$$\frac{dE}{dt} = \frac{2rB}{\mu_0\epsilon_0 R^2} = \frac{2(6.0 \times 10^{-3}\,\text{m})(2.0 \times 10^{-7}\,\text{T})}{(4\pi \times 10^{-7}\,\text{H/m})(8.85 \times 10^{-12}\,\text{Fm})(3.0 \times 10^{-3}\,\text{m})^2} = 2.4 \times 10^{13}\,\text{V/m}\cdot\text{s}\,.$$

13

The displacement current is given by

$$i_d = \epsilon_0 A \frac{dE}{dt}\,,$$

where $A$ is the area of a plate and $E$ is the magnitude of the electric field between the plates. The field between the plates is uniform, so $E = V/d$, where $V$ is the potential difference across the plates and $d$ is the plate separation. Thus

$$i_d = \frac{\epsilon_0 A}{d}\frac{dV}{dt}.$$

Now $\epsilon_0 A/d$ is the capacitance $C$ of a parallel-plate capacitor without a dielectric, so

$$i_d = C\frac{dV}{dt}.$$

**21**

(a) For a parallel-plate capacitor, the charge $q$ on the positive plate is given by $q = (\epsilon_0 A/d)V$, where $A$ is the plate area, $d$ is the plate separation, and $V$ is the potential difference between the plates. In terms of the electric field $E$ between the plates, $V = Ed$, so $q = \epsilon_0 AE = \epsilon_0 \Phi_E$, where $\Phi_E$ is the total electric flux through the region between the plates. The true current into the positive plate is $i = dq/dt = \epsilon_0\, d\Phi_E/dt = i_{d\text{ total}}$, where $i_{d\text{ total}}$ is the total displacement current between the plates. Thus $i_{d\text{ total}} = 2.0\text{ A}$.

(b) Since $i_{d\text{ total}} = \epsilon_0\, d\Phi_E/dt = \epsilon_0 A\, dE/dt$,

$$\frac{dE}{dt} = \frac{i_{d\text{ total}}}{\epsilon_0 A} = \frac{2.0\text{ A}}{(8.85 \times 10^{-12}\text{ F/m})(1.0\text{ m})^2} = 2.3 \times 10^{11}\text{ V/m}\cdot\text{s}.$$

(c) The displacement current is uniformly distributed over the area. If $a$ is the area enclosed by the dashed lines and $A$ is the area of a plate, then the displacement current through the dashed path is

$$i_{d\text{ enc}} = \frac{a}{A} i_{d\text{ total}} = \frac{(0.50\text{ m})^2}{(1.0\text{ m})^2}(2.0\text{ A}) = 0.50\text{ A}.$$

(d) According to Maxwell's law of induction,

$$\oint \vec{B}\cdot d\vec{s} = \mu_0 i_{d\text{ enc}} = (4\pi \times 10^{-7}\text{ H/m})(0.50\text{ A}) = 6.3 \times 10^{-7}\text{ T}\cdot\text{m}.$$

Notice that the integral is around the dashed path and the displacement current on the right side of the Maxwell's law equation is the displacement current through that path, not the total displacement current.

**35**

(a) The $z$ component of the orbital angular momentum is given by $L_{\text{orb}, z} = m_\ell h/2\pi$, where $h$ is the Planck constant. Since $m_\ell = 0$, $L_{\text{orb}, z} = 0$.

(b) The $z$ component of the orbital contribution to the magnetic dipole moment is given by $\mu_{\text{orb}, z} = -m_\ell \mu_B$, where $\mu_B$ is the Bohr magneton. Since $m_\ell = 0$, $\mu_{\text{orb}, z} = 0$.

(c) The potential energy associated with the orbital contribution to the magnetic dipole moment is given by $U = -\mu_{\text{orb}, z} B_{\text{ext}}$, where $B_{\text{ext}}$ is the $z$ component of the external magnetic field. Since $\mu_{\text{orb}, z} = 0$, $U = 0$.

(d) The $z$ component of the spin magnetic dipole moment is either $+\mu_B$ or $-\mu_B$, so the potential energy is either

$$U = -\mu_B B_{\text{ext}} = -(9.27 \times 10^{-24}\ \text{J/T})(35 \times 10^{-3}\ \text{T}) = -3.2 \times 10^{-25}\ \text{J}\,.$$

or $U = +3.2 \times 10^{-25}$ J.

(e) Substitute $m_\ell$ into the equations given above. The $z$ component of the orbital angular momentum is

$$L_{\text{orb},\, z} = \frac{m_\ell h}{2\pi} = \frac{(-3)(6.626 \times 10^{-34}\ \text{J}\cdot\text{s})}{2\pi} = -3.2 \times 10^{-34}\ \text{J}\cdot\text{s}\,.$$

(f) The $z$ component of the orbital contribution to the magnetic dipole moment is

$$\mu_{\text{orb},\, z} = -m_\ell \mu_B = -(-3)(9.27 \times 10^{-24}\ \text{J/T}) = 2.8 \times 10^{-23}\ \text{J/T}\,.$$

(g) The potential energy associated with the orbital contribution to the magnetic dipole moment is

$$U = -\mu_{\text{orb},\ z} B_{\text{ext}} = -(2.78 \times 10^{-23}\ \text{J/T})(35 \times 10^{-3}\ \text{T}) = -9.7 \times 10^{-25}\ \text{J}\,.$$

(h) The potential energy associated with spin does not depend on $m_\ell$. It is $\pm 3.2 \times 10^{-25}$ J.

**39**

The magnetization is the dipole moment per unit volume, so the dipole moment is given by $\mu = MV$, where $M$ is the magnetization and $V$ is the volume of the cylinder. Use $V = \pi r^2 L$, where $r$ is the radius of the cylinder and $L$ is its length. Thus

$$\mu = M\pi r^2 L = (5.30 \times 10^3\ \text{A/m})\pi(0.500 \times 10^{-2}\ \text{m})^2(5.00 \times 10^{-2}\ \text{m}) = 2.08 \times 10^{-2}\ \text{J/T}\,.$$

**45**

(a) The number of atoms per unit volume in states with the dipole moment aligned with the magnetic field is $N_+ = Ae^{\mu B/kT}$ and the number per unit volume in states with the dipole moment antialigned is $N_- = Ae^{-\mu B/kT}$, where $A$ is a constant of proportionality. The total number of atoms per unit volume is $N = N_+ + N_- = A\left(e^{\mu B/kT} + e^{-\mu B/kT}\right)$. Thus

$$A = \frac{N}{e^{\mu B/kT} + e^{-\mu B/kT}}\,.$$

The magnetization is the net dipole moment per unit volume. Subtract the magnitude of the total dipole moment per unit volume of the antialigned moments from the total dipole moment per unit volume of the aligned moments. The result is

$$M = \frac{N\mu e^{\mu B/kT} - N\mu e^{-\mu B/kT}}{e^{\mu B/kT} + e^{-\mu B/kT}} = \frac{N\mu\left(e^{\mu B/kT} - e^{-\mu B/kT}\right)}{e^{\mu B/kT} + e^{-\mu B/kT}} = N\mu \tanh(\mu B/kT)\,.$$

(b) If $\mu B \ll kT$, then $e^{\mu B/kT} \approx 1 + \mu B/kT$ and $e^{-\mu B/kT} \approx 1 - \mu B/kT$. (See Appendix E for the power series expansion of the exponential function.) The expression for the magnetization becomes

$$M \approx \frac{N\mu\left[(1+\mu B/kT) - (1-\mu B/kT)\right]}{(1+\mu B/kT) + (1-\mu B/kT)} = \frac{N\mu^2 B}{kT}\,.$$

(c) If $\mu B \gg kT$, then $e^{-\mu B/kT}$ is negligible compared to $e^{\mu B/kT}$ in both the numerator and denominator of the expression for $M$. Thus

$$M \approx \frac{N\mu e^{\mu B/kT}}{e^{\mu B/kT}} = N\mu\,.$$

(d) The expression for $M$ predicts that it is linear in $B/kT$ for $\mu B/kT$ small and independent of $B/kT$ for $\mu B/kT$ large. The figure agrees with these predictions.

47

(a) The field of a dipole along its axis is given by Eq. 29–27:

$$\vec{B} = \frac{\mu_0}{2\pi}\frac{\vec{\mu}}{z^3}\,,$$

where $\mu$ is the dipole moment and $z$ is the distance from the dipole. Thus the magnitude of the magnetic field is

$$B = \frac{(4\pi \times 10^{-7}\,\text{T}\cdot\text{m/A})(1.5 \times 10^{-23}\,\text{J/T})}{2\pi(10 \times 10^{-9}\,\text{m})^3} = 3.0 \times 10^{-6}\,\text{T}\,.$$

(b) The energy of a magnetic dipole with dipole moment $\vec{\mu}$ in a magnetic field $\vec{B}$ is given by $U = -\vec{\mu}\cdot\vec{B} = -\mu B\cos\phi$, where $\phi$ is the angle between the dipole moment and the field. The energy required to turn it end for end (from $\phi = 0°$ to $\phi = 180°$) is

$$\begin{aligned}\Delta U &= -\mu B(\cos 180° - \cos 0°) = 2\mu B = 2(1.5 \times 10^{-23}\,\text{J/T})(3.0 \times 10^{-6}\,\text{T})\\ &= 9.0 \times 10^{-29}\,\text{J} = 5.6 \times 10^{-10}\,\text{eV}\,.\end{aligned}$$

The mean kinetic energy of translation at room temperature is about 0.04 eV (see Eq. 19–24 or Sample Problem 32–3). Thus if dipole-dipole interactions were responsible for aligning dipoles, collisions would easily randomize the directions of the moments and they would not remain aligned.

53

(a) If the magnetization of the sphere is saturated, the total dipole moment is $\mu_{\text{total}} = N\mu$, where $N$ is the number of iron atoms in the sphere and $\mu$ is the dipole moment of an iron atom. We wish to find the radius of an iron sphere with $N$ iron atoms. The mass of such a sphere is $Nm$, where $m$ is the mass of an iron atom. It is also given by $4\pi\rho R^3/3$, where $\rho$ is the density of iron and $R$ is the radius of the sphere. Thus $Nm = 4\pi\rho R^3/3$ and

$$N = \frac{4\pi\rho R^3}{3m}\,.$$

Substitute this into $\mu_{\text{total}} = N\mu$ to obtain

$$\mu_{\text{total}} = \frac{4\pi\rho R^3 \mu}{3m} .$$

Solve for $R$ and obtain

$$R = \left[\frac{3m\mu_{\text{total}}}{4\pi\rho\mu}\right]^{1/3} .$$

The mass of an iron atom is

$$m = 56\,\text{u} = (56\,\text{u})(1.66 \times 10^{-27}\,\text{kg/u}) = 9.30 \times 10^{-26}\,\text{kg} .$$

So

$$R = \left[\frac{3(9.30 \times 10^{-26}\,\text{kg})(8.0 \times 10^{22}\,\text{J/T})}{4\pi(14 \times 10^3\,\text{kg/m}^3)(2.1 \times 10^{-23}\,\text{J/T})}\right]^{1/3} = 1.8 \times 10^5\,\text{m} .$$

(b) The volume of the sphere is

$$V_s = \frac{4\pi}{3}R^3 = \frac{4\pi}{3}(1.82 \times 10^5\,\text{m})^3 = 2.53 \times 10^{16}\,\text{m}^3$$

and the volume of Earth is

$$V_e = \frac{4\pi}{3}(6.37 \times 10^6\,\text{m})^3 = 1.08 \times 10^{21}\,\text{m}^3 ,$$

so the fraction of Earth's volume that is occupied by the sphere is

$$\frac{2.53 \times 10^{16}\,\text{m}^3}{1.08 \times 10^{21}\,\text{m}^3} = 2.3 \times 10^{-5} .$$

The radius of Earth was obtained from Appendix C.

<u>**55**</u>

(a) The horizontal and vertical directions are perpendicular to each other, so the magnitude of the field is

$$B = \sqrt{B_h^2 + B_v^2} = \frac{\mu_0\mu}{4\pi r^3}\sqrt{\cos^2\lambda_m + 4\sin^2\lambda_m} = \frac{\mu_0\mu}{4\pi r^3}\sqrt{1 - \sin^2\lambda_m + 4\sin^2\lambda_m}$$

$$= \frac{\mu_0\mu}{4\pi r^3}\sqrt{1 + 3\sin^2\lambda_m} ,$$

where the trigonometric identity $\cos^2\lambda_m = 1 - \sin^2\lambda_m$ was used.

(b) The tangent of the inclination angle is

$$\tan\phi_i = \frac{B_v}{B_h} = \left(\frac{\mu_0\mu}{2\pi r^3\sin\lambda_m}\right)\left(\frac{4\pi r^3}{\mu_0\mu\cos\lambda_m} = \frac{2\sin\lambda_m}{\cos\lambda_m}\right) = 2\tan\lambda_m ,$$

where $\tan\lambda_m = (\sin\lambda_m)/(\cos\lambda_m)$ was used.

<u>**61**</u>

(a) The $z$ component of the orbital angular momentum can have the values $L_{\text{orb},z} = m_\ell h/2\pi$, where $m_\ell$ can take on any integer value from $-3$ to $+3$, inclusive. There are seven such values ($-3$, $-2$, $-1$, 0, $+1$, $+2$, and $+3$).

(b) The $z$ component of the orbital magnetic moment is given by $\mu_{\text{orb},z} = -m_\ell eh/4\pi m$, where $m$ is the electron mass. Since there is a different value for each possible value of $m_\ell$, there are seven different values in all.

(c) The greatest possible value of $L_{\text{orb},z}$ occurs if $m_\ell = +3$ is $3h/2\pi$.

(d) The greatest value of $\mu_{\text{orb},\,z}$ is $3eh/4\pi m$.

(e) Add the orbital and spin angular momenta: $L_{\text{net},z} = L_{\text{orb},z} + L_{s,z} = (m_\ell h/2\pi) + (m_s h/2\pi)$. To obtain the maximum value, set $m_\ell$ equal to +3 and $m_s$ equal to $+\frac{1}{2}$. The result is $L_{\text{net},z} = 3.5h/2\pi$.

(f) Write $L_{\text{net},z} = Mh/2\pi$, where $M$ is half an odd integer. $M$ can take on all such values from $-3.5$ to $+3.5$. There are eight of these: $-3.5$, $-2.5$, $-1.5$, $-0.5$, $+0.5$, $+1.5$, $+2.5$, and $+3.5$.

# Chapter 33

**5**

If $f$ is the frequency and $\lambda$ is the wavelength of an electromagnetic wave, then $f\lambda = c$. The frequency is the same as the frequency of oscillation of the current in the $LC$ circuit of the generator. That is, $f = 1/2\pi\sqrt{LC}$, where $C$ is the capacitance and $L$ is the inductance. Thus

$$\frac{\lambda}{2\pi\sqrt{LC}} = c\,.$$

The solution for $L$ is

$$L = \frac{\lambda^2}{4\pi^2 C c^2} = \frac{(550\times 10^{-9}\,\mathrm{m})^2}{4\pi^2(17\times 10^{-12}\,\mathrm{F})(3.00\times 10^{8}\,\mathrm{m/s})^2} = 5.00\times 10^{-21}\,\mathrm{H}\,.$$

This is exceedingly small.

**21**

The plasma completely reflects all the energy incident on it, so the radiation pressure is given by $p_r = 2I/c$, where $I$ is the intensity. The intensity is $I = P/A$, where $P$ is the power and $A$ is the area intercepted by the radiation. Thus

$$p_r = \frac{2P}{Ac} = \frac{2(1.5\times 10^{9}\,\mathrm{W})}{(1.00\times 10^{-6}\,\mathrm{m}^2)(3.00\times 10^{8}\,\mathrm{m/s})} = 1.0\times 10^{7}\,\mathrm{Pa} = 10\,\mathrm{MPa}\,.$$

**23**

Let $f$ be the fraction of the incident beam intensity that is reflected. The fraction absorbed is $1 - f$. The reflected portion exerts a radiation pressure of $p_r = (2fI_0)/c$ and the absorbed portion exerts a radiation pressure of $p_a = (1 - f)I_0/c$, where $I_0$ is the incident intensity. The factor 2 enters the first expression because the momentum of the reflected portion is reversed. The total radiation pressure is the sum of the two contributions:

$$p_{\text{total}} = p_r + p_a = \frac{2fI_0 + (1-f)I_0}{c} = \frac{(1+f)I_0}{c}\,.$$

To relate the intensity and energy density, consider a tube with length $\ell$ and cross-sectional area $A$, lying with its axis along the propagation direction of an electromagnetic wave. The electromagnetic energy inside is $U = uA\ell$, where $u$ is the energy density. All this energy will pass through the end in time $t = \ell/c$ so the intensity is

$$I = \frac{U}{At} = \frac{uA\ell c}{A\ell} = uc\,.$$

Thus $u = I/c$. The intensity and energy density are inherently positive, regardless of the propagation direction.

For the partially reflected and partially absorbed wave, the intensity just outside the surface is $I = I_0 + fI_0 = (1+f)I_0$, where the first term is associated with the incident beam and the second is associated with the reflected beam. The energy density is, therefore,

$$u = \frac{I}{c} = \frac{(1+f)I_0}{c},$$

the same as radiation pressure.

**25**

(a) Since $c = \lambda f$, where $\lambda$ is the wavelength and $f$ is the frequency of the wave,

$$f = \frac{c}{\lambda} = \frac{3.00 \times 10^8\,\text{m/s}}{3.0\,\text{m}} = 1.0 \times 10^8\,\text{Hz}.$$

(b) The angular frequency is

$$\omega = 2\pi f = 2\pi(1.0 \times 10^8\,\text{Hz}) = 6.3 \times 10^8\,\text{rad/s}.$$

(c) The angular wave number is

$$k = \frac{2\pi}{\lambda} = \frac{2\pi}{3.0\,\text{m}} = 2.1\,\text{rad/m}.$$

(d) The magnetic field amplitude is

$$B_m = \frac{E_m}{c} = \frac{300\,\text{V/m}}{3.00 \times 10^8\,\text{m/s}} = 1.00 \times 10^{-6}\,\text{T}.$$

(e) $\vec{B}$ must be in the positive $z$ direction when $\vec{E}$ is in the positive $y$ direction in order for $\vec{E} \times \vec{B}$ to be in the positive $x$ direction (the direction of propagation).

(f) The time-averaged rate of energy flow or intensity of the wave is

$$I = \frac{E_m^2}{2\mu_0 c} = \frac{(300\,\text{V/m})^2}{2(4\pi \times 10^{-7}\,\text{H/m})(3.00 \times 10^8\,\text{m/s})} = 1.2 \times 10^2\,\text{W/m}^2.$$

(g) Since the sheet is perfectly absorbing, the rate per unit area with which momentum is delivered to it is $I/c$, so

$$\frac{dp}{dt} = \frac{IA}{c} = \frac{(119\,\text{W/m}^2)(2.0\,\text{m}^2)}{3.00 \times 10^8\,\text{m/s}} = 8.0 \times 10^{-7}\,\text{N}.$$

(h) The radiation pressure is

$$p_r = \frac{dp/dt}{A} = \frac{8.0 \times 10^{-7}\,\text{N}}{2.0\,\text{m}^2} = 4.0 \times 10^{-7}\,\text{Pa}.$$

**27**

If the beam carries energy $U$ away from the spaceship, then it also carries momentum $p = U/c$ away. Since the total momentum of the spaceship and light is conserved, this is the magnitude of the momentum acquired by the spaceship. If $P$ is the power of the laser, then the energy carried away in time $t$ is $U = Pt$. Thus $p = Pt/c$ and, if $m$ is mass of the spaceship, its speed is

$$v = \frac{p}{m} = \frac{Pt}{mc} = \frac{(10 \times 10^3\,\text{W})(1\,d)(8.64 \times 10^4\,\text{s/d})}{(1.5 \times 10^3\,\text{kg})(3.00 \times 10^8\,\text{m/s})} = 1.9 \times 10^{-3}\,\text{m/s} = 1.9\,\text{mm/s}\,.$$

**35**

Let $I_0$ be in the intensity of the unpolarized light that is incident on the first polarizing sheet. Then the transmitted intensity is $I_1 = \frac{1}{2}I_0$ and the direction of polarization of the transmitted light is $\theta_1$ (= 40°) counterclockwise from the $y$ axis in the diagram.

The polarizing direction of the second sheet is $\theta_2$ (= 20°) clockwise from the $y$ axis so the angle between the direction of polarization of the light that is incident on that sheet and the polarizing direction of the of the sheet is $40° + 20° = 60°$. The transmitted intensity is

$$I_2 = I_1 \cos^2 60° = \frac{1}{2} I_0 \cos^2 60°$$

and the direction of polarization of the transmitted light is 20° clockwise from the $y$ axis.

The polarizing direction of the third sheet is $\theta_3$ (= 40°) counterclockwise from the $y$ axis so the angle between the direction of polarization of the light incident on that sheet and the polarizing direction of the sheet is $20° + 40° = 60°$. The transmitted intensity is

$$I_3 = I_2 \cos^2 60° = \frac{1}{2} I_0 \cos^4 60° = 3.1 \times 10^{-2}\,.$$

3.1% of the light's initial intensity is transmitted.

**43**

(a) The rotation cannot be done with a single sheet. If a sheet is placed with its polarizing direction at an angle of 90° to the direction of polarization of the incident radiation, no radiation is transmitted.

It can be done with two sheets. Place the first sheet with its polarizing direction at some angle $\theta$, between 0 and 90°, to the direction of polarization of the incident radiation. Place the second sheet with its polarizing direction at 90° to the polarization direction of the incident radiation. The transmitted radiation is then polarized at 90° to the incident polarization direction. The intensity is $I_0 \cos^2\theta \cos^2(90° - \theta) = I_0 \cos^2\theta \sin^2\theta$, where $I_0$ is the incident radiation. If $\theta$ is not 0 or 90°, the transmitted intensity is not zero.

(b) Consider $n$ sheets, with the polarizing direction of the first sheet making an angle of $\theta = 90°/n$ with the direction of polarization of the incident radiation and with the polarizing direction of each successive sheet rotated $90°/n$ in the same direction from the polarizing direction of the previous sheet. The transmitted radiation is polarized with its direction of polarization making an angle of 90° with the direction of polarization of the incident radiation. The intensity is $I = I_0 \cos^{2n}(90°/n)$. You want the smallest integer value of $n$ for which this is greater than $0.60I_0$.

Start with $n = 2$ and calculate $\cos^{2n}(90°/n)$. If the result is greater than 0.60, you have obtained the solution. If it is less, increase $n$ by 1 and try again. Repeat this process, increasing $n$ by 1 each time, until you have a value for which $\cos^{2n}(90°/n)$ is greater than 0.60. The first one will be $n = 5$.

## 51

Consider a ray that grazes the top of the pole, as shown in the diagram to the right. Here $\theta_1 = 35°$, $\ell_1 = 0.50\,\text{m}$, and $\ell_2 = 1.50\,\text{m}$. The length of the shadow is $x + L$. $x$ is given by $x = \ell_1 \tan\theta_1 = (0.50\,\text{m})\tan 35° = 0.35\,\text{m}$. According to the law of refraction, $n_2 \sin\theta_2 = n_1 \sin\theta_1$. Take $n_1 = 1$ and $n_2 = 1.33$ (from Table 33–1). Then,

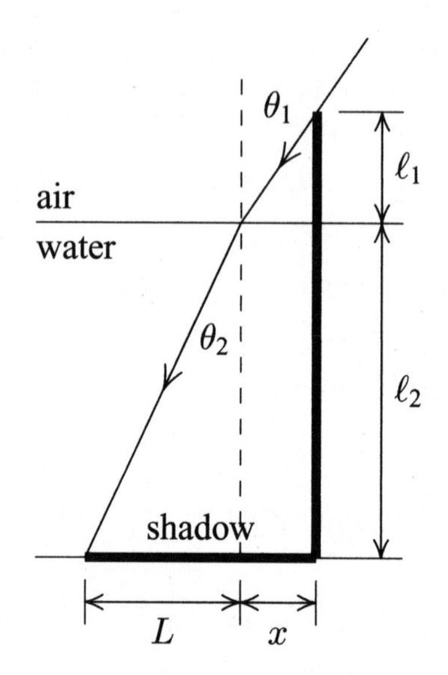


$$\theta_2 = \sin^{-1}\left(\frac{\sin\theta_1}{n_2}\right) = \sin^{-1}\left(\frac{\sin 35.0°}{1.33}\right) = 25.55°\,.$$

$L$ is given by

$$L = \ell_2 \tan\theta_2 = (1.50\,\text{m})\tan 25.55° = 0.72\,\text{m}\,.$$

The length of the shadow is $0.35\,\text{m} + 0.72\,\text{m} = 1.07\,\text{m}$.

## 55

Look at the diagram on the right. The two angles labeled $\alpha$ have the same value. $\theta_2$ is the angle of refraction. Because the dotted lines are perpendicular to the prism surface $\theta_2 + \alpha = 90°$ and $\alpha = 90° - \theta_2$. Because the interior angles of a triangle sum to 180°, $180° - 2\theta_2 + \phi = 180°$ and $\theta_2 = \phi/2$.

Now look at the next diagram and consider the triangle formed by the two normals and the ray in the interior. The two equal interior angles each have the value $\theta - \theta_2$. Because the exterior angle of a triangle is equal to the sum of the two opposite interior angles, $\psi = 2(\theta - \theta_2)$ and $\theta = \theta_2 + \psi/2$. Upon substitution for $\theta_2$ this becomes $\theta = (\phi + \psi)/2$.

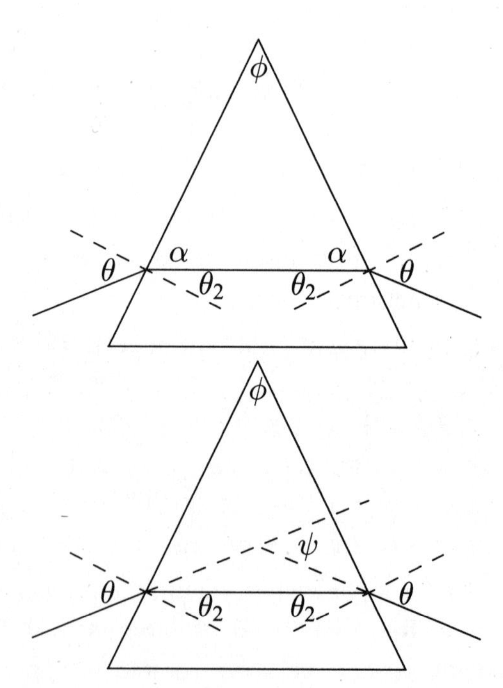


According to the law of refraction the index of refraction of the prism material is

$$n = \frac{\sin\theta}{\sin\theta_2} = \frac{\sin(\phi + \psi)/2}{\sin\phi/2}\,.$$

**65**

(a) No refraction occurs at the surface $ab$, so the angle of incidence at surface $ac$ is $90° - \phi$. For total internal reflection at the second surface, $n_g \sin(90° - \phi)$ must be greater than $n_a$. Here $n_g$ is the index of refraction for the glass and $n_a$ is the index of refraction for air. Since $\sin(90° - \phi) = \cos\phi$, you want the largest value of $\phi$ for which $n_g \cos\phi \geq n_a$. Recall that $\cos\phi$ decreases as $\phi$ increases from zero. When $\phi$ has the largest value for which total internal reflection occurs, then $n_g \cos\phi = n_a$, or

$$\phi = \cos^{-1}\left(\frac{n_a}{n_g}\right) = \cos^{-1}\left(\frac{1}{1.52}\right) = 48.9° .$$

The index of refraction for air was taken to be unity.

(b) Replace the air with water. If $n_w$ (= 1.33) is the index of refraction for water, then the largest value of $\phi$ for which total internal reflection occurs is

$$\phi = \cos^{-1}\left(\frac{n_w}{n_g}\right) = \cos^{-1}\left(\frac{1.33}{1.52}\right) = 29.0° .$$

**69**

The angle of incidence $\theta_B$ for which reflected light is fully polarized is given by Eq. 33–49 of the text. If $n_1$ is the index of refraction for the medium of incidence and $n_2$ is the index of refraction for the second medium, then $\theta_B = \tan^{-1}(n_2/n_1) = \tan^{-1}(1.53/1.33) = 63.8°$.

**73**

Let $\theta_1$ (= 45°) be the angle of incidence at the first surface and $\theta_2$ be the angle of refraction there. Let $\theta_3$ be the angle of incidence at the second surface. The condition for total internal reflection at the second surface is $n \sin\theta_3 \geq 1$. You want to find the smallest value of the index of refraction $n$ for which this inequality holds.

The law of refraction, applied to the first surface, yields $n \sin\theta_2 = \sin\theta_1$. Consideration of the triangle formed by the surface of the slab and the ray in the slab tells us that $\theta_3 = 90° - \theta_2$. Thus the condition for total internal reflection becomes $1 \leq n \sin(90° - \theta_2) = n \cos\theta_2$. Square this equation and use $\sin^2\theta_2 + \cos^2\theta_2 = 1$ to obtain $1 \leq n^2(1 - \sin^2\theta_2)$. Now substitute $\sin\theta_2 = (1/n)\sin\theta_1$ to obtain

$$1 \leq n^2\left(1 - \frac{\sin^2\theta_1}{n^2}\right) = n^2 - \sin^2\theta_1 .$$

The largest value of $n$ for which this equation is true is the value for which $1 = n^2 - \sin^2\theta_1$. Solve for $n$:

$$n = \sqrt{1 + \sin^2\theta_1} = \sqrt{1 + \sin^2 45°} = 1.22 .$$

75

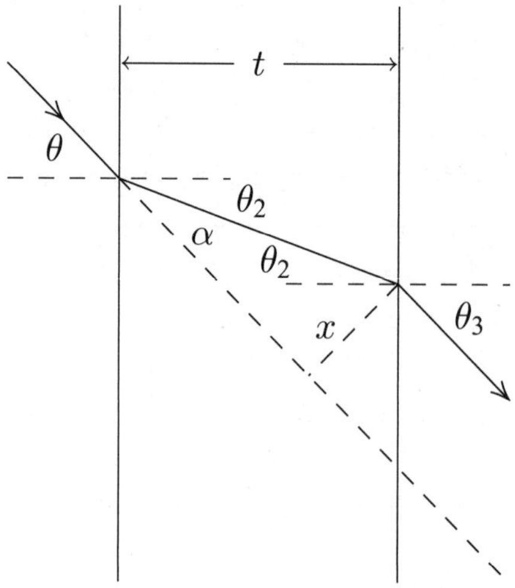

Let $\theta$ be the angle of incidence and $\theta_2$ be the angle of refraction at the left face of the plate. Let $n$ be the index of refraction of the glass. Then, the law of refraction yields $\sin\theta = n\sin\theta_2$. The angle of incidence at the right face is also $\theta_2$. If $\theta_3$ is the angle of emergence there, then $n\sin\theta_2 = \sin\theta_3$. Thus $\sin\theta_3 = \sin\theta$ and $\theta_3 = \theta$. The emerging ray is parallel to the incident ray.

You wish to derive an expression for $x$ in terms of $\theta$. If $D$ is the length of the ray in the glass, then $D\cos\theta_2 = t$ and $D = t/\cos\theta_2$. The angle $\alpha$ in the diagram equals $\theta - \theta_2$ and $x = D\sin\alpha = D\sin(\theta - \theta_2)$. Thus

$$x = \frac{t\sin(\theta - \theta_2)}{\cos\theta_2}.$$

If all the angles $\theta$, $\theta_2$, $\theta_3$, and $\theta - \theta_2$ are small and measured in radians, then $\sin\theta \approx \theta$, $\sin\theta_2 \approx \theta_2$, $\sin(\theta - \theta_2) \approx \theta - \theta_2$, and $\cos\theta_2 \approx 1$. Thus $x \approx t(\theta - \theta_2)$. The law of refraction applied to the point of incidence at the left face of the plate is now $\theta \approx n\theta_2$, so $\theta_2 \approx \theta/n$ and

$$x \approx t\left(\theta - \frac{\theta}{n}\right) = \frac{(n-1)t\theta}{n}.$$

77

The time for light to travel a distance $d$ in free space is $t = d/c$, where $c$ is the speed of light ($3.00 \times 10^8$ m/s).

(a) Take $d$ to be 150 km = $150 \times 10^3$ m. Then,

$$t = \frac{d}{c} = \frac{150 \times 10^3\ \text{m}}{3.00 \times 10^8\ \text{m/s}} = 5.00 \times 10^{-4}\ \text{s}.$$

(b) At full moon, the Moon and Sun are on opposite sides of Earth, so the distance traveled by the light is $d = (1.5 \times 10^8\ \text{km}) + 2(3.8 \times 10^5\ \text{km}) = 1.51 \times 10^8\ \text{km} = 1.51 \times 10^{11}$ m. The time taken by light to travel this distance is

$$t = \frac{d}{c} = \frac{1.51 \times 10^{11}\ \text{m}}{3.00 \times 10^8\ \text{m/s}} = 500\ \text{s} = 8.4\ \text{min}.$$

The distances are given in the problem.

(c) Take $d$ to be $2(1.3 \times 10^9\ \text{km}) = 2.6 \times 10^{12}$ m. Then,

$$t = \frac{d}{c} = \frac{2.6 \times 10^{12}\ \text{m}}{3.00 \times 10^8\ \text{m/s}} = 8.7 \times 10^3\ \text{s} = 2.4\ \text{h}.$$

(d) Take $d$ to be 6500 ly and the speed of light to be 1.00 ly/y. Then,

$$t = \frac{d}{c} = \frac{6500\,\text{ly}}{1.00\,\text{ly/y}} = 6500\,\text{y}\,.$$

The explosion took place in the year 1054 – 6500 = –5446 or B.C. 5446.

**79**

(a) The amplitude of the magnetic field is $B = E/c = (5.00\,\text{V/m})/(3.00 \times 10^8\,\text{m/s}) = 1.67 \times 10^{-8}\,\text{T}$. According to the argument of the trigonometric function in the expression for the electric field, the wave is moving in the negative $z$ direction and the electric field is parallel to the $y$ axis. In order for $\vec{E} \times \vec{B}$ to be in the negative $z$ direction, $\vec{B}$ must be in the positive $x$ direction when $\vec{E}$ is in the positive $y$ direction. Thus

$$B_x = (1.67 \times 10^{-8}\,\text{T})\sin[(1.00 \times 10^6\,\text{m}^{-1})z + \omega t]$$

is the only nonvanishing component of the magnetic field.

The angular wave number is $k = 1.00 \times 10^6\,\text{m}^{-1}$ so the angular frequency is $\omega = kc = (1.00 \times 10^6\,\text{m}^{-1})(3.00 \times 10^8\,\text{m/s}) = 3.00 \times 10^{14}\,\text{s}^{-1}$ and

$$B_x = (1.67 \times 10^{-8}\,\text{T})\sin[(1.00 \times 10^6\,\text{m}^{-1})z + (3.00 \times 10^{14}\,\text{s}^{-1})t]\,.$$

(b) The wavelength is $\lambda = 2\pi/k = 2\pi/(1.00 \times 10^6\,\text{m}^{-1}) = 6.28 \times 10^{-6}\,\text{m}$.

(c) The period is $T = 2\pi/\omega = 2\pi/(3.00 \times 10^{14}\,\text{s}^{-1}) = 2.09 \times 10^{-14}\,\text{s}$.

(d) The intensity of this wave is $I = E_m^2/2\mu_0 c = (5.00\,\text{V/m})^2/2(4\pi \times 10^{-7}\,\text{H/m})(3.00 \times 10^8\,\text{m/s} = 0.0332\,\text{W/m}^2$. (f) A wavelength of $6.28 \times 10^{-6}\,\text{m}$ places this wave in the infrared portion of the electromagnetic spectrum. See Fig. 33-1.

**83**

(a) The power is the same through any hemisphere centered at the source. The area of a hemisphere of radius $r$ is $A = 2\pi r^2$. In this case $r$ is the distance from the source to the aircraft. Thus the intensity at the aircraft is $I = P/A = P/2\pi r^2 = (180 \times 10^3\,\text{W})/2\pi(90 \times 10^3\,\text{m})^2 = 3.5 \times 10^{-6}\,\text{W/m}^2$.

(b) The power of the reflection is the product of the intensity at the aircraft and the cross section of the aircraft: $P_r = (3.5 \times 10^{-6}\,\text{W/m}^2)(0.22\,\text{m}^2) = 7.8 \times 10^{-7}\,\text{W}$.

(c) The intensity at the detector is $P_r/2\pi r^2 = (7.8 \times 10^{-7}\,\text{W})/2\pi(90 \times 10^3\,\text{m})^2 = 1.5 \times 10^{-17}\,\text{W/m}^2$.

(d) Since the intensity is given by $I = E_m^2/2\mu_0 c$,

$$E_m = \sqrt{2\mu_0 cI} = \sqrt{2(4\pi \times 10^{-7}\,\text{H/m})(3.00 \times 10^8\,\text{m/s})(1.5 \times 10^{-17}\,\text{W/m}^2)} = 1.1 \times 10^{-7}\,\text{V/m}\,.$$

(e) The rms value of the magnetic field is $B_{\text{rms}} = E_m/\sqrt{2}c = (1.1 \times 10^{-7}\,\text{V/m})/(\sqrt{2})(3.00 \times 10^8\,\text{m/s}) = 2.5 \times 10^{-16}\,\text{T}$.

**91**

The critical angle for total internal reflection is given by $\theta_c = \sin^{-1}(1/n)$. For $n = 1.456$ this angle is $\theta_c = 43.38°$ and for $n = 1.470$ it is $\theta_c = 42.86°$.

(a) An incidence angle of 42.00° is less than the critical angle for both red and blue light. The refracted light is white.

(b) An incidence angle of 43.10° is less than the critical angle for red light and greater than the critical angle for blue light. Red light is refracted but blue light is not. The refracted light is reddish.

(c) An incidence angle of 44.00° is greater than the critical angle for both red and blue light. Neither is refracted.

**103**

(a) Take the derivative of the functions given for $E$ and $B$, then substitute them into

$$\frac{\partial^2 E}{\partial t^2} = c^2 \frac{\partial^2 E}{\partial x^2} \quad \text{and} \quad \frac{\partial^2 B}{\partial t^2} = c^2 \frac{\partial^2 B}{\partial x^2}\,.$$

The derivatives of $E$ are $\partial^2 E/\partial t^2 = -\omega^2 E_m \sin(kx - \omega t)$ and $\partial^2 E/\partial x^2 = -k^2 E_m \sin(kx - \omega t)$, so the wave equation for the electric field yields $\omega^2 = c^2 k^2$. Since $\omega = ck$ the function satisfies the wave equation. Similarly, the derivatives of $B$ are $\partial^2 B/\partial t^2 = -\omega^2 B_m \sin(kx - \omega t)$ and $\partial^2 B/\partial x^2 = -k^2 B_m \sin(kx - \omega t)$ and the wave equation for the magnetic field yields $\omega^2 = c^2 k^2$. Since $\omega = ck$ the function satisfies the wave equation.

(b) Let $u = kx \pm \omega t$ and consider $f$ to be a function of $u$, which in turn is a function of $x$ and $t$. Then the chain rule of the calculus gives

$$\frac{\partial^2 E}{\partial t^2} = \frac{d^2 f}{du^2}\left(\frac{\partial u}{\partial t}\right)^2 = \frac{d^2 f}{du^2}\omega^2$$

and

$$\frac{\partial^2 E}{\partial x^2} = \frac{d^2 f}{du^2}\left(\frac{\partial u}{\partial x}\right)^2 = \frac{d^2 f}{du^2}k^2\,.$$

Substitution into the wave equation again yields $\omega^2 = c^2 k^2$, so the function obeys the wave equation. A similar analysis shows that the function for $B$ also satisfies the wave equation.

# Chapter 34

**5**

The light bulb is labeled O and its image is labeled I on the digram to the right. Consider the two rays shown on the diagram to the right. One enters the water at A and is reflected from the mirror at B. This ray is perpendicular to the water line and mirror. The second ray leaves the lightbulb at the angle $\theta$, enters the water at C, where it is refracted. It is reflected from the mirror at D and leaves the water at E. At C the angle of incidence is $\theta$ and the angle of refraction is $\theta'$. At D the angles of incidence and reflection are both $\theta'$. At E the angle of incidence is $\theta'$ and the angle of refraction is $\theta$. The dotted lines that meet at I represent extensions of the emerging rays. Light appears to come from I. We want to compute $d_3$.

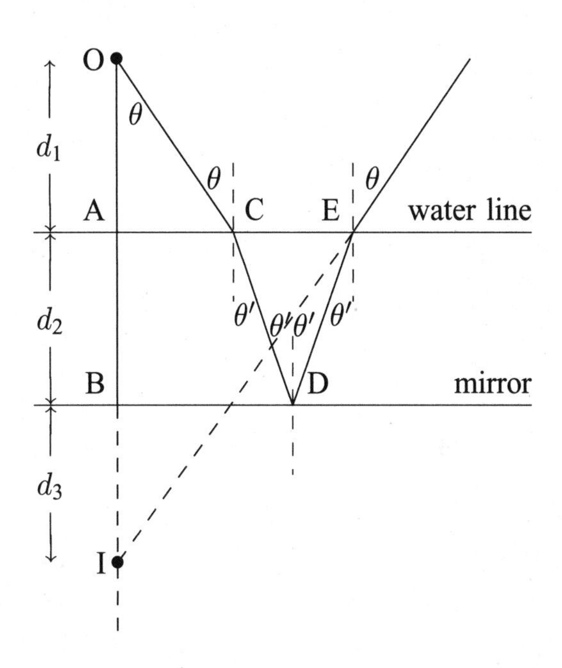


Consideration of the triangle OBE tells us that the distance $d_2 + d_3$ is $L\tan(90° - \theta) = L/\tan\theta$, where $L$ is the distance between A and E. Consideration of the triangle OBC tells us that the distance between A and C is $d_1 \tan\theta$ and consideration of the triangle CDE tells us that the distance between C and E is $2d_2 \tan\theta'$, so $L = d_1 \tan\theta + 2d_2 \tan\theta'$, $d_2 + d_3 = (d_1 \tan\theta + 2d_2 \tan\theta')/\tan\theta$, and

$$d_3 = \frac{d_1 \tan\theta + 2d_2 \tan\theta'}{\tan\theta} - d_2 .$$

Apply the law of refraction at point C: $\sin\theta = n \sin\theta'$, where $n$ is the index of refraction of water. Since the angles $\theta$ and $\theta'$ are small we may approximate their sines by their tangents and write $\tan\theta = n \tan\theta'$. Us this to substitute for $\tan\theta$ in the expression for $d_3$ to obtain

$$d_3 = \frac{nd_1 + 2d_2}{n} - d_2 = \frac{(1.33)(250\,\text{cm}) + 2(200\,\text{cm})}{1.33} - 200\,\text{cm} = 350\,\text{cm},$$

where the index of refraction of water was taken to be 1.33.

**9**

(a) The radius of curvature $r$ and focal length $f$ are positive for a concave mirror and are related by $f = r/2$, so $r = 2(+18\,\text{cm}) = +36\,\text{cm}$.

(b) Since $(1/p)+(1/i)=1/f$, where $i$ is the image distance,

$$i=\frac{fp}{p-f}=\frac{(18\text{ cm})(12\text{ cm})}{12\text{ cm}-18\text{ cm}}=-36\text{ cm}.$$

(c) The magnification is $m=-i/p=-(-36\text{ cm})/(12\text{ cm}=3.0$.

(d) The value obtained for $i$ is negative, so the image is virtual.

(e) The value obtained for the magnification is positive, so the image is not inverted.

(f) Real images are formed by mirrors on the same side as the object and virtual images are formed on the opposite side. Since the image here is virtual it is on the opposite side of the mirror from the object.

<u>11</u>

(a) The radius of curvature $r$ and focal length $f$ are positive for a concave mirror and are related by $f=r/2$, so $r=2(+12\text{ cm})=+24\text{ cm}$.

(b) Since $(1/p)+(1/i)=1/f$, where $i$ is the image distance,

$$i=\frac{fp}{p-f}=\frac{(12\text{ cm})(18\text{ cm})}{18\text{ cm}-12\text{ cm}}=36\text{ cm}.$$

(c) The magnification is $m=-i/p=-(36\text{ cm})/(18\text{ cm}=-2.0$.

(d) The value obtained for $i$ is positive, so the image is real.

(e) The value obtained for the magnification is negative, so the image is inverted.

(f) Real images are formed by mirrors on the same side as the object. Since the image here is real it is on the same side of the mirror as the object.

<u>15</u>

(a) The radius of curvature $r$ and focal length $f$ are negative for a convex mirror and are related by $f=r/2$, so $r=2(-10\text{ cm})=-20\text{ cm}$.

(b) Since $(1/p)+(1/i)=1/f$, where $i$ is the image distance,

$$i=\frac{fp}{p-f}=\frac{(-10\text{ cm})(8\text{ cm})}{(8\text{ cm})-(-10\text{ cm})}=-4.44\text{ cm}.$$

(c) The magnification is $m=-i/p=-(-4.44\text{ cm})/(8\text{ cm}=+0.56$.

(d) The value obtained for $i$ is negative, so the image is virtual.

(e) The value obtained for the magnification is positive, so the image is not inverted.

(f) Real images are formed by mirrors on the same side as the object and virtual images are formed on the opposite side. Since the image here is virtual it is on the opposite side of the mirror from the object

<u>27</u>

Since the mirror is convex the radius of curvature is negative. The focal length is $f=r/2=(-40\text{ cm})/2=-20\text{ cm}$.

Since $(1/p) + (1/i) = (1/f)$,

$$p = \frac{if}{i - f} .$$

This yields $p = +5.0\,\text{cm}$ if $i = -4.0\,\text{cm}$ and $p = -3.3\,\text{cm}$ if $i = -4.0\,\text{cm}$. Since $p$ must be positive we select $i = -4.0\,\text{cm}$ and take $p$ to be $+5.0\,\text{cm}$.

The magnification is $m = -i/p = -(-4.0\,\text{cm})/(5.0\,\text{cm}) = +0.80$. Since the image distance is negative the image is virtual and on the opposite side of the mirror from the object. Since the magnification is positive the image is not inverted.

**29**

Since the magnification $m$ is $m = -i/p$, where $p$ is the object distance and $i$ is the image distance, $i = -mp$. Use this to substitute for $i$ in $(1/p) + (1/i) = (1/f)$, where $f$ is the focal length. The solve for $p$. The result is

$$p = f\left(1 - \frac{1}{m}\right) = (\pm 30\,\text{cm})\left(1 - \frac{1}{0.20}\right) = \pm 120\,\text{cm} .$$

Since $p$ must be positive we must use the lower sign. Thus the focal length is $-30\,\text{cm}$ and the radius of curvature is $r = 2f == 60\,\text{cm}$. Since the focal length and radius of curvature are negative the mirror is convex.

The object distance is 1.2 m and the image distance is $i = -mp = -(0.20)(120\,\text{cm}) = -24\,\text{cm}$.

Since the image distance is negative the image is virtual and on the opposite side of the mirror from the object. Since the magnification is positive the image is not inverted.

**35**

Solve

$$\frac{n_1}{p} + \frac{n_2}{i} = \frac{n_2 - n_1}{r}$$

for $r$. the result is

$$r = \frac{ip(n_2 - n_1)}{n_1 i + n_2 p} = \frac{(-13\,\text{cm})(+10\,\text{cm})}{((1.0)(-13\,\text{cm}) + (1.5)(+10\,\text{cm})} = -33\,\text{cm} .$$

Since the image distance is negative the image is virtual and appears on the same side of the surface as the object.

**37**

Solve

$$\frac{n_1}{p} + \frac{n_2}{i} = \frac{n_2 - n_1}{r}$$

for $r$. the result is

$$i = \frac{n_2 r p}{(n_2 - n_1)p - n_1 r} = \frac{(1.0)(+30\,\text{cm})(+70\,\text{cm})}{(1.0 - 1.5)(+70\,\text{cm}) - (1.5)(+30\,\text{cm})} = -26\,\text{cm} .$$

Since the image distance is negative the image is virtual and appears on the same side of the surface as the object.

**41**

Use the lens maker's equation, Eq. 34–10:

$$\frac{1}{f} = (n - 1)\left(\frac{1}{r_1} - \frac{1}{r_2}\right),$$

where $f$ is the focal length, $n$ is the index of refraction, $r_1$ is the radius of curvature of the first surface encountered by the light and $r_2$ is the radius of curvature of the second surface. Since one surface has twice the radius of the other and since one surface is convex to the incoming light while the other is concave, set $r_2 = -2r_1$ to obtain

$$\frac{1}{f} = (n - 1)\left(\frac{1}{r_1} + \frac{1}{2r_1}\right) = \frac{3(n-1)}{2r_1}.$$

Solve for $r_1$:

$$r_1 = \frac{3(n-1)f}{2} = \frac{3(1.5-1)(60\,\text{mm})}{2} = 45\,\text{mm}.$$

The radii are 45 mm and 90 mm.

**47**

The object distance $p$ and image distance $i$ obey $(1/p) + (1/i) = (1/f)$, where $f$ is the focal length. In addition, $p + i = L$, where $L$ (= 44 cm) is the distance from the slide to the screen. Use $i = L - p$ to substitute for $i$ in the first equation and obtain $p^2 - pL + Lf = 0$. The solution is

$$p = \frac{L \pm \sqrt{L^2 - 4Lf}}{2} = \frac{(44\,\text{cm}) \pm \sqrt{4(44\,\text{cm})(11\,\text{cm})}}{2} = 22\,\text{cm}.$$

**51**

The lens is diverging, so the focal length is negative. Solve $(1/p) + (1/i) = (1/f)$ for $i$. The result is

$$i = \frac{pf}{p - f} = \frac{(+8.0\,\text{cm})(-12\,\text{cm})}{(8.0\,\text{cm}) - (-12\,\text{cm})} = -4.8\,\text{cm}.$$

The magnification is $m = -i/p = -(-4.8\,\text{cm})/(+8.0\,\text{cm}) = 0.60$. Since the image distance is negative the image is virtual and appears on the same side of the lens as the object. Since the magnification is positive the image is not inverted.

**55**

The lens is converging, so the focal length is positive. Solve $(1/p) + (1/i) = (1/f)$ for $i$. The result is

$$i = \frac{pf}{p - f} = \frac{(+45\,\text{cm})(+20\,\text{cm})}{(45\,\text{cm}) - (+20\,\text{cm})} = +36\,\text{cm}.$$

The magnification is $m = -i/p = -(36\,\text{cm})/(45\,\text{cm}) = -0.80$. Since the image distance is positive the image is real and appears on the opposite side of the lens from the object. Since the magnification is negative the image is inverted.

<u>**61**</u>

The focal length is

$$f = \frac{r_1 r_2}{(n-1)(r_2 - r_1)} = \frac{(+30\,\text{cm})(-42\,\text{cm})}{(1.55-1)[(-42\,\text{cm}) - (+30\,\text{cm})]} = +31.8\,\text{cm}\,.$$

Solve $(1/p) + (1/i) = (1/f)$ for $i$. the result is

$$i = \frac{pf}{p-f} = \frac{(+75\,\text{cm})(+31.8\,\text{cm})}{(+75\,\text{cm}) - (+31.8\,\text{cm})} = 55\,\text{cm}\,.$$

The magnification is $m = -i/p = -(55\,\text{cm})/(75\,\text{cm}) = -0.73$.

Since the image distance is positive the image is real and on the opposite side of the lends from the object. Since the magnification is negative the image is inverted.

<u>**75**</u>

Since $m = -i/p$, $i = -mp = -(+1.25)(+16\,\text{cm}) = -20\,\text{cm}$. Solve $(1/p) + (1/i) = (1/f)$ for $f$. The result is

$$f = \frac{pi}{p+i} = \frac{(+16\,\text{cm})(-20\,\text{cm})}{(+16\,\text{cm}) + (-20\,\text{cm})} = +80\,\text{cm}\,.$$

Since $f$ is positive the lens is a converging lens. Since the image distance is negative the image is virtual and appears on the same side of the lens as the object. Since the magnification is positive the image is not inverted.

<u>**79**</u>

The image is on the same side of the lens as the object. This means that the image is virtual and the image distance is negative. Solve $(1/p) + (1/i) = (1/f)$ for $i$. The result is

$$i = \frac{pf}{p-f}\,.$$

and the magnification is

$$m = -\frac{i}{p} = -\frac{f}{p-f}\,.$$

Since the magnification is less than 1.0, $f$ must be negative and the lens must be a diverging lens. The image distance is

$$i = \frac{(+5.0\,\text{cm})(-10\,\text{cm})}{(5.0\,\text{cm}) - (-10\,\text{cm})} = -3.3\,\text{cm}\,.$$

and the magnification is $m = -i/p = -(-3.3\,\text{cm})/(5.0\,\text{cm}) = 0.66\,\text{cm}$.

Since the magnification is positive the image is not inverted.

**81**

Lens 1 is converging and so has a positive focal length. Solve $(1/p_1)+(1/i_1)=(1/f_1)$ for the image distance $i_1$ associated with the image produced by this lens. The result is

$$i_1=\frac{p_1 f_1}{p_1-f_1}=\frac{(20\text{ cm})(+9.0\text{ cm})}{(20\text{ cm})-(9.0\text{ cm})}=16.4\text{ cm}.$$

This image is the object for lens 2. The object distance is $d-p_2=(8.0\text{ cm})-(16.4\text{ cm})=-8.4\text{ cm}$. The negative sign indicates that the image is behind the second lens. The lens equation is still valid. The second lens has a positive focal length and the image distance for the image it forms is

$$i_2=\frac{p_2 f_2}{p_2-f_2}=\frac{(-8.4\text{ cm})(5.0\text{ cm})}{(-8.4\text{ cm})-(5.0\text{ cm})}=+3.1\text{ cm}.$$

The overall magnification is the product of the individual magnifications:

$$m=m_1m_2=\left(-\frac{i_1}{p_1}\right)\left(-\frac{i_2}{p_2}\right)=\left(-\frac{16.4\text{ cm}}{20\text{ cm}}\right)\left(-\frac{3.1\text{ cm}}{-8.4\text{ cm}}\right)=-0.30.$$

Since the final image distance is positive the final image is real and on the opposite side of lens 2 from the object. Since the magnification is negative the image is inverted.

**89**

(a) If $L$ is the distance between the lenses, then according to Fig. 34–20, the tube length is $s=L-f_{ob}-f_{ey}=25.0\text{ cm}-4.00\text{ cm}-8.00\text{ cm}=13.0\text{ cm}$.

(b) Solve $(1/p)+(1/i)=(1/f_{ob})$ for $p$. The image distance is $i=f_{ob}+s=4.00\text{ cm}+13.0\text{ cm}=17.0\text{ cm}$, so

$$p=\frac{i f_{ob}}{i-f_{ob}}=\frac{(17.0\text{ cm})(4.00\text{ cm})}{17.0\text{ cm}-4.00\text{ cm}}=5.23\text{ cm}.$$

(c) The magnification of the objective is

$$m=-\frac{i}{p}=-\frac{17.0\text{ cm}}{5.23\text{ cm}}=-3.25.$$

(d) The angular magnification of the eyepiece is

$$m_\theta=\frac{25\text{ cm}}{f_{ey}}=\frac{25\text{ cm}}{8.00\text{ cm}}=3.13.$$

(e) The overall magnification of the microscope is

$$M=mm_\theta=(-3.25)(3.13)=-10.2.$$

**93**

(a) When the eye is relaxed, its lens focuses far-away objects on the retina, a distance $i$ behind the lens. Set $p=\infty$ in the thin lens equation to obtain $1/i=1/f$, where $f$ is the focal length of the relaxed effective lens. Thus $i = f = 2.50\,\text{cm}$. When the eye focuses on closer objects, the image distance $i$ remains the same but the object distance and focal length change. If $p$ is the new object distance and $f'$ is the new focal length, then

$$\frac{1}{p} + \frac{1}{i} = \frac{1}{f'} .$$

Substitute $i = f$ and solve for $f'$. You should obtain

$$f' = \frac{pf}{f + p} = \frac{(40.0\,\text{cm})(2.50\,\text{cm})}{40.0\,\text{cm} + 2.50\,\text{cm}} = 2.35\,\text{cm} .$$

(b) Consider the lensmaker's equation

$$\frac{1}{f} = (n - 1)\left(\frac{1}{r_1} - \frac{1}{r_2}\right) ,$$

where $r_1$ and $r_2$ are the radii of curvature of the two surfaces of the lens and $n$ is the index of refraction of the lens material. For the lens pictured in Fig. 34–46, $r_1$ and $r_2$ have about the same magnitude, $r_1$ is positive, and $r_2$ is negative. Since the focal length decreases, the combination $(1/r_1) - (1/r_2)$ must increase. This can be accomplished by decreasing the magnitudes of either or both radii.

**103**

For a thin lens, $(1/p) + (1/i) = (1/f)$, where $p$ is the object distance, $i$ is the image distance, and $f$ is the focal length. Solve for $i$:

$$i = \frac{fp}{p - f} .$$

Let $p = f + x$, where $x$ is positive if the object is outside the focal point and negative if it is inside. Then

$$i = \frac{f(f + x)}{x} .$$

Now let $i = f + x'$, where $x'$ is positive if the image is outside the focal point and negative if it is inside. Then

$$x' = i - f = \frac{f(f + x)}{x} - f = \frac{f^2}{x}$$

and $xx' = f^2$.

**105**

Place an object far away from the composite lens and find the image distance $i$. Since the image is at a focal point, $i = f$, the effective focal length of the composite. The final image is produced by two lenses, with the image of the first lens being the object for the second. For the first lens, $(1/p_1) + (1/i_1) = (1/f_1)$, where $f_1$ is the focal length of this lens and $i_1$ is the image distance for the image it forms. Since $p_1 = \infty$, $i_1 = f_1$.

The thin lens equation, applied to the second lens, is $(1/p_2) + (1/i_2) = (1/f_2)$, where $p_2$ is the object distance, $i_2$ is the image distance, and $f_2$ is the focal length. If the thicknesses of the lenses can be ignored, the object distance for the second lens is $p_2 = -i_1$. The negative sign must be used since the image formed by the first lens is beyond the second lens if $i_1$ is positive. This means the object for the second lens is virtual and the object distance is negative. If $i_1$ is negative, the image formed by the first lens is in front of the second lens and $p_2$ is positive. In the thin lens equation, replace $p_2$ with $-f_1$ and $i_2$ with $f$ to obtain

$$-\frac{1}{f_1} + \frac{1}{f} = \frac{1}{f_2}.$$

The solution for $f$ is

$$f = \frac{f_1 f_2}{f_1 + f_2}.$$

**107**

(a) and (b) Since the height of the image is twice the height of the fly and since the fly and its image have the same orientation the magnification of the lens is $m = +2.0$. Since $m = -i/p$, where $p$ is the object distance and $i$ is the image distance, $i = -2p$. Now $|p+i| = d$, so $|-p| = d$ and $p = d = 20$ cm. The image distance is $-40$ cm.
Solve $(1/p) + (1/i) = (1/f)$ for $f$. the result is

$$f = \frac{pi}{p+i} = \frac{(20\text{ cm})(-40\text{ cm})}{(20\text{ cm}) + (-40\text{ cm})} = +40\text{ cm}.$$

(c) and (d) Now $m = +0.5$ and $i = -0.5p$. Since $|p+i| = d$, $0.5p = d$ and $p = 2d = 40$ cm. The image distance is $-20$ cm and the focal length is

$$f = \frac{pi}{p+i} = \frac{(40\text{ cm})(-20\text{ cm})}{(40\text{ cm}) + (-20\text{ cm})} = -40\text{ cm}.$$

# Chapter 35

<u>5</u>

(a) Take the phases of both waves to be zero at the front surfaces of the layers. The phase of the first wave at the back surface of the glass is given by $\phi_1 = k_1 L - \omega t$, where $k_1$ $(= 2\pi/\lambda_1)$ is the angular wave number and $\lambda_1$ is the wavelength in glass. Similarly, the phase of the second wave at the back surface of the plastic is given by $\phi_2 = k_2 L - \omega t$, where $k_2$ $(= 2\pi/\lambda_2)$ is the angular wave number and $\lambda_2$ is the wavelength in plastic. The angular frequencies are the same since the waves have the same wavelength in air and the frequency of a wave does not change when the wave enters another medium. The phase difference is

$$\phi_1 - \phi_2 = (k_1 - k_2)L = 2\pi \left( \frac{1}{\lambda_1} - \frac{1}{\lambda_2} \right) L .$$

Now $\lambda_1 = \lambda_{\text{air}}/n_1$, where $\lambda_{\text{air}}$ is the wavelength in air and $n_1$ is the index of refraction of the glass. Similarly, $\lambda_2 = \lambda_{\text{air}}/n_2$, where $n_2$ is the index of refraction of the plastic. This means that the phase difference is $\phi_1 - \phi_2 = (2\pi/\lambda_{\text{air}})(n_1 - n_2)L$. The value of $L$ that makes this 5.65 rad is

$$L = \frac{(\phi_1 - \phi_2)\lambda_{\text{air}}}{2\pi(n_1 - n_2)} = \frac{5.65(400 \times 10^{-9}\,\text{m})}{2\pi(1.60 - 1.50)} = 3.60 \times 10^{-6}\,\text{m} .$$

(b) 5.65 rad is less than $2\pi$ rad (= 6.28 rad), the phase difference for completely constructive interference, and greater than $\pi$ rad (= 3.14 rad), the phase difference for completely destructive interference. The interference is therefore intermediate, neither completely constructive nor completely destructive. It is, however, closer to completely constructive than to completely destructive.

<u>15</u>

Interference maxima occur at angles $\theta$ such that $d \sin\theta = m\lambda$, where $d$ is the separation of the sources, $\lambda$ is the wavelength, and $m$ is an integer. Since $d = 2.0$ m and $\lambda = 0.50$ m, this means that $\sin\theta = 0.25m$. You want all values of $m$ (positive and negative) for which $|0.25m| \leq 1$. These are $-4$, $-3$, $-2$, $-1$, 0, +1, +2, +3, and +4. For each of these except $-4$ and +4, there are two different values for $\theta$. A single value of $\theta$ ($-90°$) is associated with $m = -4$ and a single value ($-90°$) is associated with $m = +4$. There are sixteen different angles in all and therefore sixteen maxima.

<u>17</u>

The angular positions of the maxima of a two-slit interference pattern are given by $d \sin\theta = m\lambda$, where $d$ is the slit separation, $\lambda$ is the wavelength, and $m$ is an integer. If $\theta$ is small, $\sin\theta$ may be approximated by $\theta$ in radians. Then $d\theta = m\lambda$. The angular separation of two adjacent maxima is $\Delta\theta = \lambda/d$. Let $\lambda'$ be the wavelength for which the angular separation is 10.0% greater. Then $1.10\lambda/d = \lambda'/d$ or $\lambda' = 1.10\lambda = 1.10(589\,\text{nm}) = 648\,\text{nm}$.

**19**

The condition for a maximum in the two-slit interference pattern is $d\sin\theta = m\lambda$, where $d$ is the slit separation, $\lambda$ is the wavelength, $m$ is an integer, and $\theta$ is the angle made by the interfering rays with the forward direction. If $\theta$ is small, $\sin\theta$ may be approximated by $\theta$ in radians. Then $d\theta = m\lambda$ and the angular separation of adjacent maxima, one associated with the integer $m$ and the other associated with the integer $m+1$, is given by $\Delta\theta = \lambda/d$. The separation on a screen a distance $D$ away is given by $\Delta y = D\,\Delta\theta = \lambda D/d$. Thus

$$\Delta y = \frac{(500\times10^{-9}\,\text{m})(5.40\,\text{m})}{1.20\times10^{-3}\,\text{m}} = 2.25\times10^{-3}\,\text{m} = 2.25\,\text{mm}\,.$$

**29**

The phasor diagram is shown to the right. Here $E_1 = 1.00$, $E_2 = 2.00$, and $\phi = 60°$. The resultant amplitude $E_m$ is given by the trigonometric law of cosines:

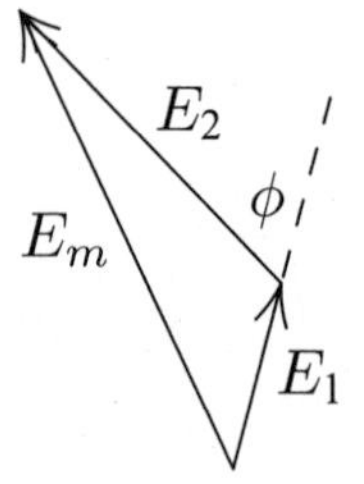


$$E_m^2 = E_1^2 + E_2^2 - 2E_1E_2\cos(180° - \phi)\,,$$

so

$$E_m = \sqrt{(1.00)^2 + (2.00)^2 - 2(1.00)(2.00)\cos 120°} = 2.65\,.$$

**39**

For complete destructive interference, you want the waves reflected from the front and back of the coating to differ in phase by an odd multiple of $\pi$ rad. Each wave is incident on a medium of higher index of refraction from a medium of lower index, so both suffer phase changes of $\pi$ rad on reflection. If $L$ is the thickness of the coating, the wave reflected from the back surface travels a distance $2L$ farther than the wave reflected from the front. The phase difference is $2L(2\pi/\lambda_c)$, where $\lambda_c$ is the wavelength in the coating. If $n$ is the index of refraction of the coating, $\lambda_c = \lambda/n$, where $\lambda$ is the wavelength in vacuum, and the phase difference is $2nL(2\pi/\lambda)$. Solve

$$2nL\left(\frac{2\pi}{\lambda}\right) = (2m+1)\pi$$

for $L$. Here $m$ is an integer. The result is

$$L = \frac{(2m+1)\lambda}{4n}\,.$$

To find the least thickness for which destructive interference occurs, take $m = 0$. Then

$$L = \frac{\lambda}{4n} = \frac{600\times10^{-9}\,\text{m}}{4(1.25)} = 1.2\times10^{-7}\,\text{m}\,.$$

<u>41</u>

Since $n_1$ is greater than $n_2$ there is no change in phase on reflection from the first surface. Since $n_2$ is less than $n_3$ there is a change in phase of $\pi$ rad on reflection from the second surface. One wave travels a distance $2L$ further than the other, so the difference in the phases of the two waves is $4\pi L/\lambda_2 + \pi$, where $\lambda_2$ is the wavelength in medium 2. Since interference produces a minimum the phase difference must be an odd multiple of $\pi$. Thus $4\pi L/\lambda_2 + \pi = (2m+1)\pi$, where $m$ is an integer or zero. Replace $\lambda_2$ with $\lambda/n_2$, where $\lambda$ is the wavelength in air, and solve for $\lambda$. The result is

$$\lambda = \frac{4Ln_2}{2m} = \frac{2(380\,\text{nm})(1.1.34)}{m} = \frac{1018\,\text{nm}}{m}\,.$$

For $m = 1$, $\lambda = 1018\,\text{nm}$ and for $m = 2$, $\lambda = (1018\,\text{nm})/2 = 509\,\text{nm}$. Other wavelengths are shorter. Only $\lambda = 509\,\text{nm}$ is in the visible range.

<u>47</u>

There is a phase shift on reflection of $\pi$ for both waves and one wave travels a distance $2L$ further than the other, so the phase difference of the reflected waves is $4\pi L/\lambda_2$, where $\lambda_2$ is the wavelength in medium 2. Since the result of the interference is a minimum of intensity the phase difference must be an odd multiple of $\pi$. Thus $4\pi L/\lambda_2 = (2m+1)\pi$, where $m$ is an integer or zero. Replace $\lambda_2$ with $\lambda/n_2$, where $\lambda$ is the wavelength in air, and solve for $\lambda$. The result is

$$\lambda = \frac{4Ln_2}{2m+1} = \frac{4(210\,\text{nm})(1.46)}{2m+1} = \frac{1226\,\text{nm}}{2m+1}\,.$$

For $m = 1$, $\lambda = (1226\,\text{nm})/3 = 409\,\text{nm}$. This is in the visible range. Other values of $m$ are associated with wavelengths that are not in the visible range.

<u>53</u>

(a) Oil has a greater index of refraction than air and water has a still greater index of refraction. There is a change of phase of $\pi$ rad at each reflection. One wave travels a distance $2L$ further than the other, where $L$ is the thickness of the oil. The phase difference of the two reflected waves is $4\pi L/\lambda_o$, where $\lambda$ is the wavelength in oil, and this must be equal to a multiple of $2\pi$ for a bright reflection. Thus $4\pi L/\lambda_o = 2m\pi$, where $m$ is an integer. Use $\lambda = n_o\lambda_o$, where $n_o$ is the index of refraction for oil, to find the wavelength in air. The result is

$$\lambda = \frac{2n_oL}{m} = \frac{2(1.20)(460\,\text{nm})}{m} = \frac{1104\,\text{nm}}{m}\,.$$

For $m = 1$, $\lambda = 1104\,\text{nm}$; for $m = 2$, $\lambda = (1104\,\text{nm})/2 = 552\,\text{nm}$; and for $m = 3$, $\lambda = (1104\,\text{nm})/3 = 368\,\text{nm}$. Other wavelengths are shorter. Only $\lambda = 552\,\text{nm}$ is in the visible range.

(b) A maximum in transmission occurs for wavelengths for which the reflection is a minimum. The phases of the two reflected waves then differ by an odd multiple of $\pi$ rad. This means $4\pi L/\lambda_o = (2m+1)\pi$ and

$$\lambda = \frac{4n_oL}{2m+1} = \frac{4(1.20)(460\,\text{nm})}{2m+1} = \frac{2208\,\text{nm}}{2m+1}\,.$$

For $m = 0$, $\lambda = 2208\,\text{nm}$; for $m = 1$, $\lambda = (2208\,\text{nm})/3 = 736\,\text{nm}$; and for $m = 3$, $\lambda = (2208\,\text{nm})/5 = 442\,\text{nm}$. Other wavelengths are shorter. Only $\lambda = 442\,\text{nm}$ falls in the visible range.

**63**

One wave travels a distance $2L$ further than the other. This wave is reflected twice, once from the back surface and once from the front surface. Since $n_2$ is greater than $n_3$ there is no change in phase at the back-surface reflection. Since $n_1$ is greater than $n_2$ there is a phase change of $\pi$ at the front-surface reflection. Thus the phase difference of the two waves as they exit material 2 is $4\pi L/\lambda_2 + \pi$, where $\lambda_2$ is the wavelength in material 2. For a maximum in intensity the phase difference is a multiple of $2\pi$. Thus $4\pi L/\lambda_2 + \pi = 2m\pi$, where $m$ is an integer. The solution for $\lambda_2$ is

$$\lambda_2 = \frac{4L}{2m-1} = \frac{4(415\,\text{nm})}{2m-1} = \frac{1660\,\text{nm}}{2m-1}\,.$$

The wavelength in air is

$$\lambda = n_2\lambda_2 = \frac{(1.59)(1660\,\text{nm})}{2m-1} = \frac{2639\,\text{nm}}{2m-1}\,.$$

For $m = 1$, $\lambda = 2639\,\text{nm}$; for $m = 2$, $\lambda = 880\,\text{nm}$; for $m = 3$, $\lambda = 528\,\text{nm}$; and for $m = 4$, $\lambda = 377\,\text{nm}$. Other wavelengths are shorter. Only $\lambda = 528\,\text{nm}$ is in the visible range.

**71**

Consider the interference of waves reflected from the top and bottom surfaces of the air film. The wave reflected from the upper surface does not change phase on reflection but the wave reflected from the bottom surface changes phase by $\pi$ rad. At a place where the thickness of the air film is $L$, the condition for fully constructive interference is $2L = (m + \frac{1}{2})\lambda$, where $\lambda$ ($= 683\,\text{nm}$) is the wavelength and $m$ is an integer. This is satisfied for $m = 140$:

$$L = \frac{(m + \frac{1}{2})\lambda}{2} = \frac{(140.5)(683 \times 10^{-9}\,\text{m})}{2} = 4.80 \times 10^{-5}\,\text{m} = 0.048\,\text{mm}\,.$$

At the thin end of the air film, there is a bright fringe. It is associated with $m = 0$. There are, therefore, 140 bright fringes in all.

**75**

Consider the interference pattern formed by waves reflected from the upper and lower surfaces of the air wedge. The wave reflected from the lower surface undergoes a $\pi$-rad phase change while the wave reflected from the upper surface does not. At a place where the thickness of the wedge is $d$, the condition for a maximum in intensity is $2d = (m + \frac{1}{2})\lambda$, where $\lambda$ is the wavelength in air and $m$ is an integer. Thus $d = (2m+1)\lambda/4$. As the geometry of Fig. 35–47 shows, $d = R - \sqrt{R^2 - r^2}$, where $R$ is the radius of curvature of the lens and $r$ is the radius of a Newton's ring. Thus $(2m+1)\lambda/4 = R - \sqrt{R^2 - r^2}$. Solve for $r$. First rearrange the terms so the equation becomes

$$\sqrt{R^2 - r^2} = R - \frac{(2m+1)\lambda}{4}\,.$$

Now square both sides and solve for $r^2$. When you take the square root, you should get

$$r = \sqrt{\frac{(2m+1)R\lambda}{2} - \frac{(2m+1)^2\lambda^2}{16}}\,.$$

If $R$ is much larger than a wavelength, the first term dominates the second and

$$r = \sqrt{\frac{(2m+1)R\lambda}{2}}\,.$$

**81**

Let $\phi_1$ be the phase difference of the waves in the two arms when the tube has air in it and let $\phi_2$ be the phase difference when the tube is evacuated. These are different because the wavelength in air is different from the wavelength in vacuum. If $\lambda$ is the wavelength in vacuum, then the wavelength in air is $\lambda/n$, where $n$ is the index of refraction of air. This means

$$\phi_1 - \phi_2 = 2L\left[\frac{2\pi n}{\lambda} - \frac{2\pi}{\lambda}\right] = \frac{4\pi(n-1)L}{\lambda}\,,$$

where $L$ is the length of the tube. The factor 2 arises because the light traverses the tube twice, once on the way to a mirror and once after reflection from the mirror.

Each shift by one fringe corresponds to a change in phase of $2\pi$ rad, so if the interference pattern shifts by $N$ fringes as the tube is evacuated,

$$\frac{4\pi(n-1)L}{\lambda} = 2N\pi$$

and

$$n = 1 + \frac{N\lambda}{2L} = 1 + \frac{60(500\times10^{-9}\,\text{m})}{2(5.0\times10^{-2}\,\text{m})} = 1.00030\,.$$

**87**

Suppose the wave that goes directly to the receiver travels a distance $L_1$ and the reflected wave travels a distance $L_2$. Since the index of refraction of water is greater than that of air this last wave suffers a phase change on reflection of half a wavelength. To obtain constructive interference at the receiver the difference $L_2 - L_2$ in the distances traveled must be an odd multiple of a half wavelength.

Look at the diagram on the right. The right triangle on the left, formed by the vertical line from the water to the transmitter T, the ray incident on the water, and the water line, gives $D_a = a/\tan\theta$ and the right triangle on the right, formed by the vertical line from the water to the receiver R, the reflected ray, and the water line gives $D_b = x/\tan\theta$. Since $D_a + D_b = D$,

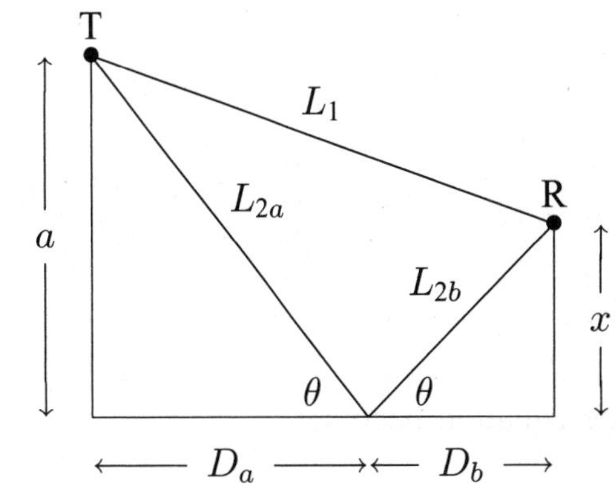


$$\tan\theta = \frac{a+x}{D}\,.$$

Use the identity $\sin^2\theta = \tan^2\theta/(1+\tan^2\theta)$ to show that $\sin\theta = (a+x)/\sqrt{D^2+(a+x)^2}$. This means

$$L_{2a} = \frac{a}{\sin\theta} = \frac{a\sqrt{D^2+(a+x)^2}}{a+x}$$

and

$$L_{2b} = \frac{x}{\sin\theta} = \frac{x\sqrt{D^2+(a+x)^2}}{a+x},$$

so

$$L_2 = L_{2a} + L_{2b} = \frac{(a+x)\sqrt{D^2+(a+x)^2}}{a+x} = \sqrt{D^2+(a+x)^2}\,.$$

Use the binomial theorem, with $D^2$ large and $a^2 + x^2$ small, to approximate this expression: $L_2 \approx D + (a+x)^2/2D$.

The distance traveled by the direct wave is $L_1 = \sqrt{D^2+(a-x)^2}$. Use the binomial theorem to approximate this expression: $L_1 \approx D + (a-x)^2/2D$. Thus

$$L_2 - L_1 \approx D + \frac{a^2+2ax+x^2}{2D} - D - \frac{a^2-2ax+x^2}{2D} = \frac{2ax}{D}\,.$$

Set this equal to $(m+\frac{1}{2})\lambda$, where $m$ is zero or a positive integer. Solve for $x$. The result is $x = (m+\frac{1}{2})(D/2a)\lambda$.

**89**

Bright fringes occur at an angle $\theta$ such that $d\sin\theta = m\lambda$, where $d$ is the slit separation, $\lambda$ is the wavelength in the medium of propagation, and $m$ is an integer. Near the center of the pattern the angles are small and $\sin\theta$ can be approximated by $\theta$ in radians. Thus $\theta = m\lambda/d$ and the angular separation of two adjacent bright fringes is $\Delta\theta = \lambda/d$. When the arrangement is immersed in water the angular separation of the fringes becomes $\Delta\theta' = \lambda_w/d$, where $\lambda_w$ is the wavelength in water. Since $\lambda_w = \lambda/n_w$, where $n_w$ is the index of refraction of water, $\Delta\theta' = \lambda/n_w d = (\Delta\theta)/n_w$. Since the units of the angles cancel from this equation we may substitute the angles in degrees and obtain $\Delta\theta' = 0.30°/1.33 = 0.23°$.

**93**

(a) For wavelength $\lambda$ dark bands occur where the path difference is an odd multiple of $\lambda/2$. That is, where the path difference is $(2m+1)\lambda/2$, where $m$ is an integer. The fourth dark band from the central bright fringe is associated with $m = 3$ and is $7\lambda/2 = 7(500\,\text{nm})/2 = 1750\,\text{nm}$.

(b) The angular position $\theta$ of the first bright band on either side of the central band is given by $\sin\theta = \lambda/d$, where $d$ is the slit separation. The distance on the screen is given by $\Delta y = D\tan\theta$, where $D$ is the distance from the slits to the screen. Because $\theta$ is small its sine and tangent are very nearly equal and $\Delta y = D\sin\theta = D\lambda/d$.

Dark bands have angular positions that are given by $\sin\theta = (m+\frac{1}{2})\lambda/d$ and, for the fourth dark band, $m = 3$ and $\sin\theta_4 = (7/2)\lambda/d$. Its distance on the screen from the central fringe is $\Delta y_4 = D\tan\theta_4 = D\sin\theta_4 = 7D\lambda/2d$. This means that $D\lambda/d = 2\Delta y_4/7 = 2(1.68\,\text{cm})/7 = 0.48\,\text{cm}$. Note that this is $\Delta y$.

**97**

(a) If $I$ is the incident intensity then the radiation pressure for total absorption is

$$p_r = \frac{I}{c} = \frac{1.4 \times 10^3\,\mathrm{W/m^2}}{3.00 \times 10^8\,\mathrm{m/s}} = 4.67 \times 10^{-6}\,\mathrm{Pa}\,.$$

(b) The ratio is

$$\text{ratio} = \frac{4.57 \times 10^{-6}\,\mathrm{Pa}}{1.0 \times 10^5\,\mathrm{Pa}} = 4.7 \times 10^{-11}\,.$$

**99**

Minima occur at angles $\theta$ for which $\sin\theta = (m + \frac{1}{2})\lambda/d$, where $\lambda$ is the wavelength, $d$ is the slit separation, and $m$ is an integer. For the first minimum, $m = 0$ and $\sin\theta_1 = \lambda/2d$. For the tenth minimum, $m = 9$ and $\sin\theta_{10} = 19\lambda/2d$.

The distance on the screen from the central fringe to a minimum is $y = D\tan\theta$, where $D$ is the distance from the slits to the screen. Since the angle is small we may approximate its tangent with its sine and write $y = D\sin\theta = D(m + \frac{1}{2})\lambda/d$. Thus the separation of the first and tenth minima is

$$\Delta y = \frac{D}{d}\left(\frac{19\lambda}{2} - \frac{\lambda}{2}\right) = \frac{9D\lambda}{d}$$

and

$$\lambda = \frac{d\,\Delta y}{9D} = \frac{(0.150 \times 10^{-3}\,\mathrm{m})(18.0 \times 10^{-3}\,\mathrm{m})}{9(50.0 \times 10^{-2}\,\mathrm{m})} = 6.00 \times 10^{-7}\,\mathrm{m}\,.$$

**103**

The difference in the path lengths of the two beams is $2x$, so their difference in phase when they reach the detector is $\phi = 4\pi x/\lambda$, where $\lambda$ is the wavelength. Assume their amplitudes are the same. According to Eq. 35–22 the intensity associated with the addition of two waves is proportional to the square of the cosine of half their phase difference. Thus the intensity of the light observed in the interferometer is proportional to $\cos^2(2\pi x/\lambda)$. Since the intensity is maximum when $x = 0$ (and the arms have equal lengths), the constant of proportionality is the maximum intensity $I_m$ and $I = I_m\cos^2(2\pi x/\lambda)$.

# Chapter 36

**9**

The condition for a minimum of intensity in a single-slit diffraction pattern is $a \sin\theta = m\lambda$, where $a$ is the slit width, $\lambda$ is the wavelength, and $m$ is an integer. To find the angular position of the first minimum to one side of the central maximum, set $m = 1$:

$$\theta_1 = \sin^{-1}\left(\frac{\lambda}{a}\right) = \sin^{-1}\left(\frac{589 \times 10^{-9}\,\text{m}}{1.00 \times 10^{-3}\,\text{m}}\right) = 5.89 \times 10^{-4}\,\text{rad}\,.$$

If $D$ is the distance from the slit to the screen, the distance on the screen from the center of the pattern to the minimum is $y_1 = D\tan\theta_1 = (3.00\,\text{m})\tan(5.89 \times 10^{-4}\,\text{rad}) = 1.767 \times 10^{-3}\,\text{m}$.

To find the second minimum, set $m = 2$:

$$\theta_2 = \sin^{-1}\left[\frac{2(589 \times 10^{-9}\,\text{m})}{1.00 \times 10^{-3}\,\text{m}}\right] = 1.178 \times 10^{-3}\,\text{rad}\,.$$

The distance from the pattern center to the minimum is $y_2 = D\tan\theta_2 = (3.00\,\text{m})\tan(1.178 \times 10^{-3}\,\text{rad}) = 3.534 \times 10^{-3}\,\text{m}$. The separation of the two minima is $\Delta y = y_2 - y_1 = 3.534\,\text{mm} - 1.767\,\text{mm} = 1.77\,\text{mm}$.

**17**

(a) The intensity for a single-slit diffraction pattern is given by

$$I = I_m \frac{\sin^2\alpha}{\alpha^2}\,,$$

where $\alpha = (\pi a/\lambda)\sin\theta$, $a$ is the slit width and $\lambda$ is the wavelength. The angle $\theta$ is measured from the forward direction. You want $I = I_m/2$, so

$$\sin^2\alpha = \frac{1}{2}\alpha^2\,.$$

(b) Evaluate $\sin^2\alpha$ and $\alpha^2/2$ for $\alpha = 1.39\,\text{rad}$ and compare the results. To be sure that 1.39 rad is closer to the correct value for $\alpha$ than any other value with three significant digits, you should also try 1.385 rad and 1.395 rad.

(c) Since $\alpha = (\pi a/\lambda)\sin\theta$,

$$\theta = \sin^{-1}\left(\frac{\alpha\lambda}{\pi a}\right)\,.$$

Now $\alpha/\pi = 1.39/\pi = 0.442$, so

$$\theta = \sin^{-1}\left(\frac{0.442\lambda}{a}\right)\,.$$

The angular separation of the two points of half intensity, one on either side of the center of the diffraction pattern, is

$$\Delta\theta = 2\theta = 2\sin^{-1}\left(\frac{0.442\lambda}{a}\right).$$

(d) For $a/\lambda = 1.0$,

$$\Delta\theta = 2\sin^{-1}(0.442/1.0) = 0.916\,\text{rad} = 52.5°.$$

(e) For $a/\lambda = 5.0$,

$$\Delta\theta = 2\sin^{-1}(0.442/5.0) = 0.177\,\text{rad} = 10.1°.$$

(f) For $a/\lambda = 10$,

$$\Delta\theta = 2\sin^{-1}(0.442/10) = 0.0884\,\text{rad} = 5.06°.$$

**21**

(a) Use the Rayleigh criteria. To resolve two point sources, the central maximum of the diffraction pattern of one must lie at or beyond the first minimum of the diffraction pattern of the other. This means the angular separation of the sources must be at least $\theta_R = 1.22\lambda/d$, where $\lambda$ is the wavelength and $d$ is the diameter of the aperture. For the headlights of this problem,

$$\theta_R = \frac{1.22(550 \times 10^{-9}\,\text{m})}{5.0 \times 10^{-3}\,\text{m}} = 1.3 \times 10^{-4}\,\text{rad}.$$

(b) If $D$ is the distance from the headlights to the eye when the headlights are just resolvable and $\ell$ is the separation of the headlights, then $\ell = D\tan\theta_R \approx D\theta_R$, where the small angle approximation $\tan\theta_R \approx \theta_R$ was made. This is valid if $\theta_R$ is measured in radians. Thus $D = \ell/\theta_R = (1.4\,\text{m})/(1.34 \times 10^{-4}\,\text{rad}) = 1.0 \times 10^4\,\text{m} = 10\,\text{km}$.

**25**

(a) Use the Rayleigh criteria: two objects can be resolved if their angular separation at the observer is greater than $\theta_R = 1.22\lambda/d$, where $\lambda$ is the wavelength of the light and $d$ is the diameter of the aperture (the eye or mirror). If $D$ is the distance from the observer to the objects, then the smallest separation $\ell$ they can have and still be resolvable is $\ell = D\tan\theta_R \approx D\theta_R$, where $\theta_R$ is measured in radians. The small angle approximation $\tan\theta_R \approx \theta_R$ was made. Thus

$$\ell = \frac{1.22D\lambda}{d} = \frac{1.22(8.0 \times 10^{10}\,\text{m})(550 \times 10^{-9}\,\text{m})}{5.0 \times 10^{-3}\,\text{m}} = 1.1 \times 10^7\,\text{m} = 1.1 \times 10^4\,\text{km}.$$

This distance is greater than the diameter of Mars. One part of the planet's surface cannot be resolved from another part.

(b) Now $d = 5.1$ m and

$$\ell = \frac{1.22(8.0 \times 10^{10}\,\text{m})(550 \times 10^{-9}\,\text{m})}{5.1\,\text{m}} = 1.1 \times 10^4\,\text{m} = 11\,\text{km}.$$

## 29

(a) The first minimum in the diffraction pattern is at an angular position $\theta$, measured from the center of the pattern, such that $\sin\theta = 1.22\lambda/d$, where $\lambda$ is the wavelength and $d$ is the diameter of the antenna. If $f$ is the frequency, then the wavelength is

$$\lambda = \frac{c}{f} = \frac{3.00\times 10^{8}\,\text{m/s}}{220\times 10^{9}\,\text{Hz}} = 1.36\times 10^{-3}\,\text{m}\,.$$

Thus

$$\theta = \sin^{-1}\left(\frac{1.22\lambda}{d}\right) = \sin^{-1}\left(\frac{1.22(1.36\times 10^{-3}\,\text{m})}{55.0\times 10^{-2}\,\text{m}}\right) = 3.02\times 10^{-3}\,\text{rad}\,.$$

The angular width of the central maximum is twice this, or $6.04\times 10^{-3}$ rad (0.346°).

(b) Now $\lambda = 1.6$ cm and $d = 2.3$ m, so

$$\theta = \sin^{-1}\left(\frac{1.22(1.6\times 10^{-2}\,\text{m})}{2.3\,\text{m}}\right) = 8.5\times 10^{-3}\,\text{rad}\,.$$

The angular width of the central maximum is $1.7\times 10^{-2}$ rad (0.97°).

## 39

(a) The angular positions $\theta$ of the bright interference fringes are given by $d\sin\theta = m\lambda$, where $d$ is the slit separation, $\lambda$ is the wavelength, and $m$ is an integer. The first diffraction minimum occurs at the angle $\theta_1$ given by $a\sin\theta_1 = \lambda$, where $a$ is the slit width. The diffraction peak extends from $-\theta_1$ to $+\theta_1$, so you want to count the number of values of $m$ for which $-\theta_1 < \theta < +\theta_1$, or what is the same, the number of values of $m$ for which $-\sin\theta_1 < \sin\theta < +\sin\theta_1$. This means $-1/a < m/d < 1/a$ or $-d/a < m < +d/a$. Now $d/a = (0.150\times 10^{-3}\,\text{m})/(30.0\times 10^{-6}\,\text{m}) = 5.00$, so the values of $m$ are $m = -4, -3, -2, -1, 0, +1, +2, +3$, and $+4$. There are nine fringes.

(b) The intensity at the screen is given by

$$I = I_m\left(\cos^2\beta\right)\left(\frac{\sin\alpha}{\alpha}\right)^2,$$

where $\alpha = (\pi a/\lambda)\sin\theta$, $\beta = (\pi d/\lambda)\sin\theta$, and $I_m$ is the intensity at the center of the pattern. For the third bright interference fringe, $d\sin\theta = 3\lambda$, so $\beta = 3\pi$ rad and $\cos^2\beta = 1$. Similarly, $\alpha = 3\pi a/d = 3\pi/5.00 = 0.600\pi$ rad and

$$\left(\frac{\sin\alpha}{\alpha}\right)^2 = \left(\frac{\sin 0.600\pi}{0.600\pi}\right)^2 = 0.255\,.$$

The intensity ratio is $I/I_m = 0.255$.

## 45

The ruling separation is $d = 1/(400\,\text{mm}^{-1}) = 2.5\times 10^{-3}$ mm. Diffraction lines occur at angles $\theta$ such that $d\sin\theta = m\lambda$, where $\lambda$ is the wavelength and $m$ is an integer. Notice that for a given order, the line associated with a long wavelength is produced at a greater angle than the line associated with a shorter wavelength. Take $\lambda$ to be the longest wavelength in the visible spectrum (700 nm) and find the greatest integer value of $m$ such that $\theta$ is less than 90°. That is, find the greatest integer value of $m$ for which $m\lambda < d$. Since $d/\lambda = (2.5 \times 10^{-6}\,\text{m})/(700 \times 10^{-9}\,\text{m}) = 3.57$, that value is $m = 3$. There are three complete orders on each side of the $m = 0$ order. The second and third orders overlap.

**51**

(a) Maxima of a diffraction grating pattern occur at angles $\theta$ given by $d \sin\theta = m\lambda$, where $d$ is the slit separation, $\lambda$ is the wavelength, and $m$ is an integer. The two lines are adjacent, so their order numbers differ by unity. Let $m$ be the order number for the line with $\sin\theta = 0.2$ and $m+1$ be the order number for the line with $\sin\theta = 0.3$. Then $0.2d = m\lambda$ and $0.3d = (m+1)\lambda$. Subtract the first equation from the second to obtain $0.1d = \lambda$, or $d = \lambda/0.1 = (600 \times 10^{-9}\,\text{m})/0.1 = 6.0 \times 10^{-6}\,\text{m}$.

(b) Minima of the single-slit diffraction pattern occur at angles $\theta$ given by $a \sin\theta = m\lambda$, where $a$ is the slit width. Since the fourth-order interference maximum is missing, it must fall at one of these angles. If $a$ is the smallest slit width for which this order is missing, the angle must be given by $a \sin\theta = \lambda$. It is also given by $d \sin\theta = 4\lambda$, so $a = d/4 = (6.0 \times 10^{-6}\,\text{m})/4 = 1.5 \times 10^{-6}\,\text{m}$.

(c) First, set $\theta = 90°$ and find the largest value of $m$ for which $m\lambda < d \sin\theta$. This is the highest order that is diffracted toward the screen. The condition is the same as $m < d/\lambda$ and since $d/\lambda = (6.0 \times 10^{-6}\,\text{m})/(600 \times 10^{-9}\,\text{m}) = 10.0$, the highest order seen is the $m = 9$ order.

(d) and (e) The fourth and eighth orders are missing so the observable orders are $m = 0, 1, 2, 3, 5, 6, 7$, and $9$. The second highest order is the $m = 7$ order and the third highest order is the $m = 6$ order.

**61**

If a grating just resolves two wavelengths whose average is $\lambda_{\text{avg}}$ and whose separation is $\Delta\lambda$, then its resolving power is defined by $R = \lambda_{\text{avg}}/\Delta\lambda$. The text shows this is $Nm$, where $N$ is the number of rulings in the grating and $m$ is the order of the lines. Thus $\lambda_{\text{avg}}/\Delta\lambda = Nm$ and

$$N = \frac{\lambda_{\text{avg}}}{m\,\Delta\lambda} = \frac{656.3\,\text{nm}}{(1)(0.18\,\text{nm})} = 3.65 \times 10^3\ \text{rulings}\,.$$

**73**

We want the reflections to obey the Bragg condition $2d \sin\theta = m\lambda$, where $\theta$ is the angle between the incoming rays and the reflecting planes, $\lambda$ is the wavelength, and $m$ is an integer. Solve for $\theta$:

$$\theta = \sin^{-1}\left[\frac{m\lambda}{2d}\right] = \sin^{-1}\left[\frac{(0.125 \times 10^{-9}\,\text{m})m}{2(0.252 \times 10^{-9}\,\text{m})}\right] = \sin^{-1}(0.2480m)\,.$$

For $m = 1$ this gives $\theta = 14.4°$. The crystal should be turned $45° - 14.4° = 30.6°$ clockwise.

For $m = 2$ it gives $\theta = 29.7°$. The crystal should be turned $45° - 29.7° = 15.3°$ clockwise.

For $m = 3$ it gives $\theta = 48.1°$. The crystal should be turned $48.1° - 45° = 3.1°$ counterclockwise.

For $m = 4$ it gives $\theta = 82.8°$. The crystal should be turned $82.8° - 45° = 37.8°$ counterclockwise. There are no intensity maxima for $m > 4$ as you can verify by noting that $m\lambda/2d$ is greater than 1 for $m$ greater than 4. For clockwise turns the smaller value is 15.3° and the larger value is 30.6°. For counterclockwise turns the smaller value is 3.1° and the larger value is 37.8°.

**77**

Intensity maxima occur at angles $\theta$ such that $d \sin\theta = m\lambda$, where $d$ is the separation of adjacent rulings and $\lambda$ is the wavelength. Here the ruling separation is $1/(200\,\text{mm}^{-1}) = 5.00 \times 10^{-3}\,\text{mm} = 5.00 \times 10^{-6}\,\text{m}$. Thus

$$\lambda = \frac{d \sin\theta}{m} = \frac{(5.00 \times 10^{-6}\,\text{m}) \sin 30.0°}{m} = \frac{2.50 \times 10^{-6}\,\text{m}}{m}.$$

For $m = 1$, $\lambda = 2500\,\text{nm}$; for $m = 2$, $\lambda = 1250\,\text{nm}$; for $m - 3$, $\lambda = 833\,\text{nm}$; for $m = 4$, $\lambda = 625\,\text{nm}$; for $m = 5$, $\lambda = 500\,\text{nm}$, and for $m = 6$, $\lambda = 417\,\text{nm}$. Only the last three are in the visible range, so the longest wavelength in the visible range is 625 nm, the next longest is 500 nm, and the third longest is 417 nm.

**79**

Suppose $m_o$ is the order of the minimum for orange light, with wavelength $\lambda_o$, and $m_{bg}$ is the order of the minimum for blue-green light, with wavelength $\lambda_{bg}$. Then $a \sin\theta = m_o\lambda_o$ and $a \sin\theta = m_{bg}\lambda_{bg}$. Thus $m_o\lambda_o = m_{bg}\lambda_{bg}$ and $m_{bg}/m_o = \lambda_o/\lambda_{bg} = (600\,\text{nm})/(500\,\text{nm}) = 6/5$. The smallest two integers with this ratio are $m_{bg} = 6$ and $m_o = 5$. The slit width is

$$a = \frac{m_o\lambda_o}{\sin\theta} = \frac{5(600 \times 10^{-9}\,\text{m}}{\sin(1.00 \times 10^{-3}\,\text{rad})} = 3.0 \times 10^{-3}\,\text{m}.$$

Other values for $m_o$ and $m_{bg}$ are possible but these are associated with a wider slit.

**81**

(a) Since the first minimum of the diffraction pattern occurs at the angle $\theta$ such that $\sin\theta = \lambda/a$, where $\lambda$ is the wavelength and $a$ is the slit width, the central maximum extends from $\theta_1 = -\sin^{-1}(\lambda/a)$ to $\theta_2 = +\sin^{-1}(\lambda/a)$. Maxima of the two-slit interference pattern are at angles $\theta$ such that $\sin\theta = m\lambda/d$, where $d$ is the slit separation and $m$ is an integer. We wish to know the number of values of $m$ such that $\sin^{-1}(m\lambda/d)$ lies between $-\sin^{-1}(\lambda/a)$ and $+\sin(\lambda/a)$ or, what is the same, the number of values of $m$ such that $m/d$ lies between $-1/a$ and $+1/a$. The greatest $m$ can be is the greatest integer that is smaller than $d/a = (14\,\mu\text{m})/(2.0\,\mu\text{m}) = 7$. (The $m = 7$ maximum does not appear since it coincides with a minimum of the diffraction pattern.) There are 13 such values: 0, ±1, ±2, ±3; ±4; ±5, and ±6. Thus 13 interference maxima appear in the central diffraction envelope.

(b) The first diffraction envelope extends from $\theta_1 = \sin^{-1}(\lambda/a)$ to $\theta_2 = \sin^{-1}(2\lambda/a)$. Thus we wish to know the number of values of $m$ such that $m/d$ is greater than $1/a$ and less than $2/a$. Since $d = 7.0a$, $m$ can be 8, 9, 10, 11, 12, or 13. That is, there are 6 interference maxima in the first diffraction envelope.

**93**

If you divide the original slit into $N$ strips and represent the light from each strip, when it reaches the screen, by a phasor, then at the central maximum in the diffraction pattern you add $N$ phasors, all in the same direction and each with the same amplitude. The intensity there is proportional to $N^2$. If you double the slit width, you need $2N$ phasors if they are each to have the amplitude of the phasors you used for the narrow slit. The intensity at the central maximum is proportional to $(2N)^2$ and is, therefore, four times the intensity for the narrow slit. The energy reaching the screen per unit time, however, is only twice the energy reaching it per unit time when the narrow slit is in place. The energy is simply redistributed. For example, the central peak is now half as wide and the integral of the intensity over the peak is only twice the analogous integral for the narrow slit.

**95**

(a) Since the resolving power of a grating is given by $R = \lambda/\Delta\lambda$ and by $Nm$, the range of wavelengths that can just be resolved in order $m$ is $\Delta\lambda = \lambda/Nm$. Here $N$ is the number of rulings in the grating and $\lambda$ is the average wavelength. The frequency $f$ is related to the wavelength by $f\lambda = c$, where $c$ is the speed of light. This means $f\,\Delta\lambda + \lambda\,\Delta f = 0$, so

$$\Delta\lambda = -\frac{\lambda}{f}\,\Delta f = -\frac{\lambda^2}{c}\,\Delta f\,,$$

where $f = c/\lambda$ was used. The negative sign means that an increase in frequency corresponds to a decrease in wavelength. We may interpret $\Delta f$ as the range of frequencies that can be resolved and take it to be positive. Then

$$\frac{\lambda^2}{c}\,\Delta f = \frac{\lambda}{Nm}$$

and

$$\Delta f = \frac{c}{Nm\lambda}\,.$$

(b) The difference in travel time for waves traveling along the two extreme rays is $\Delta t = \Delta L/c$, where $\Delta L$ is the difference in path length. The waves originate at slits that are separated by $(N-1)d$, where $d$ is the slit separation and $N$ is the number of slits, so the path difference is $\Delta L = (N-1)d\sin\theta$ and the time difference is

$$\Delta t = \frac{(N-1)d\sin\theta}{c}\,.$$

If $N$ is large, this may be approximated by $\Delta t = (Nd/c)\sin\theta$. The lens does not affect the travel time.

(c) Substitute the expressions you derived for $\Delta t$ and $\Delta f$ to obtain

$$\Delta f\,\Delta t = \left(\frac{c}{Nm\lambda}\right)\left(\frac{Nd\sin\theta}{c}\right) = \frac{d\sin\theta}{m\lambda} = 1\,.$$

The condition $d\sin\theta = m\lambda$ for a diffraction line was used to obtain the last result.

**101**

The dispersion of a grating is given by $D = d\theta/d\lambda$, where $\theta$ is the angular position of a line associated with wavelength $\lambda$. The angular position and wavelength are related by $\ell \sin\theta = m\lambda$, where $\ell$ is the slit separation and $m$ is an integer. Differentiate this with respect to $\theta$ to obtain $(d\theta/d\lambda)\,\ell\cos\theta = m$ or

$$D = \frac{\ell\theta}{\ell\lambda} = \frac{m}{\ell\cos\theta}\,.$$

Now $m = (\ell/\lambda)\sin\theta$, so

$$D = \frac{\ell\sin\theta}{\ell\lambda\cos\theta} = \frac{\tan\theta}{\lambda}\,.$$

The trigonometric identity $\tan\theta = \sin\theta/\cos\theta$ was used.

# Chapter 37

## 11

(a) The rest length $L_0$ (= 130 m) of the spaceship and its length $L$ as measured by the timing station are related by $L = L_0/\gamma = L_0\sqrt{1-\beta^2}$, where $\gamma = 1/\sqrt{1-\beta^2}$ and $\beta = v/c$. Thus $L = (130\,\text{m})\sqrt{1-(0.740)^2} = 87.4\,\text{m}$.

(b) The time interval for the passage of the spaceship is

$$\Delta t = \frac{L}{v} = \frac{87.4\,\text{m}}{(0.740)(2.9979\times 10^8\,\text{m/s})} = 3.94\times 10^{-7}\,\text{s}\,.$$

## 19

The proper time is not measured by clocks in either frame $S$ or frame $S'$ since a single clock at rest in either frame cannot be present at the origin and at the event. The full Lorentz transformation must be used:

$$x' = \gamma[x - vt]$$

$$t' = \gamma[t - \beta x/c]\,,$$

where $\beta = v/c = 0.950$ and $\gamma = 1/\sqrt{1-\beta^2} = 1/\sqrt{1-(0.950)^2} = 3.2026$. Thus

$$\begin{aligned} x' &= (3.2026)\left[100\times 10^3\,\text{m} - (0.950)(2.9979\times 10^8\,\text{m/s})(200\times 10^{-6}\,\text{s}\right] \\ &= 1.38\times 10^5\,\text{m} = 138\,\text{km} \end{aligned}$$

and

$$t' = (3.2026)\left[200\times 10^{-6}\,\text{s} - \frac{(0.950)(100\times 10^3\,\text{m})}{2.9979\times 10^8\,\text{m/s}}\right] = -3.74\times 10^{-4}\,\text{s} = -374\,\mu\text{s}\,.$$

## 29

Use Eq. 37–29 with $u' = 0.40c$ and $v = 0.60c$. Then

$$u = \frac{0.40c + 0.60c}{1 + (0.40c)(0.60c)/c^2} = 0.81c\,.$$

## 33

Calculate the speed of the micrometeorite relative to the spaceship. Let $S'$ be the reference frame for which the data is given and attach frame $S$ to the spaceship. Suppose the micrometeorite is going in the positive $x$ direction and the spaceship is going in the negative $x$ direction, both as viewed from $S'$. Then, in Eq. 37–29, $u' = 0.82c$ and $v = 0.82c$. Notice that $v$ in the equation is the velocity of $S'$ relative to $S$. Thus the velocity of the micrometeorite in the frame of the spaceship is

$$u = \frac{u' + v}{1 + u'v/c^2} = \frac{0.82c + 0.82c}{1 + (0.82c)(0.82c)/c^2} = 0.9806c\,.$$

The time for the micrometeorite to pass the spaceship is

$$\Delta t = \frac{L}{u} = \frac{350\,\text{m}}{(0.9806)(2.9979 \times 10^8\,\text{m/s})} = 1.19 \times 10^{-6}\,\text{s}\,.$$

**37**

The spaceship is moving away from Earth, so the frequency received is given by

$$f = f_0\sqrt{\frac{1-\beta}{1+\beta}}\,,$$

where $f_0$ is the frequency in the frame of the spaceship, $\beta = v/c$, and $v$ is the speed of the spaceship relative to Earth. See Eq. 37–31. Thus

$$f = (100\,\text{MHz})\sqrt{\frac{1 - 0.9000}{1 + 0.9000}} = 22.9\,\text{MHz}\,.$$

**39**

The spaceship is moving away from Earth, so the frequency received is given by

$$f = f_0\sqrt{\frac{1-\beta}{1+\beta}}\,,$$

where $f_0$ is the frequency in the frame of the spaceship, $\beta = v/c$, and $v$ is the speed of the spaceship relative to Earth. See Eq. 37–31. The frequency $f$ and wavelength $\lambda$ are related by $f\lambda = c$, so if $\lambda_0$ is the wavelength of the light as seen on the spaceship and $\lambda$ is the wavelength detected on Earth, then

$$\lambda = \lambda_0\sqrt{\frac{1+\beta}{1-\beta}} = (450\,\text{nm})\sqrt{\frac{1+0.20}{1-0.20}} = 550\,\text{nm}\,.$$

This is in the yellow-green portion of the visible spectrum.

**43**

Use the two expressions for the total energy: $E = mc^2 + K$ and $E = \gamma mc^2$, where $m$ is the mass of an electron, $K$ is the kinetic energy, and $\gamma = 1/\sqrt{1-\beta^2}$. Thus $mc^2 + K = \gamma mc^2$ and

$$\gamma = 1 + \frac{K}{mc^2} = 1 + \frac{(100.000 \times 10^6\,\text{eV})(1.602\,176\,462\,\text{J/eV})}{(9.109\,381\,88 \times 10^{-31}\,\text{kg})(2.997\,924\,58 \times 10^8\,\text{m/s})^2} = 196.695\,.$$

Now $\gamma^2 = 1/(1-\beta^2)$, so

$$\beta = \sqrt{1 - \frac{1}{\gamma^2}} = \sqrt{1 - \frac{1}{(196.695)^2}} = 0.999\,987\,.$$

**53**

The energy equivalent of one tablet is $mc^2 = (320 \times 10^{-6}\,\text{kg})(2.9979 \times 10^8\,\text{m/s})^2 = 2.88 \times 10^{13}\,\text{J}$. This provides the same energy as $(2.88 \times 10^{13}\,\text{J})/(3.65 \times 10^7\,\text{J/L}) = 7.89 \times 10^5\,\text{L}$ of gasoline. The distance the car can go is $d = (7.89 \times 10^5\,\text{L})(12.75\,\text{km/L}) = 1.01 \times 10^7\,\text{km}$.

**71**

The energy of the electron is given by $E = mc^2/\sqrt{1-(v/c)^2}$, which yields

$$v = \sqrt{1 - \left[\frac{mc^2}{E}\right]^2}\,c = \sqrt{1 - \left[\frac{(9.11 \times 10^{-31}\,\text{kg})(2.9979 \times 10^8\,\text{m/s})^2}{(1533\,\text{MeV})(1.602 \times 10{-}13\,\text{J/MeV})}\right]^2} = 0.99999994c \approx c$$

for the speed $v$ of the electron. In the rest frame of Earth the trip took time $t = 26\,\text{y}$. A clock traveling with the electron records the proper time of the trip, so the trip in the rest frame of the electron took time $t' = t/\gamma$. Now

$$\gamma = \frac{E}{mc^2} = \frac{1533\,\text{MeV})(1.602 \times 10^{-13}\,\text{J/MeV})}{(9.11 \times 10^{-31}\,\text{kg})(2.9979 \times 10^8\,\text{m/s})} = 3.0 \times 10^3$$

and $t' = (26\,\text{y})/(3.0 \times 10^3) = 8.7 \times 10^{-3}\,\text{y}$. The distance traveled is $8.7 \times 10^{-3}\,\text{ly}$.

**73**

Start with $(pc)^2 = K^2 + 2Kmc^2$, where $p$ is the momentum of the particle, $K$ is its kinetic energy, and $m$ is its mass. For an electron $mc^2 = 0.511\,\text{MeV}$, so

$$pc = \sqrt{K^2 + 2Kmc^2} = \sqrt{(2.00\,\text{MeV})^2 + 2(2.00\,\text{MeV})(0.511\,\text{MeV})} = 2.46\,\text{MeV}\,.$$

Thus $p = 2.46\,\text{MeV}/c$.

**75**

The work required is the increased in the energy of the proton. The energy is given by $E = mc^2/[1-(v/c)^2]$. Let $v_1$ be the initial speed and $v_2$ be the final speed. Then the work is

$$W = \frac{mc^2}{\sqrt{1-(v_2/c)^2}} - \frac{mc^2}{\sqrt{1-(v/c)^2}} = \frac{938\,\text{MeV}}{\sqrt{1-(0.9860)^2}} - \frac{938\,\text{MeV}}{\sqrt{1-(0.9850)^2}} = 189\,\text{MeV}\,,$$

where $mc^2 = 938\,\text{MeV}$ was used.

**77**

(a) Let $v$ be the speed of either satellite, relative to Earth. According to the Galilean velocity transformation equation the relative speed is $v_{\text{rel}} = 2v = 2(2.7 \times 10^4\,\text{km/h} = 5.4 \times 10^4\,\text{km/h}$.

(b) The correct relativistic transformation equation is

$$v_{\text{rel}} = \frac{2v}{1 + \frac{v^2}{c^2}}.$$

The fractional error is

$$\text{fract err} = \frac{2v - v_{\text{rel}}}{2v} = 1 - \frac{1}{1 + \frac{v^2}{c^2}}.$$

The speed of light is $1.08 \times 10^9$ km/h, so

$$\text{fract err} = \frac{1}{1 + \frac{(2.7 \times 10^4\,\text{km/h})^2}{(1.08 \times 10^9\,\text{km/h})^2}} = 6.3 \times 10^{-10}.$$

# Chapter 38

7

(a) Let $R$ be the rate of photon emission (number of photons emitted per unit time) and let $E$ be the energy of a single photon. Then the power output of a lamp is given by $P = RE$ if all the power goes into photon production. Now $E = hf = hc/\lambda$, where $h$ is the Planck constant, $f$ is the frequency of the light emitted, and $\lambda$ is the wavelength. Thus $P = Rhc/\lambda$ and $R = \lambda P/hc$. The lamp emitting light with the longer wavelength (the 700-nm lamp) emits more photons per unit time. The energy of each photon is less so it must emit photons at a greater rate.

(b) Let $R$ be the rate of photon production for the 700 nm lamp Then

$$R = \frac{\lambda P}{hc} = \frac{(700 \times 10^{-9}\,\text{m})(400\,\text{J/s})}{(6.626 \times 10^{-34}\,\text{J}\cdot\text{s})(2.9979 \times 10^{8}\,\text{m/s})} = 1.41 \times 10^{21}\,\text{photon/s}\,.$$

17

The energy of an incident photon is $E = hf = hc/\lambda$, where $h$ is the Planck constant, $f$ is the frequency of the electromagnetic radiation, and $\lambda$ is its wavelength. The kinetic energy of the most energetic electron emitted is $K_m = E - \Phi = (hc/\lambda) - \Phi$, where $\Phi$ is the work function for sodium. The stopping potential $V_0$ is related to the maximum kinetic energy by $eV_0 = K_m$, so $eV_0 = (hc/\lambda) - \Phi$ and

$$\lambda = \frac{hc}{eV_0 + \Phi} = \frac{(6.626 \times 19^{-34}\,\text{J}\cdot\text{s})(2.9979 \times 10^{8}\,\text{m/s})}{(5.0\,\text{eV} + 2.2\,\text{eV})(1.602 \times 10^{-19}\,\text{J/eV})} = 1.7 \times 10^{-7}\,\text{m}\,.$$

Here $eV_0 = 5.0\,\text{eV}$ was used.

21

(a) The kinetic energy $K_m$ of the fastest electron emitted is given by $K_m = hf - \Phi = (hc/\lambda) - \Phi$, where $\Phi$ is the work function of aluminum, $f$ is the frequency of the incident radiation, and $\lambda$ is its wavelength. The relationship $f = c/\lambda$ was used to obtain the second form. Thus

$$K_m = \frac{(6.626 \times 10^{-34}\,\text{J}\cdot\text{s})(2.9979 \times 10^{8}\,\text{m/s})}{(200 \times 10^{-9}\,\text{m})(1.602 \times 10^{-19}\,\text{J/eV})} - 4.20\,\text{eV} = 2.00\,\text{eV}\,.$$

(b) The slowest electron just breaks free of the surface and so has zero kinetic energy.

(c) The stopping potential $V_0$ is given by $K_m = eV_0$, so $V_0 = K_m/e = (2.00\,\text{eV})/e = 2.00\,\text{V}$.

(d) The value of the cutoff wavelength is such that $K_m = 0$. Thus $hc/\lambda = \Phi$ or

$$\lambda = \frac{hc}{\Phi} = \frac{(6.626 \times 10^{-34}\,\text{J}\cdot\text{s})(2.9979 \times 10^{8}\,\text{m/s})}{(4.2\,\text{eV})(1.602 \times 10^{-19}\,\text{J/eV})} = 2.95 \times 10^{-7}\,\text{m}\,.$$

If the wavelength is longer, the photon energy is less and a photon does not have sufficient energy to knock even the most energetic electron out of the aluminum sample.

**29**

(a) When a photon scatters from an electron initially at rest, the change in wavelength is given by $\Delta\lambda = (h/mc)(1 - \cos\phi)$, where $m$ is the mass of an electron and $\phi$ is the scattering angle. Now $h/mc = 2.43 \times 10^{-12}\,\text{m} = 2.43\,\text{pm}$, so $\Delta\lambda = (2.43\,\text{pm})(1 - \cos 30°) = 0.326\,\text{pm}$. The final wavelength is $\lambda' = \lambda + \Delta\lambda = 2.4\,\text{pm} + 0.326\,\text{pm} = 2.73\,\text{pm}$.

(b) Now $\Delta\lambda = (2.43\,\text{pm})(1 - \cos 120°) = 3.645\,\text{pm}$ and $\lambda' = 2.4\,\text{pm} + 3.645\,\text{pm} = 6.05\,\text{pm}$.

**43**

Since the kinetic energy $K$ and momentum $p$ are related by $K = p^2/2m$, the momentum of the electron is $p = \sqrt{2mK}$ and the wavelength of its matter wave is $\lambda = h/p = h/\sqrt{2mK}$. Replace $K$ with $eV$, where $V$ is the accelerating potential and $e$ is the fundamental charge, to obtain

$$\lambda = \frac{h}{\sqrt{2meV}} = \frac{6.626 \times 10^{-34}\,\text{J}\cdot\text{s}}{\sqrt{2(9.109 \times 10^{-31}\,\text{kg})(1.602 \times 10^{-19}\,\text{C})(25.0 \times 10^{3}\,\text{V})}}$$
$$= 7.75 \times 10^{-12}\,\text{m} = 7.75\,\text{pm}\,.$$

**47**

(a) The kinetic energy acquired is $K = qV$, where $q$ is the charge on an ion and $V$ is the accelerating potential. Thus $K = (1.602 \times 10^{-19}\,\text{C})(300\,\text{V}) = 4.80 \times 10^{-17}\,\text{J}$. The mass of a single sodium atom is, from Appendix F, $m = (22.9898\,\text{g/mol})/(6.02 \times 10^{23}\,\text{atom/mol}) = 3.819 \times 10^{-23}\,\text{g} = 3.819 \times 10^{-26}\,\text{kg}$. Thus the momentum of an ion is

$$p = \sqrt{2mK} = \sqrt{2(3.819 \times 10^{-26}\,\text{kg})(4.80 \times 10^{-17}\,\text{J})} = 1.91 \times 10^{-21}\,\text{kg}\cdot\text{m/s}\,.$$

(b) The de Broglie wavelength is

$$\lambda = \frac{h}{p} = \frac{6.63 \times 10^{-34}\,\text{J}\cdot\text{s}}{1.91 \times 10^{-21}\,\text{kg}\cdot\text{m/s}} = 3.47 \times 10^{-13}\,\text{m}\,.$$

**49**

Since the kinetic energy $K$ and momentum $p$ are related by $K = p^2/2m$, the momentum of the electron is $p = \sqrt{2mK}$ and the wavelength of its matter wave is $\lambda = h/p = h/\sqrt{2mK}$. Thus

$$K = \frac{1}{2m}\left(\frac{h}{\lambda}\right)^2 = \frac{1}{2(9.11 \times 10^{-31}\,\text{kg})}\left(\frac{6.626 \times 10^{-34}\,\text{J}\cdot\text{s})}{590 \times 10^{-9}\,\text{m})}\right)^2$$
$$= 6.92 \times 10^{-25}\,\text{J} = 4.33 \times 10^{-6}\,\text{eV}\,.$$

**59**

The angular wave number $k$ is related to the wavelength $\lambda$ by $k = 2\pi/\lambda$ and the wavelength is related to the particle momentum $p$ by $\lambda = h/p$, so $k = 2\pi p/h$. Now the kinetic energy $K$ and the momentum are related by $K = p^2/2m$, where $m$ is the mass of the particle. Thus $p = \sqrt{2mK}$ and

$$k = \frac{2\pi\sqrt{2mK}}{h}\,.$$

**61**

For $U = U_0$, Schrodinger's equation becomes

$$\frac{d^2\psi}{dx^2} + \frac{8\pi^2 m}{h^2}\,[E - U_0]\,\psi = 0\,.$$

Substitute $\psi = \psi_0 e^{ikx}$. The second derivative is $d^2\psi/dx^2 = -k^2\psi_0 e^{ikx} = -k^2\psi$. The result is

$$-k^2\psi + \frac{8\pi^2 m}{h^2}\,[E - U_0]\,\psi = 0\,.$$

Solve for $k$ and obtain

$$k = \sqrt{\frac{8\pi^2 m}{h^2}\,[E - U_0]} = \frac{2\pi}{h}\sqrt{2m\,[E - U_0]}\,.$$

**67**

(a) If $m$ is the mass of the particle and $E$ is its energy, then the transmission coefficient for a barrier of height $U$ and width $L$ is given by

$$T = e^{-2kL}\,,$$

where

$$k = \sqrt{\frac{8\pi^2 m(U - E)}{h^2}}\,.$$

If the change $\Delta U$ in $U$ is small (as it is), the change in the transmission coefficient is given by

$$\Delta T = \frac{dT}{dU}\,\Delta U = -2LT\,\frac{dk}{dU}\,\Delta U\,.$$

Now

$$\frac{dk}{dU} = \frac{1}{2\sqrt{U - E}}\sqrt{\frac{8\pi^2 m}{h^2}} = \frac{1}{2(U - E)}\sqrt{\frac{8\pi^2 m(U - E)}{h^2}} = \frac{k}{2(U - E)}\,.$$

Thus

$$\Delta T = -LTk\,\frac{\Delta U}{U - E}\,.$$

For the data of Sample Problem 38–7, $2kL = 10.0$, so $kL = 5.0$ and

$$\frac{\Delta T}{T} = -kL\frac{\Delta U}{U - E} = -(5.0)\frac{(0.010)(6.8\,\text{eV})}{6.8\,\text{eV} - 5.1\,\text{eV}} = -0.20\,.$$

There is a 20% decrease in the transmission coefficient.

(b) The change in the transmission coefficient is given by

$$\Delta T = \frac{dT}{dL}\,\Delta L = -2ke^{-2kL}\,\Delta L = -2kT\,\Delta L$$

and

$$\frac{\Delta T}{T} = -2k\,\Delta L = -2(6.67\times 10^{9}\,\text{m}^{-1})(0.010)(750\times 10^{-12}\,\text{m}) = -0.10\,.$$

There is a 10% decrease in the transmission coefficient.

(c) The change in the transmission coefficient is given by

$$\Delta T = \frac{dT}{dE}\,\Delta E = -2Le^{-2kL}\frac{dk}{dE}\,\Delta E = -2LT\frac{dk}{dE}\,\Delta E\,.$$

Now $dk/dE = -dk/dU = -k/2(U - E)$, so

$$\frac{\Delta T}{T} = kL\frac{\Delta E}{U - E} = (5.0)\frac{(0.010)(5.1\,\text{eV})}{6.8\,\text{eV} - 5.1\,\text{eV}} = 0.15\,.$$

There is a 15% increase in the transmission coefficient.

**79**

The uncertainty in the momentum is $\Delta p = m\,\Delta v = (0.50\,\text{kg})(1.0\,\text{m/s}) = 0.50\,\text{kg}\cdot\text{m/s}$, where $\Delta v$ is the uncertainty in the velocity. Solve the uncertainty relationship $\Delta x\,\Delta p \geq \hbar$ for the minimum uncertainty in the coordinate $x$: $\Delta x = \hbar/\Delta p = (0.60\,\text{J}\cdot\text{s})/2\pi(0.50\,\text{kg}\cdot\text{m/s}) = 0.19\,\text{m}$.

# Chapter 39

## 13

The probability that the electron is found in any interval is given by $P = \int |\psi|^2\,dx$, where the integral is over the interval. If the interval width $\Delta x$ is small, the probability can be approximated by $P = |\psi|^2\,\Delta x$, where the wave function is evaluated for the center of the interval, say. For an electron trapped in an infinite well of width $L$, the ground state probability density is

$$|\psi|^2 = \frac{2}{L}\sin^2\left(\frac{\pi x}{L}\right),$$

so

$$P = \left(\frac{2\,\Delta x}{L}\right)\sin^2\left(\frac{\pi x}{L}\right).$$

(a) Take $L = 100\,\text{pm}$, $x = 25\,\text{pm}$, and $\Delta x = 5.0\,\text{pm}$. Then

$$P = \left[\frac{2(5.0\,\text{pm})}{100\,\text{pm}}\right]\sin^2\left[\frac{\pi(25\,\text{pm})}{100\,\text{pm}}\right] = 0.050\,.$$

(b) Take $L = 100\,\text{pm}$, $x = 50\,\text{pm}$, and $\Delta x = 5.0\,\text{pm}$. Then

$$P = \left[\frac{2(5.0\,\text{pm})}{100\,\text{pm}}\right]\sin^2\left[\frac{\pi(50\,\text{pm})}{100\,\text{pm}}\right] = 0.10\,.$$

(c) Take $L = 100\,\text{pm}$, $x = 90\,\text{pm}$, and $\Delta x = 5.0\,\text{pm}$. Then

$$P = \left[\frac{2(5.0\,\text{pm})}{100\,\text{pm}}\right]\sin^2\left[\frac{\pi(90\,\text{pm})}{100\,\text{pm}}\right] = 0.0095\,.$$

## 25

The energy levels are given by

$$E_{nx\ ny} = \frac{h^2}{8m}\left[\frac{n_x^2}{L_x^2} + \frac{n_y^2}{L_y^2}\right] = \frac{h^2}{8mL^2}\left[n_x^2 + \frac{n_y^2}{4}\right],$$

where the substitutions $L_x = L$ and $L_y = 2L$ were made. In units of $h^2/8mL^2$, the energy levels are given by $n_x^2 + n_y^2/4$. The lowest five levels are $E_{1,1} = 1.25$, $E_{1,2} = 2.00$, $E_{1,3} = 3.25$, $E_{2,1} = 4.25$, and $E_{2,2} = E_{1,4} = 5.00$. A little thought should convince you that there are no other possible values for the energy less than 5.

The frequency of the light emitted or absorbed when the electron goes from an initial state $i$ to a final state $f$ is $f = (E_f - E_i)/h$ and in units of $h/8mL^2$ is simply the difference in the values of $n_x^2 + n_y^2/4$ for the two states. The possible frequencies are 0.75 (1,2 $\longrightarrow$ 1,1), 2.00 (1,3 $\longrightarrow$ 1,1), 3.00 (2,1 $\longrightarrow$ 1,1), 3.75 (2,2 $\longrightarrow$ 1,1), 1.25 (1,3 $\longrightarrow$ 1,2), 2.25 (2,1 $\longrightarrow$ 1,2), 3.00 (2,2 $\longrightarrow$ 1,2), 1.00 (2,1 $\longrightarrow$ 1,3), 1.75 (2,2 $\longrightarrow$ 1,3), 0.75 (2,2 $\longrightarrow$ 2,1), all in units of $h/8mL^2$.

There are 8 different frequencies in all. In units of $h/8mL^2$ the lowest is 0.75, the second lowest is 1.00, and the third lowest is 1.25. The highest is 3.75, the second highest is 3.00, and the third highest is 2.25.

**33**

If kinetic energy is not conserved some of the neutron's initial kinetic energy is used to excite the hydrogen atom. The least energy that the hydrogen atom can accept is the difference between the first excited state ($n = 2$) and the ground state ($n = 1$). Since the energy of a state with principal quantum number $n$ is $-(13.6\,\text{eV})/n^2$, the smallest excitation energy is $13.6\,\text{eV} - (13.6\,\text{eV})/(2)^2 = 10.2\,\text{eV}$. The neutron does not have sufficient kinetic energy to excite the hydrogen atom, so the hydrogen atom is left in its ground state and all the initial kinetic energy of the neutron ends up as the final kinetic energies of the neutron and atom. The collision must be elastic.

**37**

The energy $E$ of the photon emitted when a hydrogen atom jumps from a state with principal quantum number $u$ to a state with principal quantum number $\ell$ is given by

$$E = A\left(\frac{1}{\ell^2} - \frac{1}{u^2}\right),$$

where $A = 13.6\,\text{eV}$. The frequency $f$ of the electromagnetic wave is given by $f = E/h$ and the wavelength is given by $\lambda = c/f$. Thus

$$\frac{1}{\lambda} = \frac{f}{c} = \frac{E}{hc} = \frac{A}{hc}\left(\frac{1}{\ell^2} - \frac{1}{u^2}\right).$$

The shortest wavelength occurs at the series limit, for which $u = \infty$. For the Balmer series, $\ell = 2$ and the shortest wavelength is $\lambda_B = 4hc/A$. For the Lyman series, $\ell = 1$ and the shortest wavelength is $\lambda_L = hc/A$. The ratio is $\lambda_B/\lambda_L = 4$.

**43**

The proposed wave function is

$$\psi = \frac{1}{\sqrt{\pi}a^{3/2}}e^{-r/a},$$

where $a$ is the Bohr radius. Substitute this into the right side of Schrodinger's equation and show that the result is zero. The derivative is

$$\frac{d\psi}{dr} = -\frac{1}{\sqrt{\pi}a^{5/2}}e^{-r/a},$$

so

$$r^2\frac{d\psi}{dr} = -\frac{r^2}{\sqrt{\pi}a^{5/2}}e^{-r/a}$$

and

$$\frac{1}{r^2}\frac{d}{dr}\left(r^2\frac{d\psi}{dr}\right) = \frac{1}{\sqrt{\pi}a^{5/2}}\left[-\frac{2}{r}+\frac{1}{a}\right]e^{-r/a} = \frac{1}{a}\left[-\frac{2}{r}+\frac{1}{a}\right]\psi\,.$$

Now the energy of the ground state is given by $E = -me^4/8\epsilon_0^2h^2$ and the Bohr radius is given by $a = h^2\epsilon_0/\pi me^2$, so $E = -e^2/8\pi\epsilon_0 a$. The potential energy is given by $U = -e^2/4\pi\epsilon_0 r$, so

$$\frac{8\pi^2 m}{h^2}[E-U]\psi = \frac{8\pi^2 m}{h^2}\left[-\frac{e^2}{8\pi\epsilon_0 a}+\frac{e^2}{4\pi\epsilon_0 r}\right]\psi = \frac{8\pi^2 m}{h^2}\frac{e^2}{8\pi\epsilon_0}\left[-\frac{1}{a}+\frac{2}{r}\right]\psi$$

$$= \frac{\pi me^2}{h^2\epsilon_0}\left[-\frac{1}{a}+\frac{2}{r}\right]\psi = \frac{1}{a}\left[-\frac{1}{a}+\frac{2}{r}\right]\psi\,.$$

The two terms in Schrodinger's equation obviously cancel and the proposed function $\psi$ satisfies that equation.

<u>47</u>

The radial probability function for the ground state of hydrogen is $P(r) = (4r^2/a^3)e^{-2r/a}$, where $a$ is the Bohr radius. (See Eq. 39–44.) You want to evaluate the integral $\int_0^\infty P(r)\,dr$. Eq. 15 in the integral table of Appendix E is an integral of this form. Set $n = 2$ and replace $a$ in the given formula with $2/a$ and $x$ with $r$. Then

$$\int_0^\infty P(r)\,dr = \frac{4}{a^3}\int_0^\infty r^2 e^{-2r/a}\,dr = \frac{4}{a^3}\frac{2}{(2/a)^3} = 1\,.$$

<u>49</u>

(a) $\psi_{210}$ is real. Simply square it to obtain the probability density:

$$|\psi_{210}|^2 = \frac{r^2}{32\pi a^5}e^{-r/a}\cos^2\theta\,.$$

(b) Each of the other functions is multiplied by its complex conjugate, obtained by replacing $i$ with $-i$ in the function. Since $e^{i\phi}e^{-i\phi} = e^0 = 1$, the result is the square of the function without the exponential factor:

$$|\psi_{21+1}|^2 = \frac{r^2}{64\pi a^5}e^{-r/a}\sin^2\theta$$

$$|\psi_{21-1}|^2 = \frac{r^2}{64\pi a^5}e^{-r/a}\sin^2\theta\,.$$

The last two functions lead to the same probability density.

(c) For $m_\ell = 0$ the radial probability density decreases strongly with distance from the nucleus, is greatest along the $z$ axis, and for a given distance from the nucleus decreases in proportion to $\cos^2\theta$ for points away from the $z$ axis. This is consistent with the dot plot of Fig. 39–24 (a). For $m_\ell = \pm 1$ the radial probability density decreases strongly with distance from the nucleus, is greatest in the $x, y$ plane, and for a given distance from the nucleus decreases in proportion to $\sin^2\theta$ for points away from that plane. Thus it is consistent with the dot plot of Fig. 39-24(b).

(d) The total probability density for the three states is the sum:

$$|\psi_{210}|^2 + |\psi_{21+1}|^2 + |\psi_{21-1}|^2 = \frac{r^2}{32\pi a^5} e^{-r/a} \left[\cos^2\theta + \frac{1}{2}\sin^2\theta + \frac{1}{2}\sin^2\theta\right]$$
$$= \frac{r^2}{32\pi a^5} e^{-r/a} .$$

The trigonometric identity $\cos^2\theta + \sin^2\theta = 1$ was used. The total probability density does not depend on $\theta$ or $\phi$. It is spherically symmetric.

57

The wave function is $\psi = \sqrt{C}e^{-kx}$. Substitute this function into Schrodinger's equation,

$$-\frac{h^2}{8\pi^2 m}\frac{d^2\psi}{dx^2} + U_0\psi = E\psi .$$

Since $d^2\psi/dx^2 = \sqrt{C}k^2 e^{-kx} = k^2\psi$, the result is

$$\frac{h^2 k^2}{8\pi^2 m}\psi + U_0\psi = E\psi .$$

The solution for $k$ is

$$k = \sqrt{\frac{8\pi^2 m}{h^2}(U_0 - E)} .$$

Thus the function given for $\psi$ is a solution to Schrodinger's equation provided $k$ has the value calculated from the expression given above.

# Chapter 40

## 9

Since $L^2 = L_x^2 + L_y^2 + L_z^2$, $\sqrt{L_x^2 + L_y^2} = \sqrt{L^2 - L_z^2}$. Replace $L^2$ with $\ell(\ell+1)\hbar^2$ and $L_z$ with $m_\ell \hbar$ to obtain

$$\sqrt{L_x^2 + L_y^2} = \sqrt{\ell(\ell+1) - m_\ell^2}\,\hbar\,.$$

For a given value of $\ell$, the greatest that $m_\ell$ can be is $\ell$, so the smallest that $\sqrt{L_x^2 + L_y^2}$ can be is $\sqrt{\ell(\ell+1) - \ell^2}\,\hbar = \sqrt{\ell}\,\hbar$. The smallest possible magnitude of $m_\ell$ is zero, so the largest $\sqrt{L_x^2 + L_y^2}$ can be is $\sqrt{\ell(\ell+1)}\,\hbar$. Thus

$$\sqrt{\ell}\,\hbar \le \sqrt{L_x^2 + L_y^2} \le \sqrt{\ell(\ell+1)}\,\hbar\,.$$

## 11

(a) For $\ell = 3$, the magnitude of the orbital angular momentum is $L = \sqrt{\ell(\ell+1)}\,\hbar = \sqrt{3(3+1)}\,\hbar = \sqrt{12}\,\hbar = 3.46\hbar$.

(b) The magnitude of the orbital dipole moment is $\mu_{\text{orb}} = \sqrt{\ell(\ell+1)}\,\mu_B = \sqrt{12}\,\mu_B = 3.46\,\mu_B$.

(c) The largest possible value of $m_\ell$ is $\ell$, which is +3.

(d) The corresponding value of the $z$ component of the angular momentum is $L_z = \ell\hbar = +3\hbar$.

(e) The direction of the orbital magnetic dipole moment is opposite that of the orbital angular momentum, so the corresponding value of the $z$ component of the orbital dipole moment is $\mu_{\text{orb},\,z} = -3\mu_B$.

(f) The angle $\theta$ between $\vec{L}$ and the $z$ axis is

$$\theta = \cos^{-1}\frac{L_z}{L} = \cos^{-1}\frac{3\hbar}{3.46\hbar} = 30.0^\circ\,.$$

(g) The second largest value of $m_\ell$ is $m_\ell = \ell - 1 = 2$ and the angle is

$$\theta = \cos^{-1}\frac{L_z}{L} = \cos^{-1}\frac{2\hbar}{3.46\hbar} = 54.7^\circ\,.$$

(h) The most negative value of $m_\ell$ is $-3$ and the angle is

$$\theta = \cos^{-1}\frac{L_z}{L} = \cos^{-1}\frac{-3\hbar}{3.46\hbar} = 150^\circ\,.$$

## 15

The acceleration is

$$a = \frac{F}{M} = \frac{(\mu \cos\theta)(dB/dz)}{M},$$

where $M$ is the mass of a silver atom, $\mu$ is its magnetic dipole moment, $B$ is the magnetic field, and $\theta$ is the angle between the dipole moment and the magnetic field. Take the moment and the field to be parallel ($\cos\theta = 1$) and use the data given in Sample Problem 40–1 to obtain

$$a = \frac{(9.27 \times 10^{-24}\,\mathrm{J/T})(1.4 \times 10^{3}\,\mathrm{T/m})}{1.8 \times 10^{-25}\,\mathrm{kg}} = 7.21 \times 10^{4}\,\mathrm{m/s^2}.$$

## 25

In terms of the quantum numbers $n_x$, $n_y$, and $n_z$, the single-particle energy levels are given by

$$E_{n_x,n_y,n_z} = \frac{h^2}{8mL^2}\left(n_x^2 + n_y^2 + n_z^2\right).$$

The lowest single-particle level corresponds to $n_x = 1$, $n_y = 1$, and $n_z = 1$ and is $E_{1,1,1} = 3(h^2/8mL^2)$. There are two electrons with this energy, one with spin up and one with spin down.

The next lowest single-particle level is three-fold degenerate in the three integer quantum numbers. The energy is $E_{1,1,2} = E_{1,2,1} = E_{2,1,1} = 6(h^2/8mL^2)$. Each of these states can be occupied by a spin up and a spin down electron, so six electrons in all can occupy the states. This completes the assignment of the eight electrons to single-particle states. The ground state energy of the system is $E_{\mathrm{gr}} = (2)(3)(h^2/8mL^2) + (6)(6)(h^2/8mL^2) = (42)(h^2/8mL^2)$.

## 31

(a) All states with principal quantum number $n = 1$ are filled. The next lowest states have $n = 2$. The orbital quantum number can have the values $\ell = 0$ or 1 and of these, the $\ell = 0$ states have the lowest energy. The magnetic quantum number must be $m_\ell = 0$ since this is the only possibility if $\ell = 0$. The spin quantum number can have either of the values $m_s = -\frac{1}{2}$ or $+\frac{1}{2}$. Since there is no external magnetic field, the energies of these two states are the same. Thus, in the ground state, the quantum numbers of the third electron are either $n = 2$, $\ell = 0$, $m_\ell = 0$, $m_s = -\frac{1}{2}$ or $n = 2$, $\ell = 0$, $m_\ell = 0$, $m_s = +\frac{1}{2}$.

(b) The next lowest state in energy is an $n = 2$, $\ell = 1$ state. All $n = 3$ states are higher in energy. The magnetic quantum number can be $m_\ell = -1$, 0, or +1; the spin quantum number can be $m_s = -\frac{1}{2}$ or $+\frac{1}{2}$. If both external and internal magnetic fields can be neglected, all these states have the same energy. The possible states are (2, 1, 1, +1/2), (2, 1, 1, −1/2), (2, 1, 0, +1/2), (2, 1, 0, −1/2), (2, 1, −1, +1/2), and (2, 1, −1, −1/2).

## 37

(a) The cut-off wavelength $\lambda_{\mathrm{min}}$ is characteristic of the incident electrons, not of the target material. This wavelength is the wavelength of a photon with energy equal to the kinetic energy of an incident electron. Thus

$$\lambda = \frac{hc}{\Delta E} = \frac{(6.626 \times 10^{-34}\,\mathrm{J \cdot s})(3.00 \times 10^{8}\,\mathrm{m/s})}{(35 \times 10^{3}\,\mathrm{eV})(1.60 \times 10^{-19}\,\mathrm{J/eV})} = 3.55 \times 10^{-11}\,\mathrm{m} = 35.5\,\mathrm{pm}\,.$$

(b) A $K_\alpha$ photon results when an electron in a target atom jumps from the $L$-shell to the $K$-shell. The energy of this photon is 25.51 keV – 3.56 keV = 21.95 keV and its wavelength is

$$\lambda = \frac{hc}{\Delta E} = \frac{(6.626 \times 10^{-34}\,\mathrm{J \cdot s})(3.00 \times 10^{8}\,\mathrm{m/s})}{(21.95 \times 10^{3}\,\mathrm{eV})(1.60 \times 10^{-19}\,\mathrm{J/eV})} = 9.94 \times 10^{-11}\,\mathrm{m} = 5.65 \times 10^{-11}\,\mathrm{m} = 56.5\,\mathrm{pm}\,.$$

(c) A $K_\beta$ photon results when an electron in a target atom jumps from the $M$-shell to the $K$-shell. The energy of this photon is 25.51 keV – 0.53 keV = 24.98 keV and its wavelength is

$$\lambda = \frac{hc}{\Delta E} = \frac{(6.626 \times 10^{-34}\,\mathrm{J \cdot s})(3.00 \times 10^{8}\,\mathrm{m/s})}{(24.98 \times 10^{3}\,\mathrm{eV})(1.60 \times 10^{-19}\,\mathrm{J/eV})} = 4.96 \times 10^{-11}\,\mathrm{m} = 49.6\,\mathrm{pm}\,.$$

**41**

Since the frequency of an x-ray emission is proportional to $(Z - 1)^2$, where $Z$ is the atomic number of the target atom, the ratio of the wavelength $\lambda_{\mathrm{Nb}}$ for the $K_\alpha$ line of niobium to the wavelength $\lambda_{\mathrm{Ga}}$ for the $K_\alpha$ line of gallium is given by $\lambda_{\mathrm{Nb}}/\lambda_{\mathrm{Ga}} = (Z_{\mathrm{Ga}} - 1)^2/(Z_{\mathrm{Nb}} - 1)^2$, where $Z_{\mathrm{Nb}}$ is the atomic number of niobium (41) and the $Z_{\mathrm{Ga}}$ is the atomic number of gallium (31). Thus $\lambda_{\mathrm{Nb}}/\lambda_{\mathrm{Ga}} = (30)^2\ (40)^2 = 9/16$.

**49**

The number of atoms in a state with energy $E$ is proportional to $e^{-E/kT}$, where $T$ is the temperature on the Kelvin scale and $k$ is the Boltzmann constant. Thus the ratio of the number of atoms in the thirteenth excited state to the number in the eleventh excited state is

$$\frac{n_{13}}{n_{11}} = e^{-\Delta E/kT}\,,$$

where $\Delta E$ is the difference in the energies: $\Delta E = E_{13} - E_{11} = 2(1.2\,\mathrm{eV}) = 2.4\,\mathrm{eV}$. For the given temperature, $kT = (8.62 \times 10^{-2}\,\mathrm{eV/K})(2000\,\mathrm{K}) = 0.1724\,\mathrm{eV}$. Hence,

$$\frac{n_{13}}{n_{11}} = e^{-2.4/0.1724} = 9.0 \times 10^{-7}\,.$$

**65**

(a) The intensity at the target is given by $I = P/A$, where $P$ is the power output of the source and $A$ is the area of the beam at the target. You want to compute $I$ and compare the result with $10^8\,\mathrm{W/m^2}$.

The beam spreads because diffraction occurs at the aperture of the laser. Consider the part of the beam that is within the central diffraction maximum. The angular position of the edge is given by $\sin\theta = 1.22\lambda/d$, where $\lambda$ is the wavelength and $d$ is the diameter of the aperture (see

Problem 50). At the target, a distance $D$ away, the radius of the beam is $r = D\tan\theta$. Since $\theta$ is small, we may approximate both $\sin\theta$ and $\tan\theta$ by $\theta$, in radians. Then $r = D\theta = 1.22D\lambda/d$ and

$$I = \frac{P}{\pi r^2} = \frac{Pd^2}{\pi(1.22D\lambda)^2} = \frac{(5.0\times10^6\,\mathrm{W})(4.0\,\mathrm{m})^2}{\pi\left[1.22(3000\times10^3\,\mathrm{m})(3.0\times10^{-6}\,\mathrm{m})\right]^2}$$
$$= 2.1\times10^5\,\mathrm{W/m^2}\,,$$

not great enough to destroy the missile.

(b) Solve for the wavelength in terms of the intensity and substitute $I = 1.0\times10^8\,\mathrm{W/m^2}$:

$$\lambda = \frac{d}{1.22D}\sqrt{\frac{P}{\pi I}} = \frac{4.0\,\mathrm{m}}{1.22(3000\times10^3\,\mathrm{m})}\sqrt{\frac{5.0\times10^6\,\mathrm{W}}{\pi(1.0\times10^8\,\mathrm{W/m^2})}}$$
$$= 1.4\times10^{-7}\,\mathrm{m} = 140\,\mathrm{nm}\,.$$

**71**

(a) The length of the pulse is $L = c\,\Delta t$, where $\Delta t$ is its duration. Thus $L = (3.00\times10^8\,\mathrm{m/s})(10\times10^{-15}\,\mathrm{s} = 3.0\times10^{-6}\,\mathrm{m}$. The number of wavelengths in the pulse is $N = L/\lambda = (3.0\times10^{-6}\,\mathrm{m})/(500\times10^{-9}\,\mathrm{m}) = 6.0$.

(b) Solve for $X$:

$$X = \frac{(1\,\mathrm{s})^2}{10\times10^{-15}\,\mathrm{s}} = 1.0\times10^{14}\,\mathrm{s}\,.$$

Since 1 year contains 356 days, each day contains 24 hours, and each hour contains 3600 seconds, the value of $X$ in years is

$$\frac{1.0\times10^{14}\,\mathrm{s}}{(365.2\,d)(24\,\mathrm{h/d})(3600\,\mathrm{s/h})} = 3.2\times10^6\,\mathrm{y}\,.$$

# Chapter 41

## 1

(a) At absolute temperature $T = 0$, the probability is zero that any state with energy above the Fermi energy is occupied.

(b) The probability that a state with energy $E$ is occupied at temperature $T$ is given by

$$P(E) = \frac{1}{e^{(E-E_F)/kT} + 1},$$

where $k$ is the Boltzmann constant and $E_F$ is the Fermi energy. Now, $E - E_F = 0.062\,\text{eV}$ and $(E - E_F)/kT = (0.062\,\text{eV})/(8.62 \times 10^{-5}\,\text{eV/K})(320\,\text{K}) = 2.248$, so

$$P(E) = \frac{1}{e^{2.248} + 1} = 0.0956\,.$$

See Appendix B for the value of $k$.

## 11

The Fermi-Dirac occupation probability is given by $P_{\text{FD}} = 1/\left(e^{\Delta E/kT} + 1\right)$ and the Boltzmann occupation probability is given by $P_{\text{B}} = e^{-\Delta E/kT}$. Let $f$ be the fractional difference. Then

$$f = \frac{P_{\text{B}} - P_{\text{FD}}}{P_{\text{B}}} = \frac{e^{-\Delta E/kT} - \dfrac{1}{e^{\Delta E/kT} + 1}}{e^{-\Delta E/kT}}\,.$$

Using a common denominator and a little algebra yields

$$f = \frac{e^{-\Delta E/kT}}{e^{-\Delta E/kT} + 1}\,.$$

The solution for $e^{-\Delta E/kT}$ is

$$e^{-\Delta E/kT} = \frac{f}{1-f}\,.$$

Take the natural logarithm of both sides and solve for $T$. The result is

$$T = \frac{\Delta E}{k \ln\left(\dfrac{f}{1-f}\right)}\,.$$

(a) Put $f$ equal to 0.01 and evaluate the expression for $T$:

$$T = \frac{(1.00\,\text{eV})(1.60 \times 10^{-19}\,\text{J/eV})}{(1.38 \times 10^{-23}\,\text{J/K}) \ln\left(\dfrac{0.010}{1-0.010}\right)} = 2.50 \times 10^{3}\,\text{K}\,.$$

(b) Put $f$ equal to 0.10 and evaluate the expression for $T$:

$$T = \frac{(1.00\,\text{eV})(1.60 \times 10^{-19}\,\text{J/eV})}{(1.38 \times 10^{-23}\,\text{J/K}) \ln\left(\frac{0.10}{1 - 0.10}\right)} = 5.30 \times 10^3\,\text{K}.$$

**17**

(a) According to Appendix F the molar mass of silver is 107.870 g/mol and the density is $\rho = 10.49\,\text{g/cm}^3$. The mass of a silver atom is

$$M = \frac{107.870 \times 10^{-3}\,\text{kg/mol}}{6.022 \times 10^{23}\,\text{mol}^{-1}} = 1.791 \times 10^{-25}\,\text{kg}.$$

The number of atoms per unit volume is

$$n = \frac{\rho}{M} = \frac{10.49 \times 10^3\,\text{kg/m}^3}{1.791 \times 10^{25}\,\text{kg}} = 5.86 \times 10^{28}\,\text{m}^{-3}.$$

Since silver is monovalent this is the same as the number density of conduction electrons.

(b) The Fermi energy is

$$E_F = \frac{0.121h^2}{m} n^{2/3} = \frac{(0.121)(6.626 \times 10^{-34}\,\text{J}\cdot\text{s})^2}{9.109 \times 10^{-31}\,\text{kg}} (5.86 \times 10^{28}\,\text{m}^{-1})^{2/3}$$
$$= 8.80 \times 10^{-19}\,\text{J} = 5.49\,\text{eV}.$$

(c) Since $E_F = \frac{1}{2}mv_F^2$,

$$v_F = \sqrt{\frac{2E_F}{m}} = \sqrt{\frac{2(8.80 \times 10^{-19}\,\text{J})}{9.109 \times 10^{-31}\,\text{kg}}} = 1.39 \times 10^6\,\text{m/s}.$$

(d) The de Broglie wavelength is

$$\lambda = \frac{h}{p_F} = \frac{h}{mv_F} = \frac{6.626 \times 10^{-34}\,\text{J}\cdot\text{s}}{(9.109 \times 10^{-31}\,\text{kg})(1.39 \times 10^6\,\text{m/s})} = 5.23 \times 10^{-10}\,\text{m}.$$

**31**

(a) Since the electron jumps from the conduction band to the valence band, the energy of the photon equals the energy gap between those two bands. The photon energy is given by $hf = hc/\lambda$, where $f$ is the frequency of the electromagnetic wave and $\lambda$ is its wavelength. Thus $E_g = hc/\lambda$ and

$$\lambda = \frac{hc}{E_g} = \frac{(6.63 \times 10^{-34}\,\text{J}\cdot\text{s})(3.00 \times 10^8\,\text{m/s})}{(5.50\,\text{eV})(1.60 \times 10^{-19}\,\text{J/eV})} = 2.26 \times 10^{-7}\,\text{m} = 226\,\text{nm}.$$

Photons from other transitions have a greater energy, so their waves have shorter wavelengths.

(b) These photons are in the ultraviolet portion of the electromagnetic spectrum.

**37**

Sample Problem 41–6 gives the fraction of silicon atoms that must be replaced by phosphorus atoms. Find the number the silicon atoms in 1.0 g, then the number that must be replaced, and finally the mass of the replacement phosphorus atoms. The molar mass of silicon is 28.086 g/mol, so the mass of one silicon atom is $(28.086\,\text{g/mol})/(6.022 \times 10^{23}\,\text{mol}^{-1}) = 4.66 \times 10^{-23}\,\text{g}$ and the number of atoms in 1.0 g is $(1.0\,\text{g})/(4.66 \times 10^{-23}\,\text{g}) = 2.14 \times 10^{22}$. According to Sample Problem 41–6 one of every $5 \times 10^6$ silicon atoms is replaced with a phosphorus atom. This means there will be $(2.14 \times 10^{22})/(5 \times 10^6) = 4.29 \times 10^{15}$ phosphorus atoms in 1.0 g of silicon. The molar mass of phosphorus is 30.9758 g/mol so the mass of a phosphorus atom is $(30.9758\,\text{g/mol})/(6.022 \times 10^{-23}\,\text{mol}^{-1}) = 5.14 \times 10^{-23}\,\text{g}$. The mass of phosphorus that must be added to 1.0 g of silicon is $(4.29 \times 10^{15})(5.14 \times 10^{-23}\,\text{g}) = 2.2 \times 10^{-7}\,\text{g}$.

**39**

The energy received by each electron is exactly the difference in energy between the bottom of the conduction band and the top of the valence band (1.1 eV). The number of electrons that can be excited across the gap by a single 662-keV photon is $N = (662 \times 10^3\,\text{eV})/(1.1\,\text{eV}) = 6.0 \times 10^5$. Since each electron that jumps the gap leaves a hole behind, this is also the number of electron-hole pairs that can be created.

**49**

(a) According to Eq. 41–6

$$P(E) = \frac{1}{e^{(E-E_F)/kT} + 1} .$$

Its derivative is

$$\frac{dP(e)}{dE} = \frac{-1}{\left[e^{(E-E_F)/kT} + 1\right]^2} \frac{e^{(E-E_F)/kT}}{kT} .$$

For $E = E_F$, $e^{(E-E_F)/kT} = e^0 = 1$, so the derivative at $E + E_F$ is $-1/4kT$.

(b) Represent the tangent line by $P = A + BE$, where $A$ and $B$ are constants. We want $P = 1/2$ and $dP/dE = -1/4kT$ for $E = E_F$. This means $A + BE_F = 1/2$ and $B = -1/4kT$. The solution for $A$ is $A = (1/2) + (E_F/4kT)$. Thus $P = (1/2) - (E - E_F)/4kT$. The intercept is found by setting $P$ equal to zero and solving for $E$. The result is $E = E_F + 2kT$.

# Chapter 42

**13**

(a) The de Broglie wavelength is given by $\lambda = h/p$, where $p$ is the magnitude of the momentum. The kinetic energy $K$ and momentum are related by Eq. 37–54, which yields

$$pc = \sqrt{K^2 + 2Kmc^2} = \sqrt{(200\,\text{MeV})^2 + 2(200\,\text{MeV})(0.511\,\text{MeV})} = 200.5\,\text{MeV}\,.$$

Thus

$$\lambda = \frac{hc}{pc} = \frac{1240\,\text{eV}\cdot\text{nm}}{200.5 \times 10^6\,\text{eV}} = 6.18 \times 10^{-6}\,\text{nm} = 6.18\,\text{fm}\,.$$

(b) The diameter of a copper nucleus, for example, is about 8.6 fm, just a little larger than the de Broglie wavelength of a 200-MeV electron. To resolve detail, the wavelength should be smaller than the target, ideally a tenth of the diameter or less. 200-MeV electrons are perhaps at the lower limit in energy for useful probes.

**17**

The binding energy is given by $\Delta E_{\text{be}} = [Zm_H + (A - Z)m_n - M_{\text{Pu}}]\,c^2$, where $Z$ is the atomic number (number of protons), $A$ is the mass number (number of nucleons), $m_H$ is the mass of a hydrogen atom, $m_n$ is the mass of a neutron, and $M_{\text{Pu}}$ is the mass of a ${}^{239}_{94}\text{Pu}$ atom. In principal, nuclear masses should have been used, but the mass of the $Z$ electrons included in $ZM_H$ is canceled by the mass of the $Z$ electrons included in $M_{\text{Pu}}$, so the result is the same. First, calculate the mass difference in atomic mass units: $\Delta m = (94)(1.00783\,\text{u}) + (239 - 94)(1.00867\,\text{u}) - (239.05216\,\text{u}) = 1.94101\,\text{u}$. Since $1\,\text{u}$ is equivalent to $931.5\,\text{MeV}$, $\Delta E_{\text{be}} = (1.94101\,\text{u})(931.5\,\text{MeV/u}) = 1808\,\text{MeV}$. Since there are 239 nucleons, the binding energy per nucleon is $\Delta E_{\text{ben}} = E/A = (1808\,\text{MeV})/(239) = 7.56\,\text{MeV}$.

**19**

If a nucleus contains $Z$ protons and $N$ neutrons, its binding energy is $\Delta E_{\text{be}} = (Zm_H + Nm_n - m)c^2$, where $m_H$ is the mass of a hydrogen atom, $m_n$ is the mass of a neutron, and $m$ is the mass of the atom containing the nucleus of interest. If the masses are given in atomic mass units, then mass excesses are defined by $\Delta_H = (m_H - 1)c^2$, $\Delta_n = (m_n - 1)c^2$, and $\Delta = (m - A)c^2$. This means $m_Hc^2 = \Delta_H + c^2$, $m_nc^2 = \Delta_n + c^2$, and $mc^2 = \Delta + Ac^2$. Thus $E = (Z\Delta_H + N\Delta_n - \Delta) + (Z + N - A)c^2 = Z\Delta_H + N\Delta_n - \Delta$, where $A = Z + N$ was used. For ${}^{197}_{79}\text{Au}$, $Z = 79$ and $N = 197 - 79 = 118$. Hence

$$\Delta E_{\text{be}} = (79)(7.29\,\text{MeV}) + (118)(8.07\,\text{MeV}) - (-31.2\,\text{MeV}) = 1560\,\text{MeV}\,.$$

This means the binding energy per nucleon is $\Delta E_{\text{ben}} = (1560\,\text{MeV})/(197) = 7.92\,\text{MeV}$.

**27**

(a) The half-life $T_{1/2}$ and the disintegration constant $\lambda$ are related by $T_{1/2} = (\ln 2)/\lambda$, so $T_{1/2} = (\ln 2)/(0.0108\,\text{h}^{-1}) = 64.2\,\text{h}$.

(b) At time $t$, the number of undecayed nuclei remaining is given by

$$N = N_0\, e^{-\lambda t} = N_0\, e^{-(\ln 2)t/T_{1/2}}\,.$$

Substitute $t = 3T_{1/2}$ to obtain

$$\frac{N}{N_0} = e^{-3\ln 2} = 0.125\,.$$

In each half-life, the number of undecayed nuclei is reduced by half. At the end of one half-life, $N = N_0/2$, at the end of two half-lives, $N = N_0/4$, and at the end of three half-lives, $N = N_0/8 = 0.125N_0$.

(c) Use

$$N = N_0\, e^{-\lambda t}\,.$$

10.0 d is 240 h, so $\lambda t = (0.0108\,\text{h}^{-1})(240\,\text{h}) = 2.592$ and

$$\frac{N}{N_0} = e^{-2.592} = 0.0749\,.$$

**35**

(a) Assume that the chlorine in the sample had the naturally occurring isotopic mixture, so the average mass number was 35.453, as given in Appendix F. Then the mass of $^{226}$Ra was

$$m = \frac{226}{226 + 2(35.453)}(0.10\,\text{g}) = 76.1 \times 10^{-3}\,\text{g}\,.$$

The mass of a $^{226}$Ra nucleus is $(226\,\text{u})(1.661 \times 10^{-24}\,\text{g/u}) = 3.75 \times 10^{-22}\,\text{g}$, so the number of $^{226}$Ra nuclei present was $N = (76.1 \times 10^{-3}\,\text{g})/(3.75 \times 10^{-22}\,\text{g}) = 2.03 \times 10^{20}$.

(b) The decay rate is given by $R = N\lambda = (N\ln 2)/T_{1/2}$, where $\lambda$ is the disintegration constant, $T_{1/2}$ is the half-life, and $N$ is the number of nuclei. The relationship $\lambda = (\ln 2)/T_{1/2}$ was used. Thus

$$R = \frac{(2.03 \times 10^{20})\ln 2}{(1600\,\text{y})(3.156 \times 10^{7}\,\text{s/y})} = 2.79 \times 10^{9}\,\text{s}^{-1}\,.$$

**43**

If $N$ is the number of undecayed nuclei present at time $t$, then

$$\frac{dN}{dt} = R - \lambda N\,,$$

where $R$ is the rate of production by the cyclotron and $\lambda$ is the disintegration constant. The second term gives the rate of decay. Rearrange the equation slightly and integrate:

$$\int_{N_0}^{N} \frac{dN}{R - \lambda N} = \int_0^t dt\,,$$

where $N_0$ is the number of undecayed nuclei present at time $t = 0$. This yields

$$-\frac{1}{\lambda} \ln \frac{R - \lambda N}{R - \lambda N_0} = t\,.$$

Solve for $N$:

$$N = \frac{R}{\lambda} + \left(N_0 - \frac{R}{\lambda}\right) e^{-\lambda t}\,.$$

After many half-lives, the exponential is small and the second term can be neglected. Then $N = R/\lambda$, regardless of the initial value $N_0$. At times that are long compared to the half-life, the rate of production equals the rate of decay and $N$ is a constant.

**49**

The fraction of undecayed nuclei remaining after time $t$ is given by

$$\frac{N}{N_0} = e^{-\lambda t} = e^{-(\ln 2)t/T_{1/2}}\,,$$

where $\lambda$ is the disintegration constant and $T_{1/2}$ $(= (\ln 2)/\lambda)$ is the half-life. The time for half the original $^{238}$U nuclei to decay is $4.5 \times 10^9$ y. For $^{244}$Pu at that time

$$\frac{(\ln 2)t}{T_{1/2}} = \frac{(\ln 2)(4.5 \times 10^9\,\text{y})}{8.2 \times 10^7\,\text{y}} = 38.0$$

and

$$\frac{N}{N_0} = e^{-38.0} = 3.1 \times 10^{-17}\,.$$

For $^{248}$Cm at that time

$$\frac{(\ln 2)t}{T_{1/2}} = \frac{(\ln 2)(4.5 \times 10^9\,\text{y})}{3.4 \times 10^5\,\text{y}} = 9170$$

and

$$\frac{N}{N_0} = e^{-9170} = 3.31 \times 10^{-3983}\,.$$

For any reasonably sized sample this is less than one nucleus and may be taken to be zero. Your calculator probably cannot evaluate $e^{-9170}$ directly. Treat it as $(e^{-91.70})^{100}$.

**55**

Let $M_{\text{Cs}}$ be the mass of one atom of $^{137}_{55}$Cs and $M_{\text{Ba}}$ be the mass of one atom of $^{137}_{56}$Ba. To obtain the nuclear masses we must subtract the mass of 55 electrons from $M_{\text{Cs}}$ and the mass of 56 electrons from $M_{\text{Ba}}$. The energy released is $Q = [(M_{\text{Cs}} - 55m) - (M_{\text{Ba}} - 56m) - m]\,c^2$, where $m$ is the mass of an electron. Once cancellations have been made, $Q = (M_{\text{Cs}} - M_{\text{Ba}})c^2$ is obtained. Thus

$$Q = [136.9071\,\text{u} - 136.9058\,\text{u}]\,c^2 = (0.0013\,\text{u})c^2 = (0.0013\,\text{u})(932\,\text{MeV/u}) = 1.21\,\text{MeV}\,.$$

**59**

Since the electron has the maximum possible kinetic energy, no neutrino is emitted. Since momentum is conserved, the momentum of the electron and the momentum of the residual sulfur nucleus are equal in magnitude and opposite in direction. If $p_e$ is the momentum of the electron and $p_S$ is the momentum of the sulfur nucleus, then $p_S = -p_e$. The kinetic energy $K_S$ of the sulfur nucleus is $K_S = p_S^2/2M_S = p_e^2/2M_S$, where $M_S$ is the mass of the sulfur nucleus. Now the electron's kinetic energy $K_e$ is related to its momentum by the relativistic equation $(p_e c)^2 = K_e^2 + 2K_e mc^2$, where $m$ is the mass of an electron. See Eq. 37–54. Thus

$$K_S = \frac{(p_e c)^2}{2M_S c^2} = \frac{K_e^2 + 2K_e mc^2}{2M_S c^2} = \frac{(1.71\,\text{MeV})^2 + 2(1.71\,\text{MeV})(0.511\,\text{MeV})}{2(32\,\text{u})(931.5\,\text{MeV/u})}$$
$$= 7.83 \times 10^{-5}\,\text{MeV} = 78.3\,\text{eV}\,,$$

where $mc^2 = 0.511\,\text{MeV}$ was used.

**67**

The decay rate $R$ is related to the number of nuclei $N$ by $R = \lambda N$, where $\lambda$ is the disintegration constant. The disintegration constant is related to the half-life $T_{1/2}$ by $\lambda = (\ln 2)/T_{1/2}$, so $N = R/\lambda = RT_{1/2}/\ln 2$. Since $1\,\text{Ci} = 3.7 \times 10^{10}$ disintegrations/s,

$$N = \frac{(250\,\text{Ci})(3.7 \times 10^{10}\,\text{s}^{-1}/\text{Ci})(2.7\,\text{d})(8.64 \times 10^4\,\text{s/d})}{\ln 2} = 3.11 \times 10^{18}\,.$$

The mass of a $^{198}$Au atom is $M = (198\,\text{u})(1.661 \times 10^{-24}\,\text{g/u}) = 3.29 \times 10^{-22}\,\text{g}$ so the mass required is $NM = (3.11 \times 10^{18})(3.29 \times 10^{-22}\,\text{g}) = 1.02 \times 10^{-3}\,\text{g} = 1.02\,\text{mg}$.

**73**

A generalized formation reaction can be written X + x → Y, where X is the target nucleus, x is the incident light particle, and Y is the excited compound nucleus ($^{20}$Ne). Assume X is initially at rest. Then conservation of energy yields

$$m_X c^2 + m_x c^2 + K_x = m_Y c^2 + K_Y + E_Y\,,$$

where $m_X$, $m_x$, and $m_Y$ are masses, $K_x$ and $K_Y$ are kinetic energies, and $E_Y$ is the excitation energy of Y. Conservation of momentum yields

$$p_x = p_Y\,.$$

Now $K_Y = p_Y^2/2m_Y = p_x^2/2m_Y = (m_x/m_Y)K_x$, so

$$m_X c^2 + m_x c^2 + K_x = m_Y c^2 + (m_x/m_Y)K_x + E_Y$$

and

$$K_x = \frac{m_Y}{m_Y - m_x}\left[(m_Y - m_X - m_x)c^2 + E_Y\right]\,.$$

(a) Let $x$ represent the alpha particle and X represent the $^{16}$O nucleus. Then $(m_Y - m_X - m_x)c^2 = (19.99244\,\text{u} - 15.99491\,\text{u} - 4.00260\,\text{u})(931.5\,\text{MeV/u}) = -4.722\,\text{MeV}$ and

$$K_\alpha = \frac{19.99244\,\text{u}}{19.99244\,\text{u} - 4.00260\,\text{u}}(-4.722\,\text{MeV} + 25.0\,\text{MeV}) = 25.35\,\text{MeV}\,.$$

(b) Let $x$ represent the proton and X represent the $^{19}$F nucleus. Then $(m_Y - m_X - m_x)c^2 = (19.99244\,\text{u} - 18.99841\,\text{u} - 1.00783\,\text{u})(931.5\,\text{MeV/u}) = -12.85\,\text{MeV}$ and

$$K_\alpha = \frac{19.99244\,\text{u}}{19.99244\,\text{u} - 1.00783\,\text{u}}(-12.85\,\text{MeV} + 25.0\,\text{MeV}) = 12.80\,\text{MeV}\,.$$

(c) Let $x$ represent the photon and X represent the $^{20}$Ne nucleus. Since the mass of the photon is zero, we must rewrite the conservation of energy equation: if $E_\gamma$ is the energy of the photon, then $E_\gamma + m_X c^2 = m_Y c^2 + K_Y + E_Y$. Since $m_X = m_Y$, this equation becomes $E_\gamma = K_Y + E_Y$. Since the momentum and energy of a photon are related by $p_\gamma = E_\gamma/c$, the conservation of momentum equation becomes $E_\gamma/c = p_Y$. The kinetic energy of the compound nucleus is $K_Y = p_Y^2/2m_Y = E_\gamma^2/2m_Y c^2$. Substitute this result into the conservation of energy equation to obtain

$$E_\gamma = \frac{E_\gamma^2}{2m_Y c^2} + E_Y\,.$$

This quadratic equation has the solutions

$$E_\gamma = m_Y c^2 \pm \sqrt{(m_Y c^2)^2 - 2m_Y c^2 E_Y}\,.$$

If the problem is solved using the relativistic relationship between the energy and momentum of the compound nucleus, only one solution would be obtained, the one corresponding to the negative sign above. Since $m_Y c^2 = (19.99244\,\text{u})(931.5\,\text{MeV/u}) = 1.862 \times 10^4\,\text{MeV}$,

$$\begin{aligned} E_\gamma &= (1.862 \times 10^4\,\text{MeV}) - \sqrt{(1.862 \times 10^4\,\text{MeV})^2 - 2(1.862 \times 10^4\,\text{MeV})(25.0\,\text{MeV})} \\ &= 25.0\,\text{MeV}\,. \end{aligned}$$

The kinetic energy of the compound nucleus is very small; essentially all of the photon energy goes to excite the nucleus.

**75**

Let $A$ be the area over which fallout occurs and $a$ be the area that produces a count rate of $R = 74\,000$ counts/s. The count rate is $R = \lambda N$, where $\lambda$ is the disintegration constant and $N$ is the number of radioactive nuclei in area $a$. The number of atoms in the entire fallout is $M/m$, where $M$ is the mass of $^{90}$Sr produced and $m$ is the mass of a single nucleus of $^{90}$Sr. Thus the count rate for the area $a$ is $R = \lambda(M/m)(a/A)$. The half-life $T_{1/2}$ is related to the disintegration constant by $\lambda = (\ln 2)/T_{1/2}$, so $R = (\ln 2/T_{1/2})(M/m)(a/A)$. Solve for $a$:

$$a = AR\left(\frac{m}{M}\right)\left(\frac{T_{1/2}}{\ln 2}\right)\,.$$

The molar mass of $^{90}$Sr is 90 g/mol, so the mass of a single $^{90}$Sr nucleus is $(90^{-3}\,\text{kg/mol})/(6.02\times 10^{23}\,\text{mol}^{-1}) = 1.50\times 10^{-25}\,\text{kg}$. The half-life is $(29\,\text{y})(365\,\text{d/y})(24\,\text{h/d})(3600\,\text{s/h}) = 9.15\times 10^{8}\,\text{s}$. Therefore

$$a = (2000\times 10^{6}\,\text{m}^2)(74\,000\,\text{counts/s})\left(\frac{1.50\times 10^{-25}\,\text{kg}}{400\times 10^{-3}\,\text{kg}}\right)\left(\frac{9.14\times 10^{8}\,\text{s}}{\ln 2}\right) = 7.3\times 10^{-2}\,\text{m}^2\,.$$

**85**

The number of undecayed nuclei at time $t$ is given by $N = N_0 e^{-\lambda t}$, where $N_0$ is the number at $t = 0$ and $\lambda$ is the disintegration constant. The rate of decay is $R = -dN/dt = \lambda N_0 e^{-\lambda t} = \lambda N$ and the rate at $t =$) is $R_0 = N_0$. Thus $R/R_0 = N/N_0 = e^{-\lambda t}$. The solution for $t$ is

$$t = -\frac{1}{\lambda}\ln\frac{R}{R_0}\,.$$

The disintegration constant is related to the half-life $T_{1/2}$ by $\lambda = (\ln 2)/T_{1/2}$, so

$$t = -\frac{T_{1/2}}{\ln 2}\ln\frac{R}{R_0} = -\frac{5730\,\text{y}}{\ln 2}\ln(0.020) = 3.2\times 10^{4}\,\text{y}\,.$$

**87**

Let ${}^{A}_{Z}X$ represent the unknown nuclide. The reaction equation is

$${}^{A}_{Z}X + {}^{1}_{0}\text{n} \rightarrow {}^{0}_{-1}\text{e} + 2\,{}^{4}_{2}\text{He}\,.$$

Conservation of charge yields $Z + 0 = -1 + 4$ or $Z = 3$. Conservation of mass number yields $A + 1 = 0 + 8$ or $A = 7$. According to the periodic table in Appendix E, lithium has atomic number 3, so the nuclide must be ${}^{7}_{3}\text{Li}$.

# Chapter 43

**13**

(a) If X represents the unknown fragment, then the reaction can be written

$$^{235}_{92}\text{U} + ^{1}_{0}\text{n} \rightarrow ^{83}_{32}\text{Ge} + ^{A}_{Z}X\,,$$

where $A$ is the mass number and $Z$ is the atomic number of the fragment. Conservation of charge yields $92 + 0 = 32 + Z$, so $Z = 60$. Conservation of mass number yields $235 + 1 = 83 + A$, so $A = 153$. Look in Appendix F or G for nuclides with $Z = 60$. You should find that the unknown fragment is $^{153}_{60}\text{Nd}$.

(b) and (c) Ignore the small kinetic energy and momentum carried by the neutron that triggers the fission event. Then $Q = K_{\text{Ge}} + K_{\text{Nd}}$, where $K_{\text{Ge}}$ is the kinetic energy of the germanium nucleus and $K_{\text{Nd}}$ is the kinetic energy of the neodymium nucleus. Conservation of momentum yields $p_{\text{Ge}} + p_{\text{Nd}} = 0$, where $p_{\text{Ge}}$ is the momentum of the germanium nucleus and $p_{\text{Nd}}$ is the momentum of the neodymium nucleus. Since $p_{\text{Nd}} = -p_{\text{Ge}}$, the kinetic energy of the neodymium nucleus is

$$K_{\text{Nd}} = \frac{p_{\text{Nd}}^2}{2M_{\text{Nd}}} = \frac{p_{\text{Ge}}^2}{2M_{\text{Nd}}} = \frac{M_{\text{Ge}}}{M_{\text{Nd}}}K_{\text{Ge}}\,.$$

Thus the energy equation becomes

$$Q = K_{\text{Ge}} + \frac{M_{\text{Ge}}}{M_{\text{Nd}}}K_{\text{Ge}} = \frac{M_{\text{Nd}} + M_{\text{Ge}}}{M_{\text{Nd}}}K_{\text{Ge}}$$

and

$$K_{\text{Ge}} = \frac{M_{\text{Nd}}}{M_{\text{Nd}} + M_{\text{Ge}}}Q = \frac{153\,\text{u}}{153\,\text{u} + 83\,\text{u}}(170\,\text{MeV}) = 110\,\text{MeV}\,.$$

Similarly,

$$K_{\text{Nd}} = \frac{M_{\text{Ge}}}{M_{\text{Nd}} + M_{\text{Ge}}}Q = \frac{83\,\text{u}}{153\,\text{u} + 83\,\text{u}}(170\,\text{MeV}) = 60\,\text{MeV}\,.$$

The mass conversion factor can be found in Appendix C.

(d) The initial speed of the germanium nucleus is

$$v_{\text{Ge}} = \sqrt{\frac{2K_{\text{Ge}}}{M_{\text{Ge}}}} = \sqrt{\frac{2(110 \times 10^6\,\text{eV})(1.60 \times 10^{-19}\,\text{J/eV})}{(83\,\text{u})(1.661 \times 10^{-27}\,\text{kg/u})}} = 1.60 \times 10^7\,\text{m/s}\,.$$

(e) The initial speed of the neodymium nucleus is

$$v_{\text{Nd}} = \sqrt{\frac{2K_{\text{Nd}}}{M_{\text{Nd}}}} = \sqrt{\frac{2(60 \times 10^6\,\text{eV})(1.60 \times 10^{-19}\,\text{J/eV})}{(153\,\text{u})(1.661 \times 10^{-27}\,\text{kg/u})}} = 8.69 \times 10^6\,\text{m/s}\,.$$

**15**

(a) The energy yield of the bomb is $E = (66 \times 10^{-3}\,\text{megaton})(2.6 \times 10^{28}\,\text{MeV/megaton}) = 1.72 \times 10^{27}\,\text{MeV}$. (The energy conversion factor is given in Problem 16.) At 200 MeV per fission event, $(1.72 \times 10^{27}\,\text{MeV})/(200\,\text{MeV}) = 8.58 \times 10^{24}$ fission events take place. Since only 4.0% of the $^{235}$U nuclei originally present undergo fission, there must have been $(8.58 \times 10^{24})/(0.040) = 2.14 \times 10^{26}$ nuclei originally present. The mass of $^{235}$U originally present was $(2.14 \times 10^{26})(235\,\text{u})(1.661 \times 10^{-27}\,\text{kg/u}) = 83.7\,\text{kg}$. The mass conversion factor can be found in Appendix C.

(b) Two fragments are produced in each fission event, so the total number of fragments is $2(8.58 \times 10^{24}) = 1.72 \times 10^{25}$.

(c) One neutron produced in a fission event is used to trigger the next fission event, so the average number of neutrons released to the environment in each event is 1.5. The total number released is $(8.58 \times 10^{24})(1.5) = 1.29 \times 10^{25}$.

**23**

(a) Let $v_{ni}$ be the initial velocity of the neutron, $v_{nf}$ be its final velocity, and $v_f$ be the final velocity of the target nucleus. Then, since the target nucleus is initially at rest, conservation of momentum yields $m_n v_{ni} = m_n v_{nf} + m v_f$ and conservation of energy yields $\frac{1}{2} m_n v_{ni}^2 = \frac{1}{2} m_n v_{nf}^2 + \frac{1}{2} m v_f^2$. Solve these two equations simultaneously for $v_f$. This can be done, for example, by using the conservation of momentum equation to obtain an expression for $v_{nf}$ in terms of $v_f$ and substituting the expression into the conservation of energy equation. Solve the resulting equation for $v_f$. You should obtain $v_f = 2 m_n v_{ni}/(m + m_n)$. The energy lost by the neutron is the same as the energy gained by the target nucleus, so

$$\Delta K = \frac{1}{2} m v_f^2 = \frac{1}{2} \frac{4 m_n^2 m}{(m + m_n)^2} v_{ni}^2 .$$

The initial kinetic energy of the neutron is $K = \frac{1}{2} m_n v_{ni}^2$, so

$$\frac{\Delta K}{K} = \frac{4 m_n m}{(m + m_n)^2} .$$

(b) The mass of a neutron is 1.0 u and the mass of a hydrogen atom is also 1.0 u. (Atomic masses can be found in Appendix G.) Thus $(\Delta K)/K = 4(1.0\,\text{u})(1.0\,\text{u})/(1.0\,\text{u} + 1.0\,\text{u})^2 = 1.0$.

(c) The mass of a deuterium atom is 2.0 u, so $(\Delta K)/K = 4(1.0\,\text{u})(2.0\,\text{u})/(2.0\,\text{u} + 1.0\,\text{u})^2 = 0.89$.

(d) The mass of a carbon atom is 12 u, so $(\Delta K)/K = 4(1.0\,\text{u})(12\,\text{u})/(12\,\text{u} + 1.0\,\text{u})^2 = 0.28$.

(e) The mass of a lead atom is 207 u, so $(\Delta K)/K = 4(1.0\,\text{u})(207\,\text{u})/(207\,\text{u} + 1.0\,\text{u})^2 = 0.019$.

(f) During each collision, the energy of the neutron is reduced by the factor $1 - 0.89 = 0.11$. If $E_i$ is the initial energy, then the energy after $n$ collisions is given by $E = (0.11)^n E_i$. Take the natural logarithm of both sides and solve for $n$. The result is

$$n = \frac{\ln(E/E_i)}{\ln 0.11} = \frac{\ln(0.025\,\text{eV}/1.00\,\text{eV})}{\ln 0.11} = 7.9 .$$

The energy first falls below 0.025 eV on the eighth collision.

**25**

Let $P_0$ be the initial power output, $P$ be the final power output, $k$ be the multiplication factor, $t$ be the time for the power reduction, and $t_{\text{gen}}$ be the neutron generation time. Then according to the result of Problem 18,

$$P = P_0\, k^{t/t_{\text{gen}}} .$$

Divide by $P_0$, then take the natural logarithm of both sides of the equation and solve for $\ln k$. You should obtain

$$\ln k = \frac{t_{\text{gen}}}{t} \ln \frac{P}{P_0} .$$

Hence

$$k = e^{\alpha} ,$$

where

$$\alpha = \frac{t_{\text{gen}}}{t} \ln \frac{P}{P_0} = \frac{1.3 \times 10^{-3}\,\text{s}}{2.6000\,\text{s}} \ln \frac{350.00\,\text{MW}}{1200.0\,\text{MW}} = -6.161 \times 10^{-4} .$$

This yields $k = .99938$.

**29**

Let $t$ be the present time and $t = 0$ be the time when the ratio of $^{235}\text{U}$ to $^{238}\text{U}$ was 3.0%. Let $N_{235}$ be the number of $^{235}\text{U}$ nuclei present in a sample now and $N_{235,\,0}$ be the number present at $t = 0$. Let $N_{238}$ be the number of $^{238}\text{U}$ nuclei present in the sample now and $N_{238,\,0}$ be the number present at $t = 0$. The law of radioactive decay holds for each specie, so

$$N_{235} = N_{235,\,0}\, e^{-\lambda t}$$

and

$$N_{238} = N_{238,\,0}\, e^{-\lambda t} .$$

Divide the first equation by the second to obtain

$$r = r_0\, e^{-(\lambda - \lambda)t} ,$$

where $r = N_{235}/N_{238}$ (= 0.0072) and $r_0 = N_{235,\,0}/N_{238,\,0}$ (= 0.030). Solve for $t$:

$$t = -\frac{1}{\lambda_{235} - \lambda_{238}} \ln \frac{r}{r_0} .$$

Now use $\lambda_{235} = (\ln 2)/T_{235}$ and $\lambda_{238} = (\ln 2)/T_{238}$, where $T_{235}$ and $T_{238}$ are the half-lives, to obtain

$$t = -\frac{T_{235}T_{238}}{(T_{238} - T_{235}) \ln 2} \ln \frac{r}{r_0} = -\frac{(7.0 \times 10^8\,\text{y})(4.5 \times 10^9\,\text{y})}{(4.5 \times 10^9\,\text{y} - 7.0 \times 10^8\,\text{y}) \ln 2} \ln \frac{0.0072}{0.030} = 1.71 \times 10^9\,\text{y} .$$

**31**

The height of the Coulomb barrier is taken to be the value of the kinetic energy $K$ each deuteron must initially have if they are to come to rest when their surfaces touch (see Sample Problem 43–4). If $r$ is the radius of a deuteron, conservation of energy yields

$$2K = \frac{1}{4\pi\epsilon_0}\frac{e^2}{2r},$$

so

$$K = \frac{1}{4\pi\epsilon_0}\frac{e^2}{4r} = (8.99 \times 10^9\,\text{m/F})\frac{(1.60 \times 10^{-19}\,\text{C})^2}{4(2.1 \times 10^{-15}\,\text{m})} = 2.74 \times 10^{-14}\,\text{J}.$$

This is 170 keV.

**43**

(a) The mass of a carbon atom is $(12.0\,\text{u})(1.661 \times 10^{-27}\,\text{kg/u}) = 1.99 \times 10^{-26}\,\text{kg}$, so the number of carbon atoms in 1.00 kg of carbon is $(1.00\,\text{kg})/(1.99 \times 10^{-26}\,\text{kg}) = 5.02 \times 10^{25}$. (The mass conversion factor can be found in Appendix C.) The heat of combustion per atom is $(3.3 \times 10^7\,\text{J/kg})/(5.02 \times 10^{25}\,\text{atom/kg}) = 6.58 \times 10^{-19}\,\text{J/atom}$. This is 4.11 eV/atom.

(b) In each combustion event, two oxygen atoms combine with one carbon atom, so the total mass involved is $2(16.0\,\text{u})+(12.0\,\text{u}) = 44\,\text{u}$. This is $(44\,\text{u})(1.661\times10^{-27}\,\text{kg/u}) = 7.31\times10^{-26}\,\text{kg}$. Each combustion event produces $6.58\times10^{-19}\,\text{J}$ so the energy produced per unit mass of reactants is $(6.58 \times 10^{-19}\,\text{J})/(7.31 \times 10^{-26}\,\text{kg}) = 9.00 \times 10^6\,\text{J/kg}$.

(c) If the Sun were composed of the appropriate mixture of carbon and oxygen, the number of combustion events that could occur before the Sun burns out would be $(2.0 \times 10^{30}\,\text{kg})/(7.31 \times 10^{-26}\,\text{kg}) = 2.74 \times 10^{55}$. The total energy released would be $E = (2.74 \times 10^{55})(6.58 \times 10^{-19}\,\text{J}) = 1.80 \times 10^{37}\,\text{J}$. If $P$ is the power output of the Sun, the burn time would be $t = E/P = (1.80 \times 10^{37}\,\text{J})/(3.9 \times 10^{26}\,\text{W}) = 4.62 \times 10^{10}\,\text{s}$. This is 1460 y.

# Chapter 44

**11**

(a) The conservation laws considered so far are associated with energy, momentum, angular momentum, charge, baryon number, and the three lepton numbers. The rest energy of the muon is 105.7 MeV, the rest energy of the electron is 0.511 MeV, and the rest energy of the neutrino is zero. Thus the total rest energy before the decay is greater than the total rest energy after. The excess energy can be carried away as the kinetic energies of the decay products and energy can be conserved. Momentum is conserved if the electron and neutrino move away from the decay in opposite directions with equal magnitudes of momenta. Since the orbital angular momentum is zero, we consider only spin angular momentum. All the particles have spin $\hbar/2$. The total angular momentum after the decay must be either $\hbar$ (if the spins are aligned) or zero (if the spins are antialigned). Since the spin before the decay is $\hbar/2$, angular momentum cannot be conserved. The muon has charge $-e$, the electron has charge $-e$, and the neutrino has charge zero, so the total charge before the decay is $-e$ and the total charge after is $-e$. Charge is conserved. All the particles have baryon number zero, so baryon number is conserved. The muon lepton number of the muon is $+1$, the muon lepton number of the muon neutrino is $+1$, and the muon lepton number of the electron is 0. Muon lepton number is conserved. The electron lepton numbers of the muon and muon neutrino are 0 and the electron lepton number of the electron is $+1$. Electron lepton number is not conserved. The laws of conservation of angular momentum and electron lepton number are not obeyed and this decay does not occur..

(b) Analyze the decay in the same way. You should find that only charge is not conserved.

(c) Here you should find that energy and muon lepton number cannot be conserved.

**29**

(a) Look at Table 44–5. Since the particle is a baryon, it must consist of three quarks. To obtain a strangeness of $-2$, two of them must be s quarks. Each of these has a charge of $-e/3$, so the sum of their charges is $-2e/3$. To obtain a total charge of $e$, the charge on the third quark must be $5e/3$. There is no quark with this charge, so the particle cannot be constructed. In fact, such a particle has never been observed.

(b) Again the particle consists of three quarks (and no antiquarks). To obtain a strangeness of zero, none of them may be s quarks. We must find a combination of three u and d quarks with a total charge of $2e$. The only such combination consists of three u quarks.

**41**

(a) The mass $M$ within Earth's orbit is used to calculate the gravitational force on Earth. If $r$ is the radius of the orbit, $R$ is the radius of the new Sun, and $M_S$ is the mass of the Sun, then

$$M = \left(\frac{r}{R}\right)^3 M_S = \left(\frac{1.50 \times 10^{11}\,\text{m}}{5.90 \times 10^{12}\,\text{m}}\right)^3 (1.99 \times 10^{30}\,\text{kg}) = 3.27 \times 10^{25}\,\text{kg}\,.$$

The gravitational force on Earth is given by $GMm/r^2$, where $m$ is the mass of Earth and $G$ is the universal gravitational constant. Since the centripetal acceleration is given by $v^2/r$, where $v$ is the speed of Earth, $GMm/r^2 = mv^2/r$ and

$$v = \sqrt{\frac{GM}{r}} = \sqrt{\frac{(6.67 \times 10^{-11}\,\text{m}^3/\text{s}^2 \cdot \text{kg})(3.27 \times 10^{25}\,\text{kg})}{1.50 \times 10^{11}\,\text{m}}} = 1.21 \times 10^2\,\text{m/s}\,.$$

(b) The period of revolution is

$$T = \frac{2\pi r}{v} = \frac{2\pi(1.50 \times 10^{11}\,\text{m})}{1.21 \times 10^2\,\text{m/s}} = 7.82 \times 10^9\,\text{s}\,.$$

This is 248 y.

**45**

The energy released would be twice the rest energy of Earth, or $E = 2mc^2 = 2(5.98 \times 10^{24}\,\text{kg})(3.00 \times 10^8\,\text{m/s})^2 = 1.08 \times 10^{42}\,\text{J}$. The mass of Earth can be found in Appendix C.

**47**

(a) Since $S = -1$ the meson must contain an s quark, which has a charge quantum number of $-1/3$. To obtain a meson with charge quantum number +1, the s quark must be combined with an antiquark with strangeness 0 and charge quantum number +4/3. There is no such antiquark.

(b) Now $S = +1$, so the meson contains an $\bar{\text{s}}$ quark, which has a charge quantum number of $+1/3$. To obtain a charge quantum number of $-1$ it must also contain a quark with charge quantum number $-4/2$. There is no such quark.

**51**

(a) After time $\Delta t$ the distance between the galaxy and Earth is $r + r\alpha\,\Delta t$, where $r$ is the distance when the light is emitted. The distance when the light reaches Earth must be $c\Delta t$, so $c\Delta t = r + r\alpha\Delta t$ and $\Delta t = r/(c - r\alpha)$.

(b) The detected wavelength is longer than $\lambda$ by $\lambda\alpha\,\Delta t$, so $\Delta\lambda/\lambda = \alpha\,\Delta t = \alpha r/(c - \alpha r)$.

(c) Since $c > \alpha r$, the binomial theorem gives

$$\frac{\Delta\lambda}{\lambda} = \frac{\alpha r}{c}\left[1 - \frac{\alpha r}{c}\right]^{-1} = \frac{\alpha r}{c}\left[1 + \left(\frac{\alpha r}{c}\right) + \left(\frac{\alpha r}{c}\right)^2 + \ldots\right] = \frac{\alpha r}{c} + \left(\frac{\alpha r}{c}\right)^2 + \left(\frac{\alpha r}{c}\right)^3 + \ldots\,.$$

(d) If only the first term is retained $\Delta\lambda/\lambda = \alpha r/c$.

(e) If $v = Hr$, where $H$ is the Hubble constant, then $\Delta\lambda/\lambda = v/c = Hr/c$. Comparison with $\Delta\lambda/\lambda = \alpha r/c$ gives $\alpha = H = 0.0218\,\text{m/s}\cdot\text{ly}$.

(f) Solve $\Delta\lambda/\lambda = r\alpha/(c - r\alpha)$ for $r$. The result is

$$r = \frac{c(\Delta\lambda/\lambda)}{\alpha(1+\Delta\lambda/\lambda)} = \frac{(3.00\times10^8\,\text{m/s})(0.050)}{(0.0218\,\text{m/s}\cdot\text{ly})(1+0.050)} = 6.6\times10^8\,\text{ly}.$$

(g) According to the result of part (a)

$$\Delta t = \frac{r}{c - r\alpha} = \frac{(6.6\times10^8\,\text{ly})(9.46\times10^{15}\,\text{m/ly})}{3.00\times10^8\,\text{m/s} - (0.0218\,\text{m/s}\cdot\text{ly})(6.6\times10^8\,\text{ly})} = 2.2\times10^{16}\,\text{s}.$$

This is $6.9\times10^8\,\text{y}$.

(h) The time is $\Delta t = r/c = (6.6\times10^8\,\text{ly})/(1.00\,\text{ly/y}) = 6.6\times10^8\,\text{y}$.

(i) The distance is $c\,\Delta t = (1.00\,\text{ly/y})(6.9\times10^8\,\text{y}) = 6.9\times10^8\,\text{ly}$.

(j) Use the equation developed in part (f):

$$r = \frac{c(\Delta\lambda/\lambda)}{\alpha(1+\Delta\lambda/\lambda)} = \frac{(3.00\times10^8\,\text{m/s})(0.080)}{(0.0218\,\text{m/s}\cdot\text{ly})(1+0.080)} = 1.02\times10^9\,\text{ly}.$$

(k) The result of part (a) gives

$$\Delta t = \frac{r}{(c - r\alpha)} = \frac{(1.02\times10^9\,\text{ly})(9.49\times10^{15}\,\text{m/ly})}{3.00\times10^8\,\text{m/s} - (1.02\times10^9\,\text{ly})(0.0218\,\text{m/s}\cdot\text{ly})} = 3.5\times10^{16}\,\text{s}.$$

This is $1.1\times10^9\,\text{y}$.

(l) Galaxy B emits light $\Delta t = 1.1\times10^9\,\text{y} - 6.9\times10^8\,\text{y} = 4.1\times10^8\,\text{y}$ earlier than galaxy B. During that time the universe expands, so that the distance of galaxy B from Earth at the time A emits is

$$\begin{aligned} r_B(1+\alpha\Delta t) &= (1.02\times10^9\,\text{ly})\left[1 + \frac{0.0218\,\text{m/s}\cdot\text{ly}}{9.46\times10^{15}\,\text{m/ly}}(4.1\times10^8\,\text{y})(3.16\times10^7\,\text{s/y})\right] \\ &= 1.05\times10^9\,\text{ly}. \end{aligned}$$

The separation of the galaxies at the time A emits is $1.05\times10^9\,\text{ly} - 6.6\times10^8\,\text{ly} = 3.9\times10^8\,\text{ly}$.